Engineering Vibrations: Theory and Applications

Engineering Vibrations: Theory and Applications

Edited by
Seth Grattan

WILLFORD PRESS

www.willfordpress.com

Published by Willford Press,
118-35 Queens Blvd., Suite 400,
Forest Hills, NY 11375, USA

ISBN: 978-1-68285-641-3

Cataloging-in-Publication Data

Engineering vibrations : theory and applications / edited by Seth Grattan.
    p. cm.
Includes bibliographical references and index.
ISBN 978-1-68285-641-3
1. Vibration. 2. Engineering. I. Grattan, Seth.
TA355 .E54 2019
620.3--dc23

For information on all Willford Press publications
visit our website at www.willfordpress.com

WILLFORD PRESS

Contents

Permissions

List of Contributors

Index

Preface

Vibration is a mechanical phenomenon that is characterized by oscillations around the equilibrium point. Such oscillations can be periodic or random. Vibrating structures cause sound or pressure waves. Free, forced and damped vibrations are the different types of vibrations. The study of sound and vibration, as well as the design, analysis and control of sound, is undertaken by the field of acoustical engineering. Vibroscopes are devices that are used to trace or record vibrations. The analysis of vibrations is used in the industrial sector to detect faults in equipment and reduce their downtime. It also helps in minimizing maintenance costs. The objective of this book is to give a general view of the different areas of vibrations and their applications. It brings forth some of the most innovative concepts and elucidates the unexplored aspects of this field. As this field is emerging at a fast pace, this book will help the readers to better understand the concepts of vibrations and their engineering.

This book is a result of research of several months to collate the most relevant data in the field.

When I was approached with the idea of this book and the proposal to edit it, I was overwhelmed. It gave me an opportunity to reach out to all those who share a common interest with me in this field. I had 3 main parameters for editing this text:

1. Accuracy – The data and information provided in this book should be up-to-date and valuable to the readers.

2. Structure – The data must be presented in a structured format for easy understanding and better grasping of the readers

3. Universal Approach – This book not only targets students but also experts and innovators in the field, thus my aim was to present topics which are of use to all.

Thus, it took me a couple of months to finish the editing of this book.

I would like to make a special mention of my publisher who considered me worthy of this opportunity and also supported me throughout the editing process. I would also like to thank the editing team at the back-end who extended their help whenever required.

Editor

Forced Response of Polar Orthotropic Tapered Circular Plates Resting on Elastic Foundation

A. H. Ansari

Department of Mathematics, AlBaha University, Al Baha 1988, Saudi Arabia

Correspondence should be addressed to A. H. Ansari; ansaridranwar@rediffmail.com

Academic Editor: Mohammad Tawfik

Forced axisymmetric response of polar orthotropic circular plates of linearly varying thickness resting on Winkler type of elastic foundation has been studied on the basis of classical plate theory. An approximate solution of problem has been obtained by Rayleigh Ritz method, which employs functions based upon the static deflection of polar orthotropic circular plates. The effect of transverse loadings has been studied for orthotropic circular plate resting on elastic foundation. The transverse deflections and bending moments are presented for various values of taper parameter, rigidity ratio, foundation parameter, and flexibility parameter under different types of loadings. A comparison of results with those available in literature shows an excellent agreement.

1. Introduction

The study of vibrational characteristics of plates has been of great interest due to their increasing use in various engineering applications. The development of fibre-reinforced composite materials and their extensive use in fabrication of plate type structural components in aerospace, ocean engineering, and electronic and mechanical components has led to the study of dynamic response of anisotropic plates. The consideration of thickness variation together with anisotropy not only reduces the size and weight of components but also meets the desirability of high strength, corrosion resistance, high temperature performance, and economy. One can fabricate the orthotropic nature of plate in different ways, that is, rectangular and polar orthotropy. The consideration of polar orthotropy in circular and annular plates provides the best approximation to the results, because polar coordinate axes are also the axes for the materials symmetry. An excellent review of vibration of plates has been given by Leissa in his monograph [1] and a series of review articles [2–7]. Reddy [8] and Lal and Gupta [9] have presented an up-to-date survey analysing transverse vibrations of nonuniform rectangular orthotropic plates. In a recent study, Sayyad and Ghugal [10] presented a review of recent literatures on free vibration analysis of Laminated Composite and Sandwich Plates.

A considerable amount of work dealing with vibration of polar orthotropic circular and annular plates of uniform/nonuniform thickness in presence/absence of elastic foundation has been done by a number of workers and is reported in [11–31] to mention a few. The orthotropic plates resting on an elastic foundation find their application in foundation engineering, such as reinforced concrete pavements of high runways, foundation of deep wells and storage tanks, and slabs of buildings (Lekhnitskii [11], p. 136). In this connection, various models such as Winkler [12–17], Pasternak [18], and Vlasov [19] have been proposed in the literature. Laura et al. [13] studied the effect of Winkler Foundation on vibrations of solid circular plate of linearly varying thickness. Gupta et al. [14] analysed the effect of elastic foundation on axisymmetric vibrations of polar orthotropic circular plates of variable thickness. Further the work has been extended to asymmetric vibration of polar orthotropic plates resting on elastic foundation by Ansari and Gupta [15]. Gupta et al. [16] studied buckling and vibration of polar orthotropic circular plate attached to Winkler Foundation. Orthotropic circular/annular plates such as deck, diaphragm, and bulk heads are used as structural components in launch vehicles.

In some technological situations, a plate is exposed to transverse loads on the surfaces with a downward and/or upward thrust, that is, ship container and aeronautical structural components. The stability of these components increases with the support of elastic foundations. Thus the study of the combined effect of transverse loads and elastic foundation on vibrational characteristics of plates is of practical importance. Gupta et al. [16] analysed the complicating effect of elastic foundation on elastic properties of the plates. A number of research papers are available showing the study of transverse deflection and bending moments of polar orthotropic/isotropic plates of variable thickness resting on elastic foundation [20–31] with complicating effects.

The present work is concerned with the study of transverse loads on vibration of polar orthotropic circular plates of linearly varying thickness resting on Winkler type of elastic foundation. The support of elastic foundation provides greater stability to these structural components exposed to transverse loads. An approximate solution is obtained by Ritz method employing functions based on static deflection given by Lekhnitskii [11]. The present choice of functions has a faster rate of convergence as compared to polynomial coordinate functions employed by Laura et al. [13, 21, 26, 27].

2. Analysis

Consider a thin circular plate of radius a and variable thickness $h = h(r)$, resting on elastic foundation of modulus k_f, elastically restrained against rotation by springs of stiffness k_ϕ and subjected to $P(r) \cos \omega t$ type of excitation extending from $r = r_0$ to $r = r_1$. Let (r, θ) be the polar coordinates of any point on the neutral surface of the circular plate referred to as the centre of the plate as origin (shown in Figure 1(a)).

The maximum kinetic energy of the plate is given by

$$T_{\max} = \frac{1}{2}\rho\omega^2 \int_0^a \int_0^{2\pi} hw^2 r d\theta \, dr, \tag{1}$$

where w is the transverse deflection, ρ the mass density, and ω the frequency in radians per second.

The maximum strain energy of the plate is given by

$$U_{\max} = \frac{1}{2} \int_0^a \int_0^{2\pi} \left[D_r \left\{ \left(\frac{\partial^2 w}{\partial r^2}\right)^2 + 2v_\theta \frac{\partial^2 w}{\partial r^2}\left(\frac{1}{r}\frac{\partial w}{\partial r}\right) \right\} \right. $$
$$\left. + D_\theta \left(\frac{1}{r}\frac{\partial w}{\partial r}\right)^2 + k_f w^2 \right] r \, dr \, d\theta + \frac{1}{2} \tag{2}$$
$$\cdot ak_\phi \int_0^{2\pi} \left(\frac{\partial w(a, \theta)}{\partial r}\right)^2 d\theta,$$

where $1/k_\phi$ is the rotational flexibility of the spring and flexural rigidities of the plate are

$$D_r = \frac{E_r h^3}{12(1 - v_r v_\theta)},$$
$$\tag{3}$$
$$D_\theta = \frac{E_\theta h^3}{12(1 - v_r v_\theta)}.$$

The work done by the external force $P(r)$ acting on the plate in the direction parallel to z-axis is given by

$$V_{\max} = \int_0^{2\pi} d\theta \int_{r_0}^{r_1} wP(r) r \, dr. \tag{4}$$

3. Method of Solution: Ritz Method

Ritz method requires that the functional

$$J(w) = U_{\max} - V_{\max} - T_{\max} = \frac{1}{2}$$
$$\cdot \int_0^a \int_0^{2\pi} \left[D_r \left\{ \left(\frac{\partial^2 w}{\partial r^2}\right)^2 + 2v_\theta \frac{\partial^2 w}{\partial r^2}\left(\frac{1}{r}\frac{\partial w}{\partial r}\right) \right\} \right.$$
$$\left. + D_\theta \left(\frac{1}{r}\frac{\partial w}{\partial r}\right)^2 + k_f w^2 \right] r \, dr \, d\theta + \frac{1}{2} \tag{5}$$
$$\cdot ak_\phi \int_0^{2\pi} \left(\frac{\partial w(a, \theta)}{\partial r}\right)^2 d\theta - \int_0^{2\pi} d\theta \int_{r_0}^{r_1} wP(r)$$
$$\cdot r \, dr - \frac{1}{2}\rho\omega^2 \int_0^a \int_0^{2\pi} hw^2 r \, d\theta \, dr$$

be minimised.

Uniformly distributed load P_0 extending from r_0 to r_1 is given by

$$P(r) = q_0 = \frac{P_0}{\pi(r_1^2 - r_0^2)}\left[U(r - r_0) - U(r - r_1)\right]. \tag{6}$$

Introducing the nondimensional variables $R = r/a$, $R_1 = r_1/a$, $R_0 = r_0/a$ and assuming the deflection function as $W(R) = w/(a^4 q_0 D_{r_0})$

$$W(R) = \sum_{i=0}^m A_i F_i(R) = \sum_{i=0}^m A_i \left(1 + \alpha_i R^4 + \beta_i R^{1+p}\right) R^{2i}, \tag{7}$$

where A_i are undetermined coefficients, $p^2 = E_\theta/E_r$, and α_i, β_i are unknown constants to be determined from boundary conditions (Leissa [1], p. 14):

$$K_\phi \frac{dF_i(1)}{dR} = -(1 - \alpha)^3 \left[\frac{d^2 F_i}{dR^2} + v_\theta\left(\frac{1}{R}\frac{dF_i}{dR}\right)\right]_{R=1} \tag{8}$$
$$F_i(1) = 0.$$

Linearly thickness variation of the plate is assumed as $h = h_0(1 - \alpha R)$, where α and h_0 are taper parameter and thickness of the plate at the centre, respectively.

The choice of the function approximating the deflection of the plate $W(R)$ given in (7) is based upon the static deflection polar orthotropic plates (Lekhnitskii [11]), which has faster rate of convergence (Gupta et al. [16]) as compared to the polynomial coordinate functions used by earlier researchers [13, 21, 26, 27].

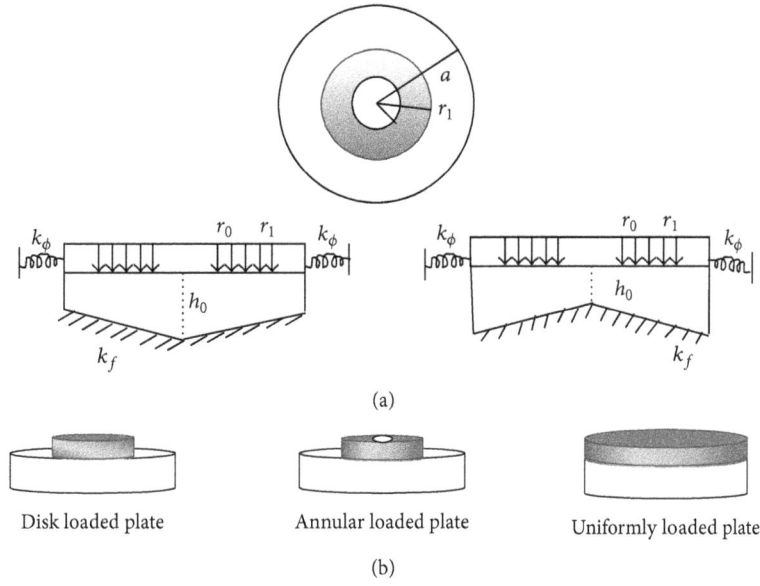

FIGURE 1: (a) Plate geometry with side and surface views. (b) Loading structure of the plate.

Introducing the nondimensional variables W and R in (5), the functional $J(W)$ becomes

$$J(W) = \pi D_{r_0} \left[\int_0^1 (1 - \alpha R)^3 \left\{ \left(\frac{\partial^2 W}{\partial R^2} \right)^2 \right. \right.$$

$$+ 2v_\theta \frac{\partial^2 W}{\partial R^2} \left(\frac{1}{R} \frac{\partial W}{\partial R} \right) + p^2 \left(\frac{1}{R} \frac{\partial W}{\partial R} \right)^2 + K_f W^2 \right\}$$

$$\cdot R\,dR + K_\phi \int_0^{2\pi} \left(\frac{\partial W(1)}{\partial R} \right)^2 d\theta - \int_{R_0}^{R_1} 2W(R) R\,dR$$

$$- \Omega^2 \int_0^1 \int_0^{2\pi} (1 - \alpha R) W^2 R\,dR \right],$$

where

$$D_{r_0} = \frac{E_r h_0{}^3}{12(1 - v_r v_\theta)},$$

$$\Omega^2 = \frac{a^4 \omega^2 \rho h_0}{D_{r_0}},$$

$$K_f = \frac{a^4 k_f}{D_{r_0}},$$

$$K_\Phi = \frac{a k_\Phi}{D_{r_0}}.$$

The minimisation of the functional $J(W)$ given by (9) requires

$$\frac{\partial J(W)}{\partial A_i} = 0, \quad i = 0, 1, 2, 3, \ldots, m. \tag{11}$$

This leads to a system of nonhomogeneous equations in A_j,

$$(a_{ij} - b_{ij}) A_j = C_i, \quad i, j = 0, 1, \ldots, m, \tag{12}$$

where $A = [a_{ij}]$ and $B = [b_{ij}]$ are square matrices of order $m + 1$ given by

$$a_{ij} = \int_0^1 (1 - \alpha R)^3 \left[F_i'' F_j'' + 2v_\theta F_i'' \left(\frac{F_j'}{R} \right) \right.$$

$$+ p^2 \left(\frac{F_i'}{R} \right) \left(\frac{F_j'}{R} \right) + K_f F_i F_j \right] R\,dR + K_\phi F_i'(1)$$

$$\cdot F_j'(1),$$

$$b_{ij} = \int_0^1 (1 - \alpha R) F_i F_j R\,dR,$$

$$C_{ij} = \int_0^1 2 F_i R\,dR.$$

The solution of system of (12) gives the value of A_i and thus the transverse deflection W and the radial and transverse bending moments

$$\frac{M_r}{a^2 q_0} = -(1 - \alpha R)^3 \left[\frac{d^2 W}{dR^2} + v_\theta \frac{1}{R} \frac{dW}{dR} \right],$$

$$\frac{M_\theta}{a^2 q_0} = -(1 - \alpha R)^3 \left[v_\theta \frac{d^2 W}{dR^2} + p^2 \frac{1}{R} \frac{dW}{dR} \right] \tag{14}$$

are computed.

4. Numerical Results

Numerical results have been calculated for different values of taper parameter α ($=\pm 0.3$), E_θ/E_r ($=5.0$), and foundation

(a) Disk loading (SS-plate)

(b) Disk loading (CL-plate)

(c) Annular loading (SS-plate)

(d) Annular loading (CL-plate)

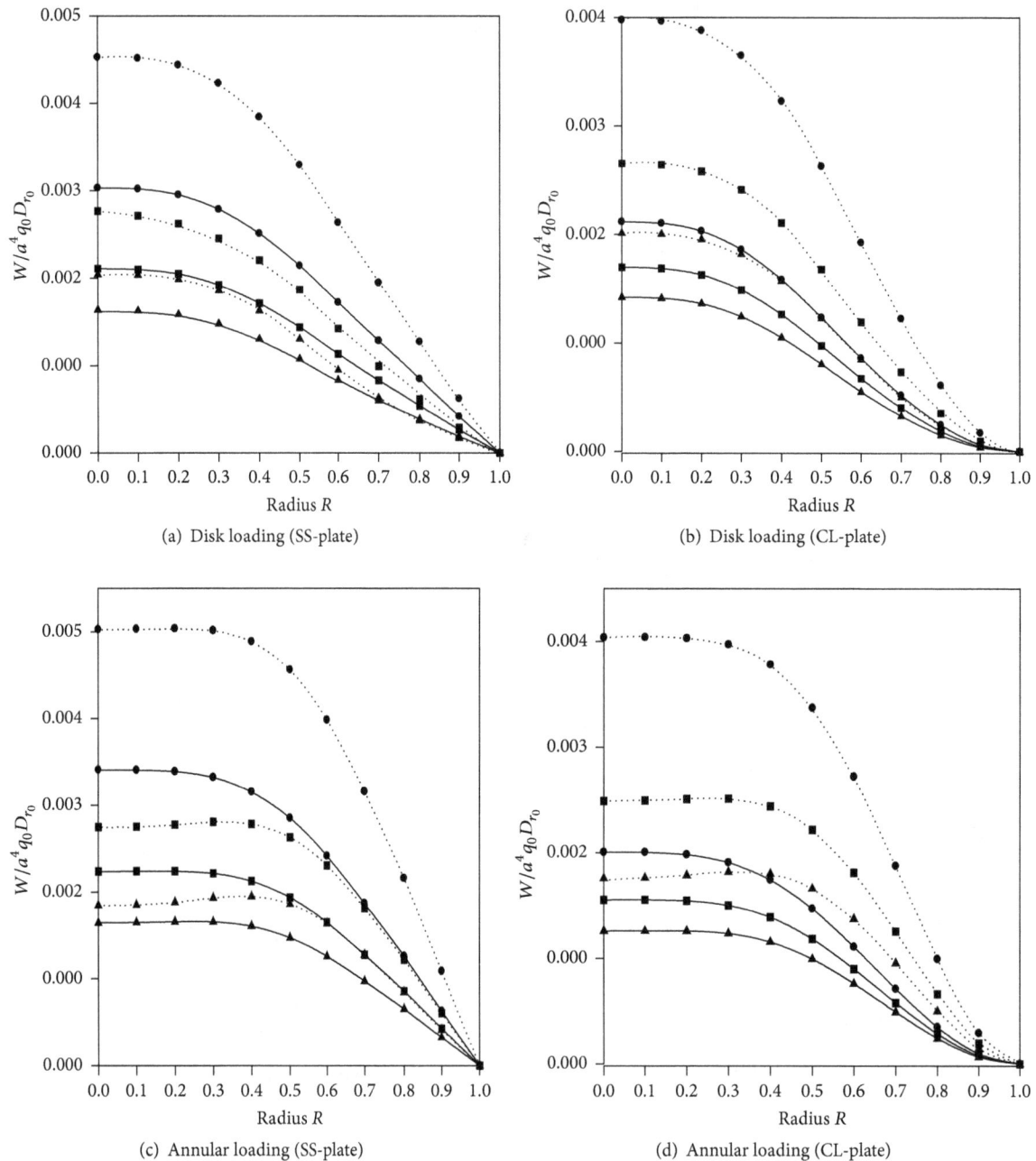

FIGURE 2: Radius vector versus deflection parameter $W/a^4 q_0 D_{r_0}$ for taper parameter $\alpha = -0.3$: — and $\alpha = 0.3$: …. Keys foundation parameter $K_f = 0.01$: •••, $K_f = 0.02$: ■■■, and $K_f = 0.03$: ▲▲▲.

modulus K_f (=0.01, 0.02, 0.03) for forced vibration of polar orthotropic circular plate with simply supported (SS-plate) and clamped (CL-plate) edges. The plate is subjected under different types of loadings such as uniform loading on entire plate, annular loading, and disk loading. The natural frequencies for free vibration are obtained by putting $P(r) = 0$. In case of forced vibration the nondimensional frequency parameter is taken as $\Omega = \eta\Omega_{00}$ for $\Omega < \Omega_{00}$ and $\Omega = \Omega_{00} + \eta(\Omega_{01} - \Omega_{00})$ for $\Omega_{00} < \Omega < \Omega_{01}$, where $\eta = 0.2$. The normalised deflection and bending moments are obtained for fixed value of Poisson's ratio as 0.3.

5. Discussion

Forced axisymmetric response of polar orthotropic circular plates of linearly varying thickness has been analysed for various values of plate parameters. Transverse deflection and bending moments are obtained for circumferentially stiffened plate, that is, $E_\theta > E_r$. Bending moments of the plate cannot be obtained for radially stiffened plate $E_\theta > E_r$, because infinite stress is developed at the centre in this case (Lekhnitskii [11], p. 372). Transverse deflection and bending moments are presented here under different types of loadings:

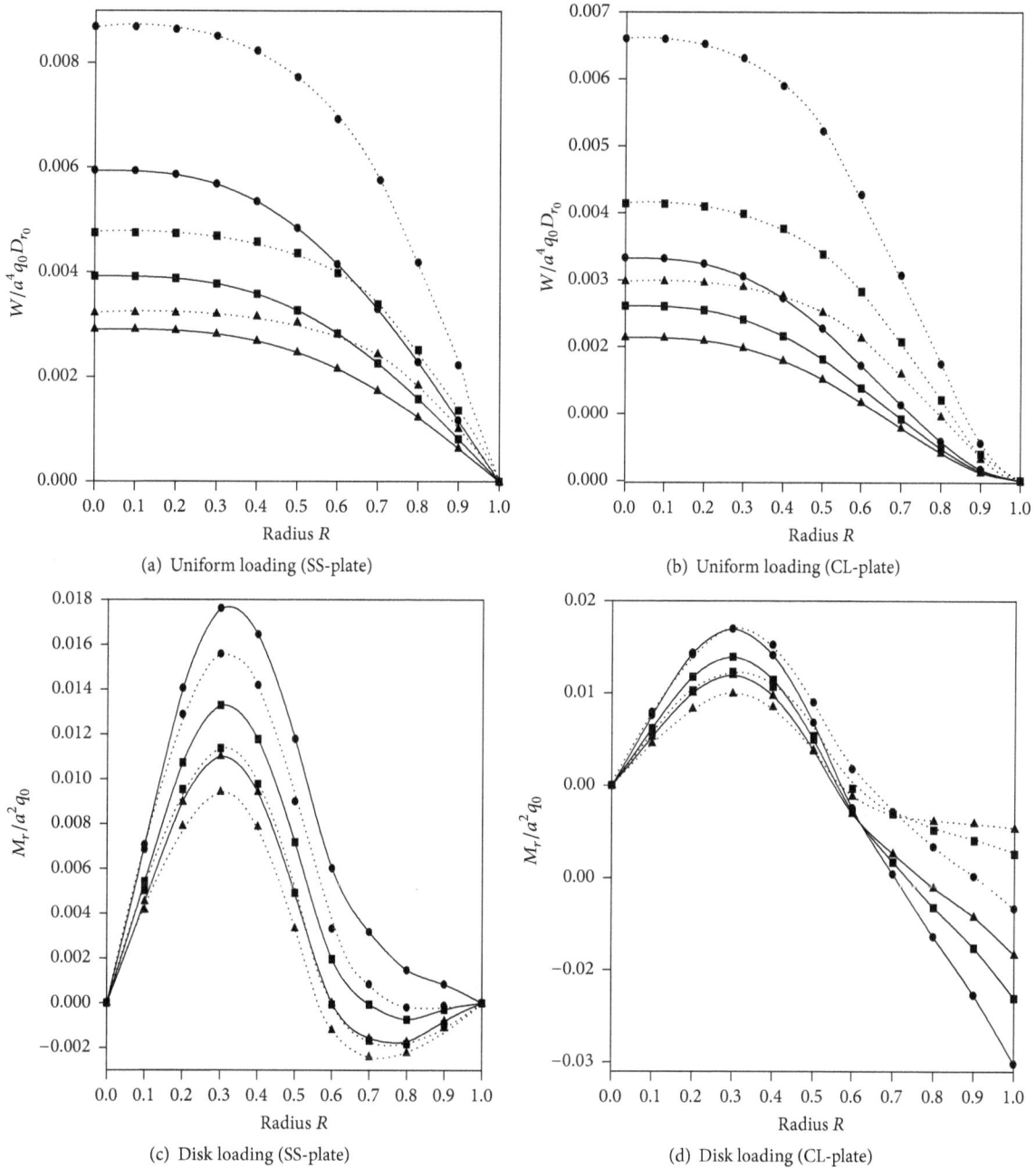

(a) Uniform loading (SS-plate)

(b) Uniform loading (CL-plate)

(c) Disk loading (SS-plate)

(d) Disk loading (CL-plate)

FIGURE 3: Radius vector versus radial bending moment $M_r/a^2 q_0$ for taper parameter $\alpha = -0.3$: — and $\alpha = 0.3$: .... Keys foundation parameter $K_f = 0.01$: •••, $K_f = 0.02$: ■■■, and $K_f = 0.03$: ▲▲▲.

(a) When load is distributed uniformly on the disk extending from $R_0 = 0.0$ to $R_1 = 0.5$ (disk loaded plate).

(b) When load is distributed uniformly on the annular region extending from $R_0 = 0.3$ to $R_1 = 0.7$ (annular loaded plate).

(c) When load is distributed uniformly on the entire region extending from $R_0 = 0.0$ to $R_1 = 1.0$ (uniformly loaded plate).

The total load on the plate is the same in all three cases (shown in Figure 1(b)).

The whole analysis of numerical results is presented in Figures 2, 3, 4, 5, 6(a), and 6(b) for $\Omega < \Omega_{00}$ and $E_\theta/E_r = 5.0$ for different values of plate parameters of simply supported (SS) and clamped (CL) edges. The transverse deflection for all three cases of loading for SS- and CL-plate is presented in Figures 2, 3(a), and 3(b). Figure 2(a) presents the transverse deflection of polar orthotropic simply supported plates along the radius vector R for taper parameter $\alpha = \pm 0.3$ and foundation parameter $K_f = 0.01, 0.02$, and 0.03 under disk loading. The transverse deflection is maximum at the centre of the plate, which decreases as the foundation parameter K_f increases, keeping other plate parameters fixed. Thus the

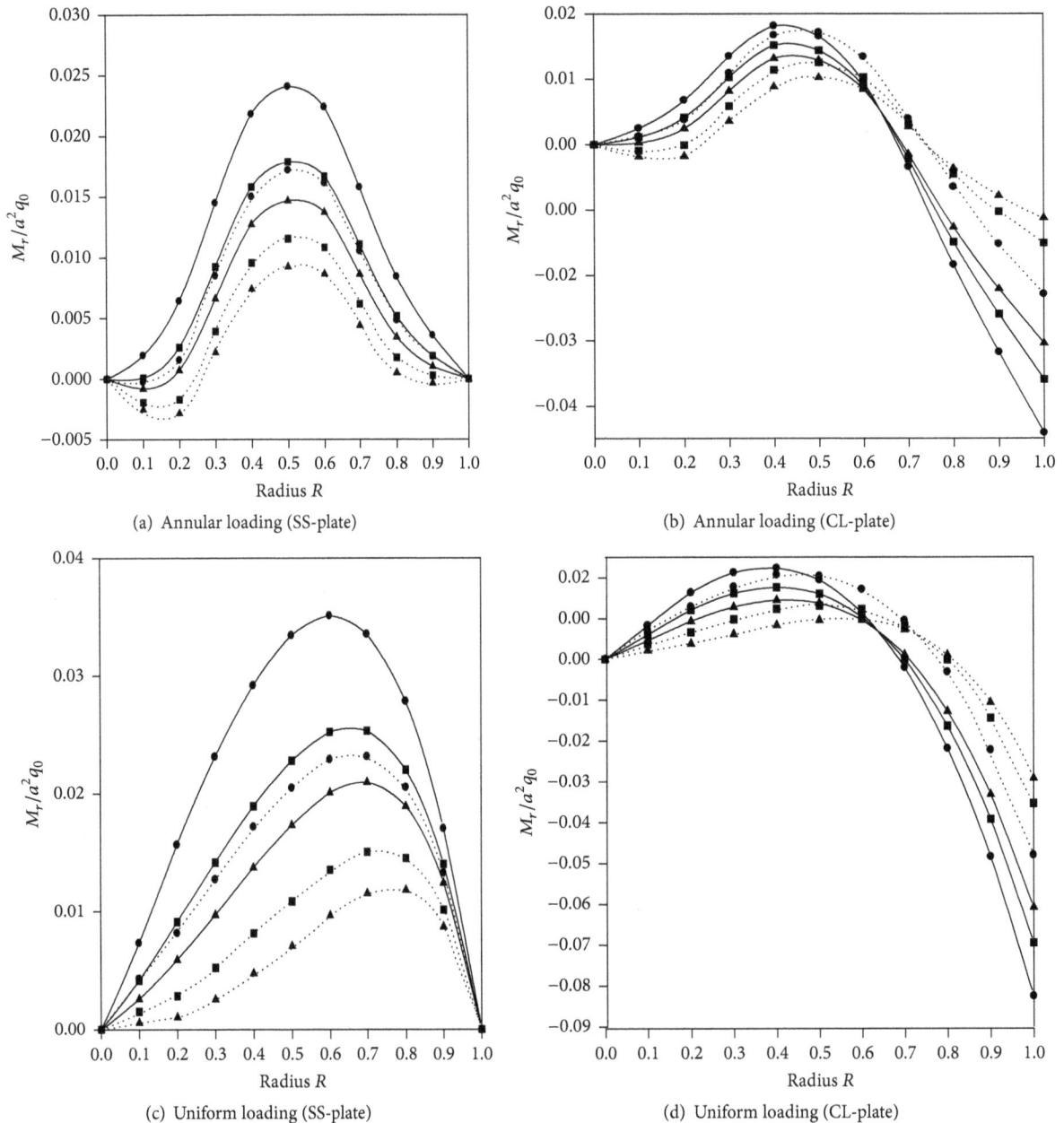

FIGURE 4: Radius vector versus radial bending moment $M_r/a^2 q_0$ for taper parameter $\alpha = -0.3$: — and $\alpha = 0.3$: .... Keys foundation parameter $K_f = 0.01$: • • •, $K_f = 0.02$: ∎∎∎, and $K_f = 0.03$: ▲▲▲.

elastic stability of the plate under loading is more, if some elastic foundation to the structural component is provided. Figure also shows that the transverse deflection for taper parameter $\alpha = 0.3$ (centrally thicker plate) is greater than that for $\alpha = -0.3$ (centrally thinner plate), the reason being greater mass attribution at the centre in case of centrally thicker plate. The centrally thicker plate having higher amplitude of deflection can be protected from being damaged using foundation. This type of structural components is highly used in civil, mechanical, aeronautical, and modern technology. The idea of structural stability in these fields is very essential for design engineer to know before. Figure 2(b) presents that transverse deflection for clamped plate under disk loading

for the same plate parameters as in Figure 2(a). It has been observed that the transverse deflection gradually decreases as the flexibility parameter K_ϕ increases for elastically restrained plate. Thus the transverse deflection for clamped plate is always less than that for simply supported plate, keeping other plate parameters fixed. The effect of foundation parameter K_f on transverse deflection for clamped plate shows that deflection at the centre decreases more rapidly for centrally thicker plate than that for centrally thinner plate, the reason being more mass attribution for centrally thicker plate. Though values are nearly close, the foundation parameter K_f increases nature of variation of transverse deflection of CL-plate which is found to be different. Figures 2(c) and

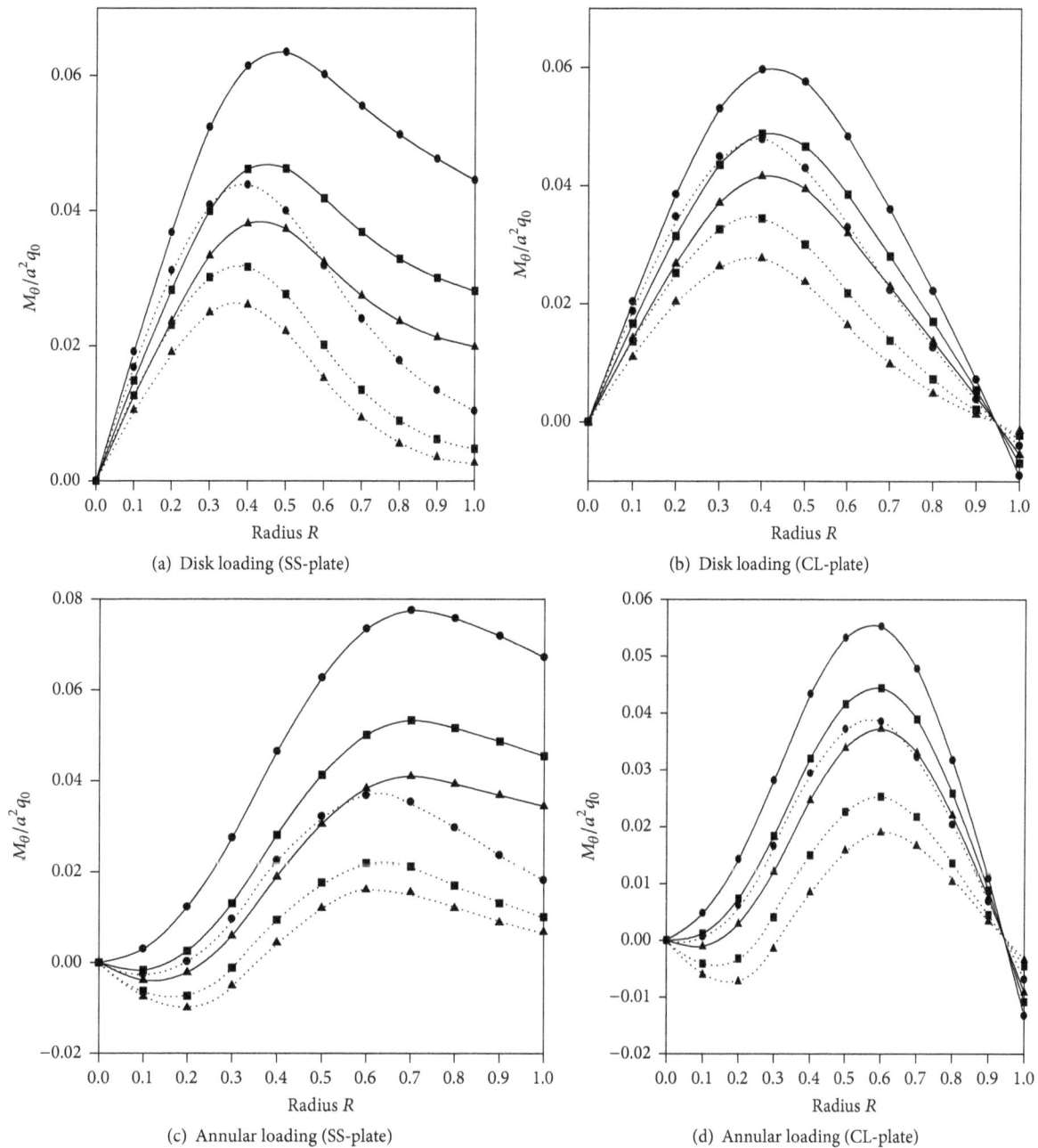

FIGURE 5: Radius vector versus tangential bending moment M_θ/a^2q_0 for taper parameter $\alpha = -0.3$: — and $\alpha = 0.3$: .... Keys foundation parameter $K_f = 0.01$: •••, $K_f = 0.02$: ■■■, and $K_f = 0.03$: ▲▲▲.

2(d) present the transverse deflection when load is annular extending from $R_0 = 0.3$ to $R_1 = 0.7$ for SS- and CL-plates, respectively. Transverse deflection for annular loaded plate is found to be greater than that for disk loaded plate; of course it depends on the position of annular region of loading. The effect on deflection for simply supported plate is more than that of clamped plate. Transverse deflection for uniformly loaded plate is observed to be maximum as compared to disk and annular loaded plates as shown in Figures 3(a) and 3(b).

The radial bending moments for disk, annular, and uniformly loaded SS- and CL-plates are presented in Figures 3(c), 3(d), 4(a), 4(b), 4(c), and 4(d). Radial bending moment for

disk loaded plate is presented in Figures 3(c) and 3(d). Figures show that radial bending moment for $\alpha = -0.3$ is greater than that for $\alpha = 0.3$, other plate parameters being fixed. It is observed that peak of the bending moment decreases as the foundation parameter K_f increases. Maximum radial bending moment occurs at $R = 0.35$ for disk loaded plate for these plate parameter values, whereas in case of annular loaded plate peak of the bending moment occurs at $R = 0.5$. Peak of the bending moment shifts towards the edge $R = 1.0$ for uniformly loaded plate (shown in Figures 4(a) and 4(b)). Radial bending moment at the edge increases as the flexibility parameter K_ϕ increases and is maximum for Cl plate. Radial

(a) Uniform loading (SS-plate)

(b) Uniform loading (CL-plate)

(c) Radius vector versus transverse deflection

(d) Radius vector versus bending moment

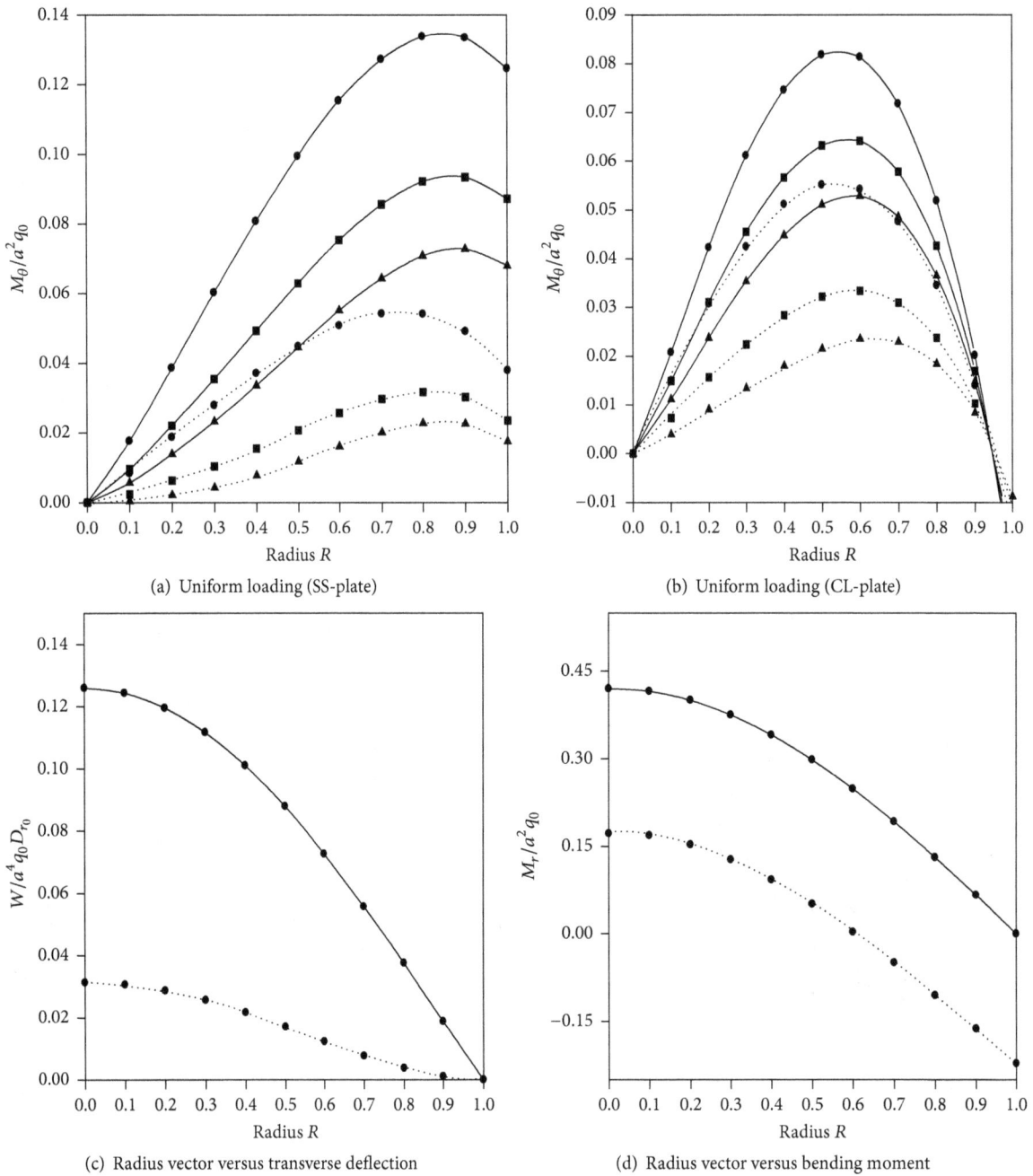

FIGURE 6: (•) values in Figures 6(c) and 6(d) are taken from [26] for SS-plate: — and CL-plate: . . ., respectively.

bending moment for $\alpha = -0.3$ is greater than that for $\alpha = 0.3$, which decreases as the foundation parameter K_f increases. Maximum bending moment at the edge occurs when the plate is uniformly loaded (Figures 4(c) and 4(d)) compared to those of disk and annular loaded plates. Tangential bending moments are also presented in Figures 5(a), 5(b), 5(c), 5(d), 6(a), and 6(b) under three types of loadings. Radial bending moment at the edge for SS orthotropic plate is zero, but the tangential bending moment is nonzero. Tangential bending moment at the edge for SS-plate is greater than that of C1 plate for corresponding value of plate parameters. The present choice of function considered in the study has faster rate

of convergence as compared to the polynomial coordinate function used by earlier researchers [13, 21, 26, 27] (see Figure 7).

Table 1 compares the deflection and radial bending moment at the centre of the plate for $\Omega < \Omega_{00}$ and $\Omega_{00} < \Omega < \Omega_{01}$ for uniform isotropic plate obtained by Laura et al. [24] using Ritz method. Table 2 compares the transverse deflection of simply supported plate of variable thickness. Figure 6(c) compares the transverse deflection with the result of Laura et al. [26] for uniformly loaded isotropic circular plate of constant thickness with simply supported and clamped edge, respectively, obtained by Glarkin's method.

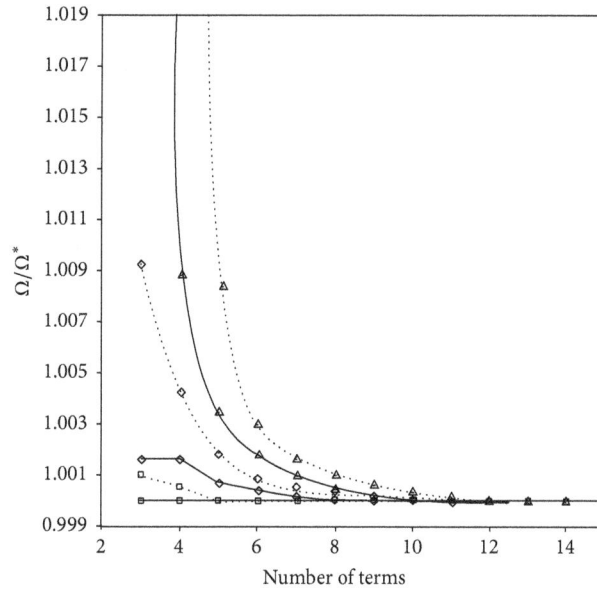

FIGURE 7: Normalised frequency parameter Ω/Ω^* for first three natural frequencies in fundamental mode: $\square\square\square$, second Mode: $\diamond\,\diamond\,\diamond$, and third mode: $\triangle\,\triangle\,\triangle$ for CL-plate with $E_\theta/E_r = 5.0$, $n = 1$, $\alpha = 0.3$. Keys present method: — and polynomial coordinate method: ..., where Ω^* is the frequency obtained using 15 terms.

TABLE 1: Comparison of displacement and radial bending moment at the centre as a function of η ($= \Omega < \Omega_{00}$) and ($= \Omega_{00} < \Omega < \Omega_{01}$) for uniform isotropic circular plate.

	η	$\Omega < \Omega_{00}$				$\Omega_{00} < \Omega < \Omega_{01}$			
		$W/a^4 q_0 D_{r_0}$		$M_r/a^2 q_0$		$W/a^4 q_0 D_{r_0}$		$M_r/a^2 q_0$	
		Reference [24]	Present	Reference [24]	Present	Reference [24]	Present	Reference [24]	Present
CL-plate $\Omega_{00} = 10.2158$ $\Omega_{01} = 39.771$	0.2	0.0160	0.01630	0.085	0.08524	0.0110	0.01181	0.084	0.08468
	0.4	0.01872	0.01873	0.099	0.09952	0.0055	0.00553	0.051	0.05121
	0.6	0.02479	0.02471	0.135	0.13533	0.0040	0.00404	0.039	0.03840
	0.8	0.0446	0.04461	0.253	0.25281	0.0044	0.00446	0.080	0.08001
SS-plate $\Omega_{00} = 4.9351$ $\Omega_{01} = 29.721$	0.2	0.0664	0.06640	0.215	0.21554	0.0226	0.02263	0.094	0.09407
	0.4	0.0760	0.07603	0.248	0.24874	0.0094	0.00946	0.052	0.05224
	0.6	0.1001	0.10010	0.331	0.33179	0.0062	0.006238	0.049	0.04948
	0.8	0.1787	0.17877	0.603	0.60332	0.0063	0.006329	0.076	0.07653

TABLE 2: Comparison of displacements for simply supported circular plate of linearly varying thickness for $\nu_\theta = 0.25$.

Taper parameter α	Reference [26]	Present
−0.5708	0.1817	0.1815
−0.0	0.06568	0.0656
−0.33333	0.1143	0.1127

The bending moments for same plate parameter values as taken by Laura et al. [26] have been also compared and presented in Figure 6(d). Comparison of these obtained results shows an excellent agreement with the result available in the literature.

6. Conclusion

The forced vibrational characteristics of polar orthotropic circular plates resting on Winkler type of elastic foundation have been studied using Ritz method. The function based on static deflection of polar orthotropic plates has been used to approximate the transverse deflection of the plate. The present choice of function has faster rate of convergence as compared to the polynomial coordinate function used by earlier researchers. Study shows that the deflection attains maximum value at the centre of the plate under uniform loading. The effect is influenced by thickness variation. Centrally thicker plate has more deflection than that of centrally thinner plate, the region being the more mass attribution at the centre in case of centrally thicker plate. While designing a structural component, stability is one of the important factors to be considered. The result analysis shows that the consideration of elastic foundation can protect the system from fatigue failure under heavy loads.

Competing Interests

The author declares that they have no competing interests.

References

[1] A. W. Leissa, "Vibration of plates," Tech. Rep. sp-160, NASA, 1969.

[2] A. W. Leissa, "Recent research in plate vibrations: classical theory," *The Shock and Vibration Digest*, vol. 9, no. 10, pp. 13–24, 1977.

[3] A. W. Leissa, "Recent research in plate vibrations 1973–1976: complicating effects," *The Shock and Vibration Digest*, vol. 10, no. 12, pp. 21–35, 1978.

[4] A. W. Leissa, "Literature review: plate vibration research, 1976–1980: classical theory," *The Shock and Vibration Digest*, vol. 13, no. 9, pp. 11–22, 1981.

[5] A. W. Leissa, "Plate vibrations research 1976–1980: complicating effects," *The Shock and Vibration Digest*, vol. 13, no. 10, pp. 19–36, 1981.

[6] A. W. Leissa, "Recent studies in plate vibrations 1981–1985, part I: classical theory," *The Shock and Vibration Digest*, vol. 19, no. 2, pp. 11–18, 1987.

[7] A. W. Leissa, "Recent studies in plate vibrations 1981–1985, part II: complicating effects," *The Shock and Vibration Digest*, vol. 19, no. 3, pp. 10–24, 1987.

[8] J. N. Reddy, *Mechanics of Laminated Composite Plates: Theory and Analysis*, CRP Press, Boca Raton, Fla, USA, 1997.

[9] R. Lal and U. S. Gupta, "Chebyshev polynomials in the study of transverse vibrations of non-uniform rectangular orthotropic plates," *The Shock and Vibration Digest*, vol. 33, no. 2, pp. 103–112, 2001.

[10] A. S. Sayyad and Y. M. Ghugal, "On the free vibration analysis of laminated composite and sandwich plates: a review of recent literature with some numerical results," *Composite Structures*, vol. 129, pp. 177–201, 2015.

[11] S. G. Lekhnitskii, *Anisotropic Plates*, Breach Science, New York, NY, USA, 1985.

[12] S. Chonan, "Random vibration of an initially stressed thick plate on an elastic foundation," *Journal of Sound and Vibration*, vol. 71, no. 1, pp. 117–127, 1980.

[13] P. A. A. Laura, R. H. Gutierrez, R. Carnicer, and H. C. Sanzi, "Free vibrations of a solid circular plate of linearly varying thickness and attached to a winkler foundation," *Journal of Sound and Vibration*, vol. 144, no. 1, pp. 149–161, 1991.

[14] U. S. Gupta, R. Lal, and S. K. Jain, "Effect of elastic foundation on axisymmetric vibrations of polar orthotropic circular plates of variable thickness," *Journal of Sound and Vibration*, vol. 139, no. 3, pp. 503–513, 1990.

[15] A. H. Ansari and U. S. Gupta, "Effect of elastic foundation on asymmetric vibration of polar orthotropic parabolically tapered circular plate," *Indian Journal of Pure and Applied Mathematics*, vol. 30, no. 10, pp. 975–990, 1999.

[16] U. S. Gupta, A. H. Ansari, and S. Sharma, "Buckling and vibration of polar orthotropic circular plate resting on Winkler foundation," *Journal of Sound and Vibration*, vol. 297, no. 3–5, pp. 457–476, 2006.

[17] K. M. Liew, J.-B. Han, Z. M. Xiao, and H. Du, "Differential quadrature method for Mindlin plates on Winkler foundations," *International Journal of Mechanical Sciences*, vol. 38, no. 4, pp. 405–421, 1996.

[18] T. M. Wang and J. E. Stephens, "Natural frequencies of Timoshenko beams on pasternak foundations," *Journal of Sound and Vibration*, vol. 51, no. 2, pp. 149–155, 1977.

[19] B. Bhattacharya, "Free vibration of plates on Vlasov's foundation," *Journal of Sound and Vibration*, vol. 54, no. 3, pp. 464–467, 1977.

[20] A. K. Upadhyay, R. Pandey, and K. K. Shukla, "Nonlinear dynamic response of laminated composite plates subjected to pulse loading," *Communications in Nonlinear Science and Numerical Simulation*, vol. 16, no. 11, pp. 4530–4544, 2011.

[21] P. A. A. Laura, G. C. Pardoen, L. E. Luisoni, and D. Ávalos, "Transverse vibrations of axisymmetric polar orthotropic circular plates elastically restrained against rotation along the edges," *Fibre Science and Technology*, vol. 15, no. 1, pp. 65–77, 1981.

[22] D. J. Gunaratnam and A. P. Bhattacharya, "Transverse vibrations and stability of polar orthotropic circular plates: high-level relationships," *Journal of Sound and Vibration*, vol. 132, no. 3, pp. 383–392, 1989.

[23] G. N. Weisensel and A. L. Schlack Jr., "Annular plate response to circumferentially moving loads with sudden radial position changes," *The International Journal of Analytical and Experimental Modal Analysis*, vol. 5, no. 4, pp. 239–250, 1990.

[24] P. A. A. Laura, D. R. Avalos, and H. A. Larrondo, "Force vibrations of circular, stepped plates," *Journal of Sound and Vibration*, vol. 136, no. 1, pp. 146–150, 1990.

[25] G. C. Pardoen, "Asymmetric vibration and stability of circular plates," *Computers and Structures*, vol. 9, no. 1, pp. 89–95, 1978.

[26] P. A. A. Laura, C. Filipich, and R. D. Santos, "Static and dynamic behavior of circular plates of variable thickness elastically restrained along the edges," *Journal of Sound and Vibration*, vol. 52, no. 2, pp. 243–251, 1977.

[27] R. H. Gutierrez, E. Romanelli, and P. A. A. Laura, "Vibrations and elastic stability of thin circular plates with variable profile," *Journal of Sound and Vibration*, vol. 195, no. 3, pp. 391–399, 1996.

[28] B. Bhushan, G. Singh, and G. Venkateswara Rao, "Asymmetric buckling of layered orthotropic circular and annular plates of varying thickness using a computationally economic semianalytical finite element approach," *Computers and Structures*, vol. 59, no. 1, pp. 21–33, 1996.

[29] U. S. Gupta and A. H. Ansari, "Asymmetric vibrations and elastic stability of polar orthotropic circular plates of linearly varying profile," *Journal of Sound and Vibration*, vol. 215, no. 2, pp. 231–250, 1998.

[30] C. S. Kim and S. M. Dickinson, "The flexural vibration of thin isotropic and polar orthotropic annular and circular plates with elastically restrained peripheries," *Journal of Sound and Vibration*, vol. 143, no. 1, pp. 171–179, 1990.

[31] S. Sharma, R. Lal, and N. Singh, "Effect of non-homogeneity on asymmetric vibrations of non-uniform circular plates," *Journal of Vibration and Control*, 2015.

2

Design of Corrugated Plates for Optimal Fundamental Frequency

Nabeel Alshabatat

Department of Mechanical Engineering, Tafila Technical University, Tafila 66110, Jordan

Correspondence should be addressed to Nabeel Alshabatat; nabeel963030@yahoo.com

Academic Editor: Marc Thomas

This paper investigates shifting the fundamental frequency of plate structures by corrugation. Creating corrugations significantly improves the flexural rigidities of plate and hence increases its natural frequencies. Two types of corrugations are investigated: sinusoidal and trapezoidal corrugations. The finite element method (FEM) is used to model the corrugated plates and extract the natural frequencies and mode shapes. The effects of corrugation geometrical parameters on simply supported plate fundamental frequency are studied. To reduce the computation time, the corrugated plates are modeled as orthotropic flat plates with equivalent rigidities. To demonstrate the validity of modeling the corrugated plates as orthotropic flat plates in studying the free vibration characteristics, a comparison between the results of finite element model and equivalent orthotropic models is made. A correspondence between the results of orthotropic models and the FE models is observed. The optimal designs of sinusoidal and trapezoidal corrugated plates are obtained based on a genetic algorithm. The optimization results show that plate corrugations can efficiently maximize plate fundamental frequency. It is found that the trapezoidal corrugation can more efficiently enhance the fundamental frequency of simply supported plate than the sinusoidal corrugation.

1. Introduction

Plate structures have been widely used in structural, naval, automobile, and aerospace engineering. However, plates exhibit poor vibration performance due to their lateral flexibility (i.e., plates can be deformed easily in the out-of-plane direction under static or dynamic loads). The vibration properties of plates can be modified by different methods. One objective of these methods is to shift the plate natural frequencies away from the frequency of the excitation force to avoid resonance. One cost effective method to improve the vibration characteristics of plates is to modify their shapes [1]. In this study, shape modification is used to optimize the vibration characteristics of plates. In particular, corrugation is used in plate construction to maximize the first plate natural frequency.

Corrugation of plate-like structures is commonly used to increase their strength for static loading conditions such as roofing in engineering structures, container walls, and bridge bulkheads. In the late 1960s, NASA developed corrugated panel structures to use them in many aerospace applications such as the space shuttle and hypersonic aircraft. The corrugated nature of these panels makes them suitable for high temperature applications because the curved sections permit panel thermal expansion without inducing a significant thermal stress [2]. In 1970, Plank et al. [3] studied the best primary structure of a Mach 8 hypersonic transport wing; and they examined the ultimate load, wing flutter, panel flutter, fatigue, and creep for different wing structures. Recently, corrugated plates are used in constructing morphing wings [4].

A precise analysis of corrugated plates can be achieved by using the finite element method. However, the finite element method does not efficiently model the corrugated plates because it requires significant computing times. A more practical method to design and analyze corrugated plates is to treat them as orthotropic flat plates with equivalent rigidities (i.e., the corrugated plate is flexible in the corrugation direction and rigid in the crosswise direction). After that, an approximate solution can be obtained by solving

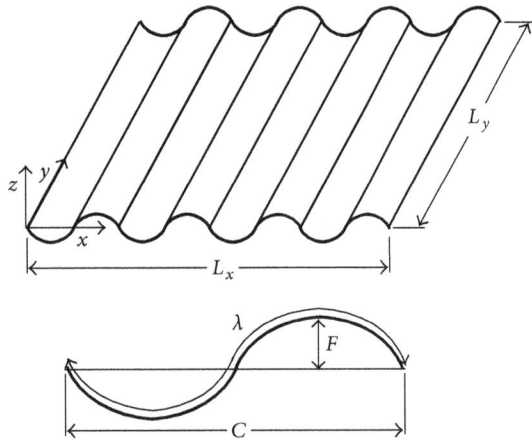

FIGURE 1: Sinusoidal corrugated plate.

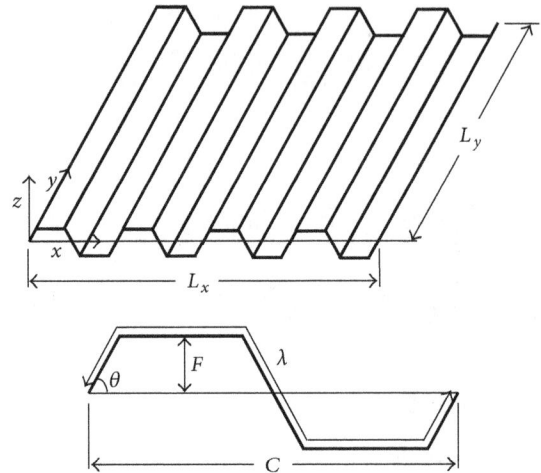

FIGURE 2: Trapezoidal corrugated plate.

the equivalent orthotropic plate problem. This approach of approximation is computationally more efficient with a little loss of precision.

The literature is rich in studies related to the behavior of corrugated plates under static loading. However, there is a limited number of papers which study the vibration of such plates. Samanta and Mukhopadhyay [5] presented formulas for the equivalent rigidities of trapezoidal corrugated plates based on energy principles; and they carried out free vibration analysis using finite element method. In addition, they carried out a nonlinear geometric analysis. Based on the nonlinear bending theory of thin shallow shells, Wang et al. [6] investigated the large amplitude vibration of corrugated circular plates with shallow sinusoidal corrugations under temperature changes. Hamilton's principle was used to derive the governing equations. Liew et al. [7] studied the free vibration of stiffened and unstiffened corrugated plates by using a mesh-free Galerkin method based on the first order shear deformation theory. Yucel and Arpaci [8] used the finite element method to study the free vibration of different types of trapezoidal and sinusoidal corrugated plates.

The objective of this study is to design corrugated plates and to optimize their fundamental frequencies. A free vibration analysis of sinusoidal and trapezoidal corrugated plates based on both the finite element method and equivalent orthotropic models is made in this paper. To demonstrate the validity of modeling the corrugated plates as orthotropic flat plates in studying the free vibration characteristics, a comparison between the results of the finite element model and equivalent orthotropic models is made. Parametric analysis and optimization examples are studied to provide a guide for design engineers. The optimal designs of the corrugated plates are found by using a genetic algorithm (GA).

2. Theoretical Background

Corrugated plates are assumed to have sinusoidal or trapezoidal corrugation in one direction as shown in Figures 1 and 2. The corrugated plate can be described by the number of corrugation N_c, the pitch C, the height F, and the trough

angle θ (for trapezoidal corrugation). In this work, FEM is used for the modal analysis of corrugated plates. In addition to FEM, the corrugated plates are studied analytically based on modeling the corrugated plates as orthotropic flat plates with the same thickness and equivalent rigidities.

2.1. Free Vibration Analysis of Orthotropic Plate. Using Love-Kirchhoff's hypotheses, the equation of the free vibration of orthotropic rectangular thin plate of length L_x and width L_y can be written as

$$D_{11}\frac{\partial^4 w}{\partial x^4} + 2H\frac{\partial^4 w}{\partial x^2\partial y^2} + D_{22}\frac{\partial^4 w}{\partial y^4} + \rho h\frac{\partial^2 w}{\partial t^2} = 0, \quad (1)$$

where w is the lateral displacement of the plate at point (x, y); D_{11} and D_{22} are the flexural rigidities in the x and y directions, respectively; H is the equivalent rigidity; h is the thickness; and ρ is the mass density of the plate. The solution of the harmonic motion of the plate has the form

$$w(x, y, t) = W(x, y)\cos(\omega t), \quad (2)$$

where $W(x, y)$ is the natural mode. Substituting (2) into (1) results in the following:

$$D_{11}\frac{\partial^4 W}{\partial x^4} + 2H\frac{\partial^4 W}{\partial x^2\partial y^2} + D_{22}\frac{\partial^4 W}{\partial y^4} - \rho h\omega^2 W = 0. \quad (3)$$

The solution of (3) depends on the boundary conditions of the plate. For a rectangular plate simply supported on all the four sides, the solution of (3), which satisfies the boundary conditions, can be written as

$$W(x, y) = \sum_{m=1}^{\infty}\sum_{n=1}^{\infty}A_{mn}\sin\left(\frac{m\pi x}{L_x}\right)\sin\left(\frac{n\pi y}{L_y}\right), \quad (4)$$

where A_{mn} is a scale factor of (m, n) mode shape. Substitution of (4) into (3) gives the circular frequency ω_{mn} which is associated to the W_{mn} mode shape as

$$\omega_{mn} = \frac{\pi^2}{\sqrt{\rho h}} \cdot \sqrt{D_{11} \left(\frac{m}{L_x}\right)^4 + 2H \left(\frac{m}{L_x}\right)^2 \left(\frac{n}{L_y}\right)^2 + D_{22} \left(\frac{n}{L_y}\right)^4}. \tag{5}$$

After calculating the equivalent rigidities of the corrugated plate, the free vibration characteristics can be estimated by solving (1) either theoretically for some boundary conditions or numerically for others.

2.2. Equivalent Orthotropic Properties of Corrugated Plates.
As mentioned previously, corrugated plates can be approximated as orthotropic flat plates. The suitable selection of the equivalent rigidities plays a significant role in the accuracy of the model. There are different equivalent rigidities available in the literature. The most complete and the most recent equivalent rigidities are adopted in this study to calculate the natural frequencies. The equivalent models are summarized here. The derivation of these models is not considered to be within the scope of this study.

The earliest complete estimation of the equivalent rigidities for sinusoidal corrugation known to the author is found in the works of Huber [9] and Seydel [10]. For many years, researchers followed the classical formulas of Huber and Seydel. Briassoulis [11] reviewed the classical formulas for the equivalent rigidities and developed more accurate expressions for the rigidities of corrugated plates:

$$D_{11} = \frac{C}{\lambda} \frac{Eh^3}{12(1-\nu^2)},$$

$$D_{22} = \frac{EhF^2}{2} + \frac{Eh^3}{12(1-\nu^2)},$$

$$D_{66} = \frac{Eh^3}{24(1+\nu)}, \tag{6}$$

$$D_{12} = \nu D_{11},$$

where C is the corrugation pitch; F is the height of corrugation (see Figure 1); λ is the developed length of unit corrugation; I is the moment of inertia of the plate along the corrugation direction; E and ν are the modulus of elasticity and Poisson's ratio of plate material, respectively. For trapezoidal corrugated plates, Samanta and Mukhopadhyay [5] presented the following bending rigidities:

$$D_{11} = \frac{C}{\lambda} \frac{Eh^3}{12},$$

$$D_{22} = \frac{EI}{C},$$

$$D_{66} = \frac{\lambda}{C} \frac{Eh^3}{24(1+\nu)},$$

$$D_{12} = 0. \tag{7}$$

Liew et al. [7] used the following formulas for flexural rigidities of trapezoidal corrugated plates:

$$D_{11} = \frac{C}{\lambda} \frac{Eh^3}{12(1-\nu^2)},$$

$$D_{22} = \frac{Eh}{C}\alpha + \frac{Eh^3}{12(1-\nu^2)},$$

$$D_{66} = \frac{Eh^3}{24(1+\nu)}, \tag{8}$$

$$D_{12} = \nu D_{11},$$

where α is a parameter that depends on the geometry of corrugation as given in [12].

Xia et al. [13] investigated the equivalent flexural rigidities for the geometry of arbitrary corrugations. The suggested bending rigidities are

$$D_{11} = \frac{C}{\lambda} \frac{Eh^3}{12(1-\nu^2)},$$

$$D_{22} = \frac{EI}{1-\nu^2} + A\frac{Eh^3}{12(1-\nu^2)},$$

$$D_{66} = \frac{\lambda}{C} \frac{Eh^3}{24(1+\nu)}, \tag{9}$$

$$D_{12} = \nu D_{11},$$

where A depends on the geometry of corrugation. For sinusoidal corrugation, A is given by

$$A = \int_{-0.5}^{0.5} \frac{1}{\sqrt{1 + ((2\pi T/C)\cos(2\pi X))^2}} dX. \tag{10}$$

For trapezoidal corrugation A is given by

$$A = C - \frac{8}{3} \frac{F}{\tan(\theta)}. \tag{11}$$

Very recently, Ye et al. [14] derived an equivalent plate model for general corrugation. For shallow sinusoidal corrugation, the bending rigidities are

$$D_{11} = \frac{C}{\lambda} \frac{Eh^3}{12(1-\nu^2)},$$

$$D_{22} = EI + A\frac{Eh^3}{12} + \nu^2 D_{11},$$

$$D_{66} = \frac{\lambda}{C} \frac{Eh^3}{12}, \tag{12}$$

$$D_{12} = \nu D_{11}.$$

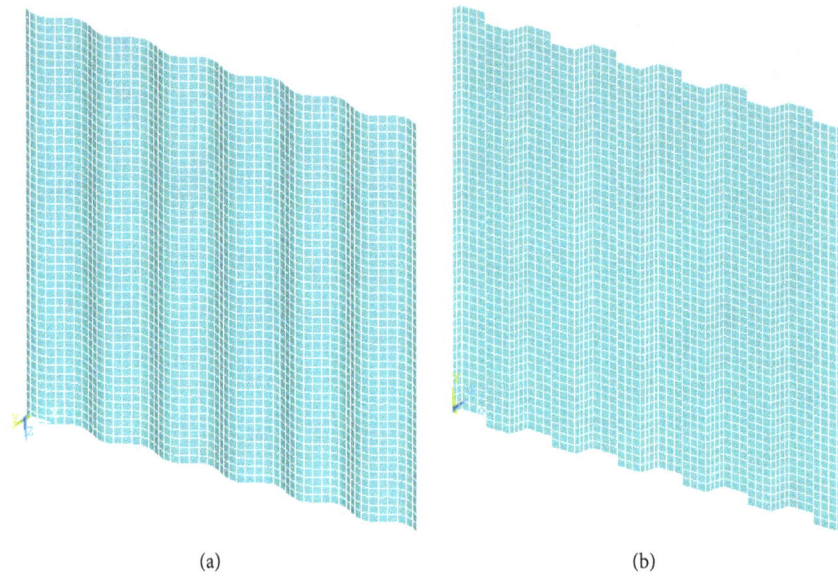

FIGURE 3: Typical finite element models for (a) sinusoidal and (b) trapezoidal corrugated plates.

Approximate values for the moment of inertia I and the developed length of a unit corrugation λ are available in various references [15, 16]. For more accuracy, the values of I and λ, in this study, are evaluated numerically. The mass density used in the equivalent orthotropic models is estimated by (λ/C) multiplied by the actual mass density of the corrugated plate.

3. Parametric Analysis

To design a corrugated plate, three independent geometrical parameters need to be determined such as the thickness of the plate h, the number of corrugation N, and the corrugation height F, or the corrugation pitch C instead of number of corrugation N, in addition to the trough angle θ for trapezoidal corrugation. The plate under consideration is a simply-supported square plate with length $L_x = L_y = 1$ m and thickness $h = 0.001$ m that is made of steel with modulus of elasticity $E = 210$ GPa, Poisson's ratio $\nu = 0.3$, and mass density $\rho = 7800$ kg/m^3.

The finite element method (FEM) is used for the modal analysis of corrugated plates. In particular, the plates are modelled by using ANSYS Parametric Design Language (APDL) in which the model can be built in terms of the geometrical parameters of corrugation. The shell element "shell63" with four nodes and six degrees of freedom per node is used to mesh the solid models. Convergence studies are performed and the finite element of 10 mm size is used throughout the parametric analysis. Typical finite element meshes of sinusoidal and trapezoidal corrugated plates are shown in Figure 3.

To gain some knowledge about the fundamental frequency of corrugated plates, it is important to study the effect of changing the corrugation geometrical parameters on the plate fundamental frequency. To study the effect of the corrugation height of sinusoidally corrugated plate on

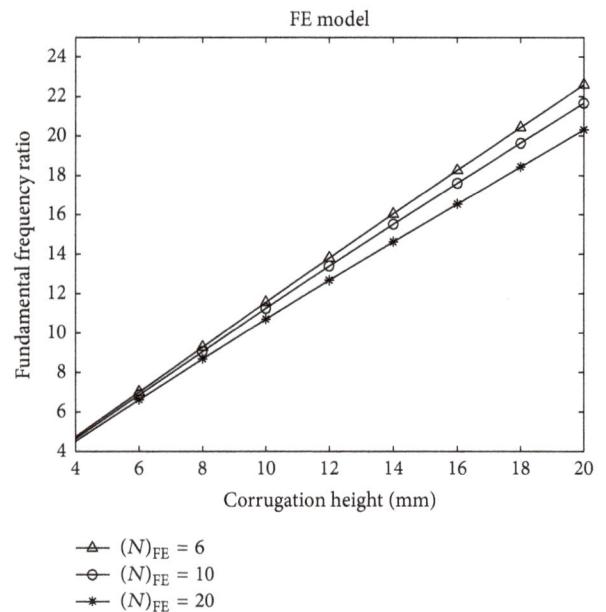

FIGURE 4: Effect of corrugation height on the fundamental frequency ratio for different number of corrugations (sinusoidal corrugation).

the fundamental frequency ratio, the number of corrugation is held constant (e.g., $N = 6$ or 10 or 20 corrugations). The corrugation height varies from 5 mm to 20 mm. The results are shown in Figure 4. The fundamental frequency ratio represents the ratio between the fundamental frequency of the corrugated plate and the fundamental frequency of the flat plate with the same lengths and thickness. Increasing the corrugation height would significantly increase the fundamental frequency ratio of the sinusoidally corrugated plate. This increase in the fundamental frequency resulted

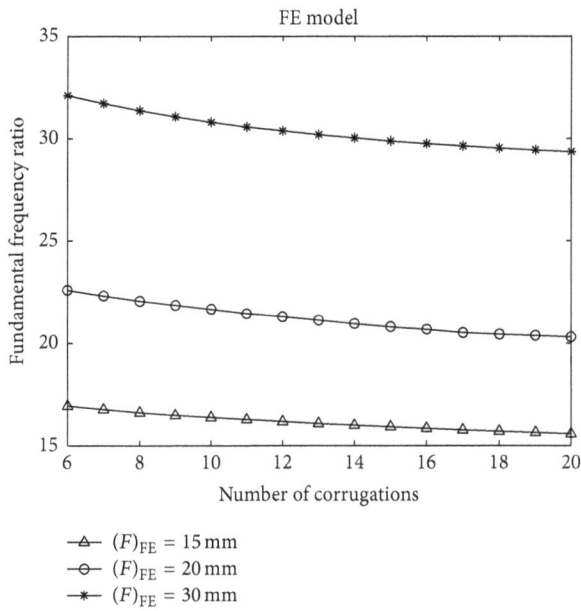

FIGURE 5: Effect of number of corrugations on the fundamental frequency ratio for different corrugation heights (sinusoidal corrugation).

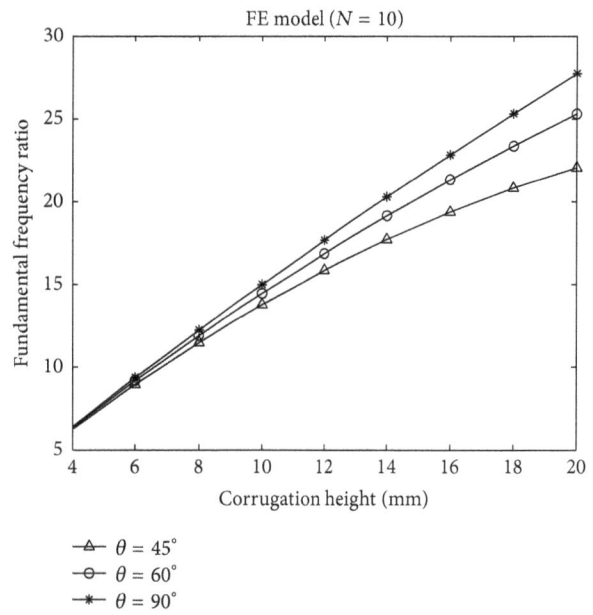

FIGURE 6: Effect of corrugation height on the fundamental frequency ratio for different trough angles (trapezoidal corrugation).

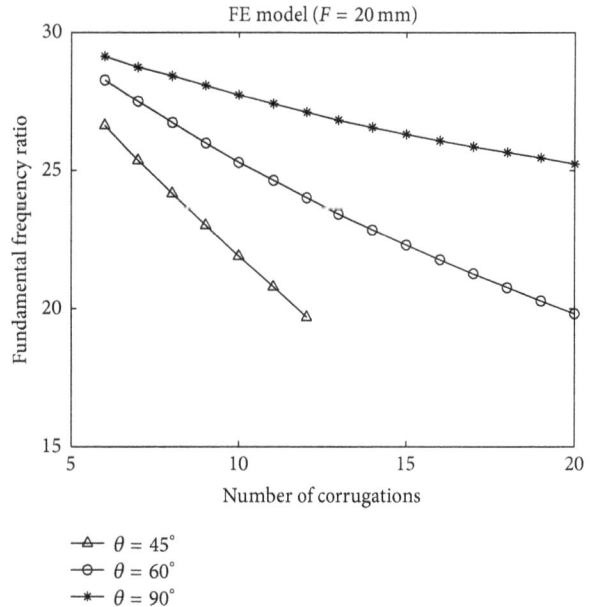

FIGURE 7: Effect of number of corrugations on the fundamental frequency ratio for different trough angles (trapezoidal corrugation).

from the obvious increase in the moment of inertia due to the corrugation. Figure 5 shows the change in the fundamental frequency ratio of sinusoidally corrugated plates with varying the number of corrugations from 6 to 20 and the height of corrugation is held constant (e.g., $F = 15$ or 20 or 30 mm). It is shown that, for small corrugation height, the frequency ratio decreases as N increases though the frequencies are still higher than those of the uncorrugated plate. For a higher corrugation number, increasing the number of corrugations ($6 \leq N \leq 14$) decreases the fundamental frequency ratio. In all cases, the significant increase in the number of corrugations has a minor effect on the fundamental frequency ratio.

The effects of changing the corrugation height, number, and the trough angle on the fundamental frequency ratio of trapezoidally corrugated plate are shown in Figures 6 and 7. Figure 6 shows that the fundamental frequency ratio of trapezoidally corrugated plate increases significantly by increasing the corrugation height and the trough angle. For small corrugation height, the trough angles have almost the same effect on the fundamental frequency ratio. As shown in Figure 7, the fundamental frequency ratio of trapezoidally corrugated plate is sensitive to the increase in the number of corrugations, especially for plates with small trough angles. By increasing the number of corrugations, the fundamental frequency ratio decreases significantly when $\theta = 45°$.

In addition to changing the natural frequencies, creating corrugations changes the plate mode shapes. For example, the first four mode shapes of a simply supported plate before and after creating 10 sinusoidal corrugations with 20 mm heights are shown in Figure 8. It is expected that this change in mode shapes results in a change in the sound radiation when the plate is subjected to an excitation force.

Modeling and vibration analysis of corrugated plates using FEM can be very computational expensive and time consuming especially for the cases of high number of corrugations (i.e., high number of corrugations per unit length requires finer mesh). Thus, using FEM in design optimization or in any rigorous analysis of corrugated plates is inefficient. Treating the corrugated plate as a flat orthotropic plate is an economic alternative for using FEM. In the optimization

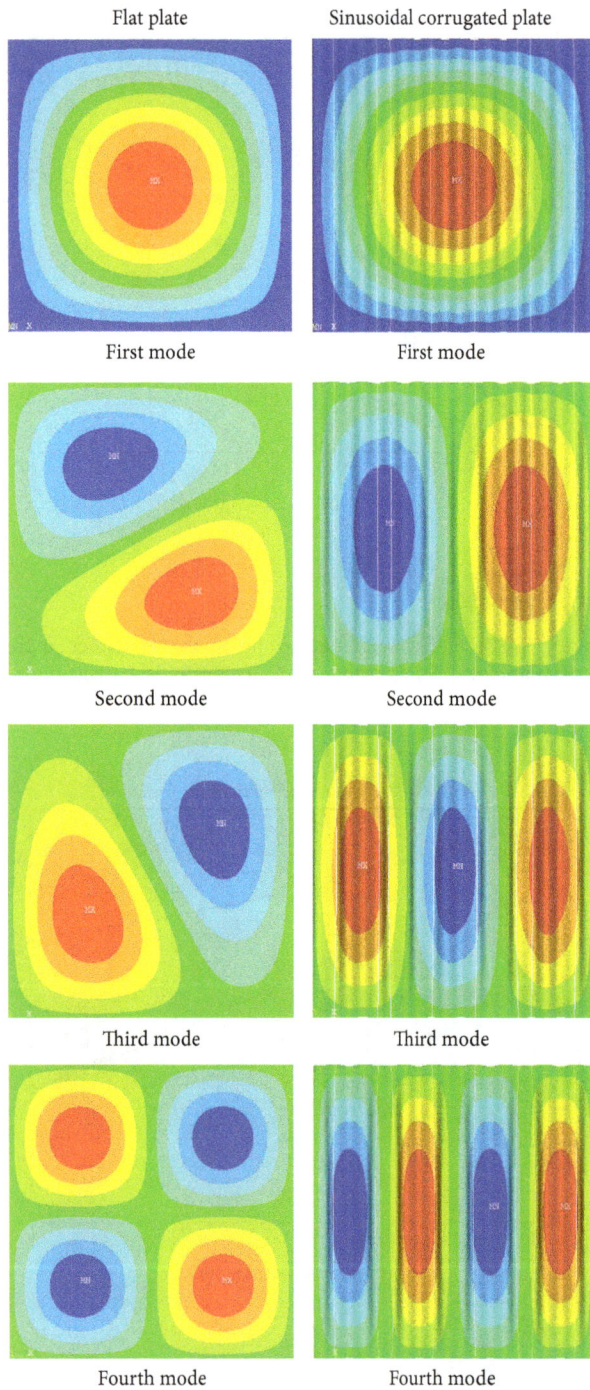

Flat plate

Sinusoidal corrugated plate

First mode

First mode

Second mode

Second mode

Third mode

Third mode

Fourth mode

Fourth mode

FIGURE 8: First four mode shapes of flat and sinusoidal corrugated simply supported plate.

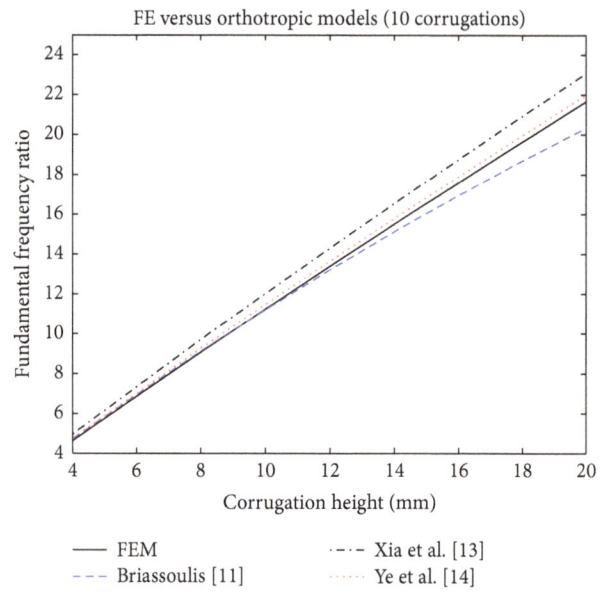

FIGURE 9: Effect of corrugation height on the fundamental frequency ratio based on finite element and equivalent orthotropic models (sinusoidal corrugation).

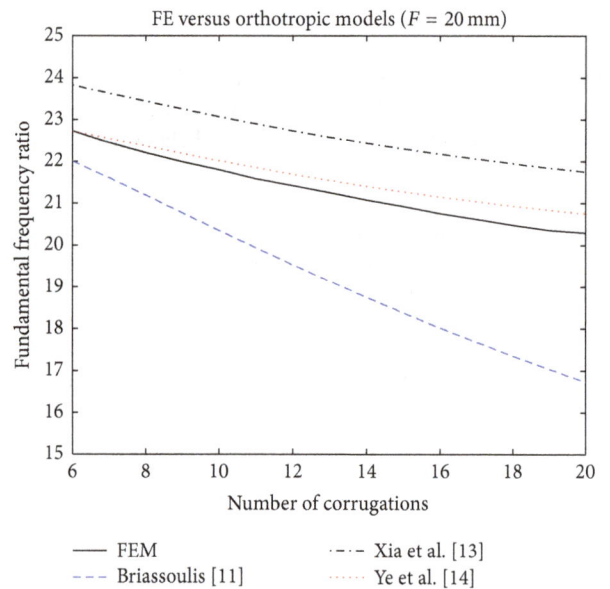

FIGURE 10: Effect of number of corrugations on the fundamental frequency ratio based on finite element and equivalent orthotropic models (sinusoidal corrugation).

section, the corrugated plates will be treated as orthotropic plates with equivalent rigidities. To select the most appropriate equivalent rigidities for the current study, the effect of geometric parameters of corrugation on plate fundamental frequency is investigated using equivalent orthotropic flat plate models. The rigidities in these models are based on the most complete and most recent equivalent rigidities ((6)–(12)).

The accuracy of the equivalent orthotropic models for corrugated plates is examined by comparing their results with those of the FE models. Comparison of the fundamental frequency ratio of the sinusoidal corrugated plates, computed by FEM and different orthotropic models based on the equivalent rigidities of Briassoulis [11], Xia et al. [13], and Ye et al. [14], is shown in Figures 9 and 10. It is clear that the orthotropic model that is based on the equivalent rigidities of Ye et al. [14] yields the most best agreement with the fundamental frequency values obtained using FEM. In

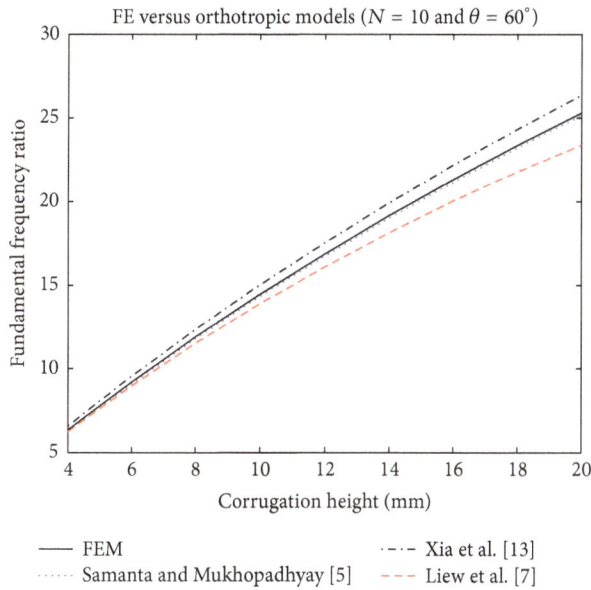

FIGURE 11: Effect of corrugation height on the fundamental frequency ratio based on finite element and equivalent orthotropic models (trapezoidal corrugation).

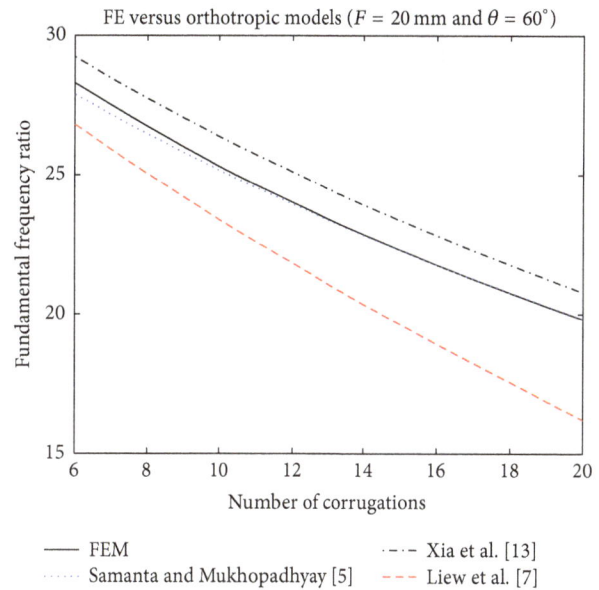

FIGURE 12: Effect of number of corrugations on the fundamental frequency ratio based on finite element and equivalent orthotropic models (trapezoidal corrugation).

general, the differences between the results of FEM and those of the equivalent orthotropic models get larger by increasing the corrugation height and number of corrugations.

For trapezoidal corrugated plates, orthotropic equivalent plates that are based on the equivalent rigidities of Samanta and Mukhopadhyay [5], Xia et al. [13], and Liew et al. [7] are compared with FE models of the actual corrugated plates in Figures 11 and 12. Like in the sinusoidal corrugated plates, a better correlation between the orthotropic models and the FE models can be obtained when the corrugation height decreases because the derivations of the equivalent rigidities are based on the assumption of shallow corrugations. Figure 12 shows that the correlation in the results of the orthotropic model that is based on equivalent rigidities of Samanta and Mukhopadhyay [5] improves with the increase in the number of corrugations.

In the optimization examples presented in the next section, the equivalent rigidities of Ye et al. [14] and Samanta and Mukhopadhyay [5] are used for sinusoidal corrugated plates and trapezoidal corrugated plates, respectively.

4. Optimization Examples

In the previous section, an insight is gained on the effect of corrugation geometric parameters on a plate fundamental frequency. In this section, the focus will be on maximizing the plate fundamental frequency by using sinusoidal or trapezoidal corrugations. We consider two optimization problems. The first is to maximize the fundamental frequency of a square corrugated plate. The second is to maximize the fundamental frequency-to-mass ratio of the same plate. Three dimensions of sinusoidal corrugated plates, including plate thickness h, height of corrugation F, and number of

corrugations N, are selected as design variables. In addition to these design variables, the trough angle θ is considered as a design variable for designing trapezoidal corrugated plates. During optimization, the number of corrugation is restricted to integer numbers. The equivalent orthotropic plate model is used to model the corrugated plates in the optimization problems.

Here, a genetic algorithm (GA) is used for optimization examples. The GAs are stochastic optimization methods that are based on the concept of biological evolution. It was introduced by Holland in 1975 [17]. In comparison with traditional optimization methods, the GAs work over a set of candidate points, which cover the entire search space instead of a single candidate point at each iteration. It, therefore, increases the probability to reach to the global optimum. GAs are not gradient-based methods. Thus, they can work with discrete design variables and do not require the objective functions to be differentiable. Also, the GAs can work with nonconvex problems. In general, GA uses a set of candidate points (called population) which cover the entire search space. Then, the objective function (fitness function) is evaluated for each candidate point. The points with relatively good objective functions are used to generate a new set of candidate points through three mechanisms known as reproduction, crossover, and mutation. In the reproduction method, the candidate point in the current generation with the best objective function is automatically passed to the next generation. The crossover method creates a new point by combining the vectors of two candidate points in the current generation. The mutation method creates a new point by randomly changing some components of a candidate point in the current generation. The algorithm continues iteration (generation) until it reaches the stopping criteria. The population size for each generation is assumed to be 70 with a crossover probability of 80% and mutation probability of 20%. The stopping criterion is chosen to be either the tolerance of less than 10^{-4} for the objective function or the maximum number of generations not to exceed 100.

4.1. Maximizing the Fundamental Frequencies of Corrugated Plates. In this section, examples are presented to show the efficacy of designing the corrugated plates to maximize their fundamental frequencies. The maximization is applied on two designs: using sinusoidal corrugations and using trapezoidal corrugations. The optimization problem is defined as follows:

Maximize: ω_1

Subject to: $F_{lb} \leq F \leq F_{ub}$

$N_{lb} \leq N \leq N_{ub}$

$h_{lb} \leq h \leq h_{ub}$

$$\left(\frac{F}{C}\right)_{lb} \leq \left(\frac{F}{C}\right) \leq \left(\frac{F}{C}\right)_{ub} \qquad (13)$$

$\theta_{lb} \leq \theta \leq \theta_{ub}$

(In trapezoidal corrugation)

$$F_{lb} \leq F \leq \left(\frac{C}{4}\right)\tan(\theta)$$

(In trapezoidal corrugation),

where lb and ub are the lower and upper bounds of design variables, respectively. Throughout this study, the bounds of design variables are selected to be $F_{lb} = 0$, $F_{ub} = 30$ mm, $N_{lb} = 6$, $N_{ub} = 30$, $h_{lb} = 0.5$ mm, $h_{ub} = 5$ mm, $(F/C)_{lb} = 0$, $(F/C)_{ub} = 0.5$, $\theta_{lb} = 0°$, and $\theta_{ub} = 90°$. The optimal design variables that maximize the fundamental frequency of sinusoidal corrugated plate are $F = 30$ mm, $N = 6$, and $h = 5$ mm. The fundamental frequency of the optimal sinusoidal corrugated plate is 165.4 Hz.

The optimal design variables that maximize the fundamental frequency of trapezoidal corrugated plate are $F = 30$ mm, $N = 6$, $h = 5$ mm, and $\theta = 90°$. The fundamental frequency of the optimal trapezoidal corrugated plate is 208.5 Hz. It can be concluded that the maximum fundamental frequency occurs at the upper bounds of the corrugation height F and trough angle θ and at the lower bound of the number of corrugations N. These results can be anticipated from the parametric study.

4.2. Maximizing the Fundamental Frequency-to-Mass Ratio of Corrugated Plates. As shown in the previous optimization example, the fundamental frequency increases when the height of corrugation, the trough angle, and the thickness increase. However, increasing the aforementioned design variables increases the mass of the plate. For example, the masses of the optimal plates in the previous example are 49.4 kg and 66.8 kg for the sinusoidal and trapezoidal corrugated plates, respectively. The fundamental frequency-to-mass ratios in the previous example are 3.35 Hz/kg and

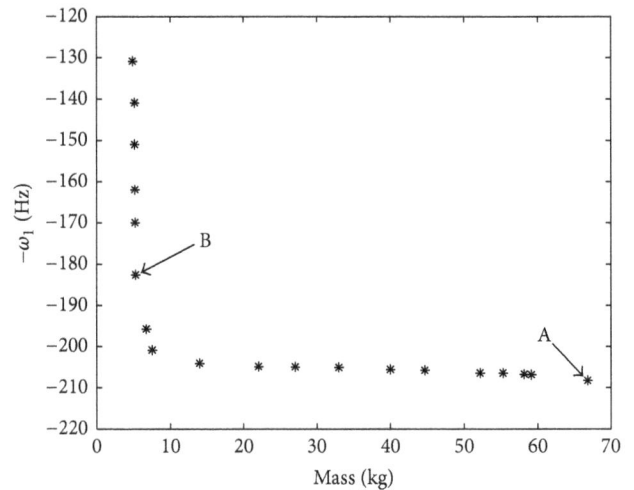

FIGURE 13: Pareto front for the fundamental frequency and the mass of the trapezoidal corrugated plates.

3.12 Hz/kg for the optimal sinusoidal and trapezoidal corrugated plates, respectively. In this example, the fundamental frequency-to-mass ratio of simply supported plate is maximized by creating sinusoidal or trapezoidal corrugations. The optimization problem is the same as the problem that is illustrated in the previous example but with maximizing ω_1/mass instead of maximizing ω_1.

The GA is used to solve the optimization problem. The design parameters of the optimal sinusoidal corrugated plate are $F = 30$ mm, $N = 6$, and $h = 0.5$ mm. The maximum ω_1/mass is 33.2 Hz/kg (i.e., ω_1 is 164 Hz and the mass is 4.94 kg). The optimal design variables, which maximize the fundamental frequency-to-mass ratio of the trapezoidal corrugated plate, are $F = 30$ mm, $N = 6$, $h = 0.5$ mm, and $\theta = 53°$. The maximum ω_1/mass is 34.5 Hz/kg (i.e., ω_1 is 183 Hz and the mass is 5.30 kg). Pareto front of the mass and the fundamental frequency of the trapezoidal corrugated plate is shown in Figure 13. In this figure, point A stands for the maximum fundamental frequency (i.e., $\omega_1 = 208.5$ Hz and $m = 66.8$ kg), and point B stands for the maximum fundamental frequency-to-mass ratio (i.e., $\omega_1 = 183$ Hz and $m = 5.30$ kg).

5. Conclusions

Designing sinusoidal and trapezoidal corrugated plates for optimal fundamental frequency was investigated. The fundamental frequencies of simply supported corrugated plates were determined numerically by using the finite element method and equivalent orthotropic models. The corrugation of the plate increases the area moment of inertia and hence increases its bending rigidities. In addition to altering the natural frequencies of plates, corrugations change the plate mode shapes.

A comparative study shows that using the equivalent orthotropic models to study the free vibration of corrugated plates significantly reduces the computational time. The parametric analysis shows the efficacy of the corrugation technique on increasing the fundamental frequency of simply supported plates. Two optimization problems were considered: the maximization of corrugated plate fundamental frequency and maximization of the corrugated plate fundamental frequency-to-mass ratio. A future extension of this work will focus on the experimental verification of the results. Future work will consider the design of corrugated plates subjected to a dynamic load with respect to the minimal sound radiation.

Competing Interests

The author declares that they have no competing interests.

References

[1] N. T. Alshabatat and K. Naghshineh, "Optimization of the natural frequencies of plates via dimpling and beading techniques," *International Journal of Modelling and Simulation*, vol. 32, no. 4, pp. 244–254, 2012.

[2] M. D. Musgrove and B. E. Greene, "Advanced beaded and tubular structural panels," *Journal of Aircraft*, vol. 11, no. 2, pp. 68–75, 1974.

[3] P. P. Plank, I. F. Sakata, G. W. Davis, and C. C. Richie, "Hypersonic cruise vehicle wing structure evaluation," Tech. Rep., NASA, Washington, DC, USA, 1970.

[4] C. Gentilinia, L. Nobilea, and K. A. Seffen, "Numerical analysis of morphing corrugated plates," *Procedia Engineering*, vol. 1, no. 1, pp. 79–82, 2009.

[5] A. Samanta and M. Mukhopadhyay, "Finite element static and dynamic analyses of folded plates," *Engineering Structures*, vol. 21, no. 3, pp. 277–287, 1999.

[6] Y. Wang, D. Gao, and X. Wang, "On the nonlinear vibration of heated corrugated circular plates with shallow sinusoidal corrugations," *International Journal of Mechanical Sciences*, vol. 50, no. 6, pp. 1082–1089, 2008.

[7] K. M. Liew, L. X. Peng, and S. Kitipornchai, "Vibration analysis of corrugated Reissner-Mindlin plates using a mesh-free Galerkin method," *International Journal of Mechanical Sciences*, vol. 51, no. 9-10, pp. 642–652, 2009.

[14] Z. Ye, V. L. Berdichevsky, and W. Yu, "An equivalent classical plate model of corrugated structures," *International Journal of Solids and Structures*, vol. 51, no. 11-12, pp. 2073–2083, 2014.

[15] S. Timoshenko and S. Woinowsky-Krieger, *Theory of Plates and Shells*, McGraw–Hill, New York, NY, USA, 1959.

[16] S. Wolford, "Sectional properties of corrugated sheets determined by formula," *Civil Engineering*, vol. 103, pp. 59–60, 1954.

[17] J. H. Holland, *Adaptation in Natural and Artificial Systems*, University of Michigan Press, Ann Arbor, Mich, USA, 1975.

[8] A. Yucel and A. Arpaci, "Theoretical and experimental vibration analyses of trapezoidal and sinusoidal corrugated plates," *Journal of Vibration and Control*, vol. 21, no. 10, pp. 2006–2026, 2015.

[9] M. T. Huber, "Die Theorie der kreuzweise bewehrten Eisenbeton-platten nebst Anwendungen auf mehrere bautechnisch wichtige Aufgaben über rechteckige Platten," *Bauingenieur*, vol. 4, pp. 354–360, 1923.

[10] E. Seydel, "Shear buckling of corrugated plates," *Jahrbuch die Deutschen Versuchsanstalt fur Luftfahrt*, vol. 9, pp. 233–245, 1931.

[11] D. Briassoulis, "Equivalent orthotropic properties of corrugated sheets," *Computers and Structures*, vol. 23, no. 2, pp. 129–138, 1986.

[12] K. Liew, L. Peng, and S. Kitipornchai, "Buckling analysis of corrugated plates using a mesh-free Galerkin method based on the first-order shear deformation theory," *Computational Mechanics*, vol. 38, no. 1, pp. 61–75, 2006.

[13] Y. Xia, M. I. Friswell, and E. I. Saavedra Flores, "Equivalent models of corrugated panels," *International Journal of Solids and Structures*, vol. 49, no. 13, pp. 1453–1462, 2012.

3

Working and Limitations of Cable Stiffening in Flexible Link Manipulators

Rahul Dixit[1] and R. Prasanth Kumar[2]

[1]*Control Systems Laboratory, Research Center Imarat, Vigyanakancha, Hyderabad 500069, India*
[2]*Department of Mechanical & Aerospace Engineering, Indian Institute of Technology Hyderabad, Kandi, Telangana 502285, India*

Correspondence should be addressed to R. Prasanth Kumar; rpkumar@iith.ac.in

Academic Editor: Emil Manoach

Rigid link manipulators (RLMs) are used in industry to move and manipulate objects in their workspaces. Flexible link manipulators (FLMs), which are much lighter and hence highly flexible compared to RLMs, have been proposed in the past as means to reduce energy consumption and increase the speed of operation. Unlike RLM, an FLM has infinite degrees of freedom actuated by finite number of actuators. Due to high flexibility affecting the precision of operation, special control algorithms are required to make them usable. Recently, a method to stiffen FLMs using cables, without adding significant inertia or adversely affecting the advantages of FLMs, has been proposed as a possible solution in a preliminary work by the authors. An FLM stiffened using cables can use existing control algorithms designed for RLMs. In this paper we discuss in detail the working principle and limitations of cable stiffening for flexible link manipulators through simulations and experiments. A systematic way of deciding the location of cable attachments to the FLM is also presented. The main result of this paper is to show the advantage of adding a second pair of cables in reducing overall link deflections.

1. Introduction

Robot manipulators used in industry spend much of their energies in moving their end effectors from one point to another. Their links are designed to have very low deflection which makes them bulky. Hence they are called rigid link manipulators (RLMs). In order to reduce wastage of energy in moving the manipulator link mass, researchers proposed flexible link manipulators (FLMs) (Figure 1) which are not only much lighter but also highly flexible because of the reduced inertia. Moving a payload at the end effector of FLM to desired location at the specified time requires special control algorithms designed taking into account the under-actuated nature of FLMs.

In order to minimize the tip vibration of FLMs, several researchers have analyzed the problem in different ways. Having a more accurate model helps in better control of FLM. Some researchers have focused on modeling of the flexible arm [1–8] so as to control it effectively. Modeling of the flexible arm has been studied for single [2, 6, 7] as well as multiple

flexible link systems [4]. Piedboeuf [9] presented six methods to model a rotating flexible beam considering foreshortening effect. Sugiyama et al. [10] presented finite segment modeling of flexible link considering discrete segments. The method explains modeling of finite segments using torsional springs between adjacent segments. Kiran et al. [11] presented bond graph technique for modeling a single link flexible space manipulator. Kinematic analysis of the FLM has been studied and suitability of coordinate frame was discussed in [8].

Different methods of control to reduce or eliminate tip oscillations of FLM have been studied which includes resonant control [12], sliding mode control [13–16], optimal control strategies [2], and other nonlinear control methods [17]. Morales et al. [18] studied behaviours of light weight single link FLM with payload variations by using disturbance observer. Baroudi et al. [19] presented their study on control of flexible manipulator using LQR technique. Ahmad and Mohamed [20] compared the control performance of LQR and PID to suppress the tip vibration. Several authors have used closed loop control methodologies by modeling flexible

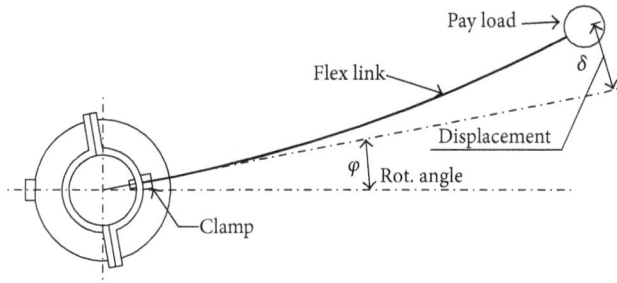

FIGURE 1: Schematic of flexible link manipulator (FLM).

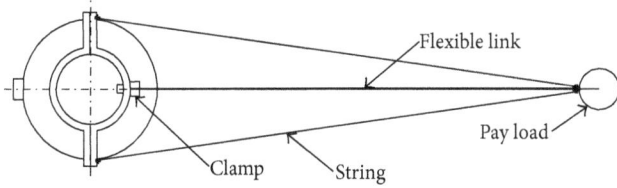

FIGURE 2: Schematic of cable stiffened FLM.

link and using sensors along the flexible link for closing the control loop. The closed loop control requires several feedback sensors and complex control algorithms. Usage of these sensors and complex control algorithms further increase the cost of system. In addition to that the accuracy of control becomes dependent on the sensor accuracy as well as the variation in control parameters. Researchers have also studied command shaping to reduce tip oscillations [21–26]. Further, force interaction with environment using FLMs remains an open problem. It has been observed that the control philosophy used for rigid link manipulators cannot be applied to FLMs directly. Despite research into special control algorithms for FLMs for more than two decades, they have not yet been adopted by industry.

Stiffening an FLM using a pair of cables attached between the flexible link and its rigid hub as shown in Figure 2 has recently been proposed by Dixit and Prasanth Kumar [27] as possible means of overcoming the disadvantages of FLMs. Mass of cables for stiffening are very small compared to the mass of FLM itself. Hence they do not add significantly to the inertia of the manipulator. An FLM so stiffened can use the same control algorithms used for RLMs. Although the tip oscillations of cable stiffened FLM are minimized, radial component of force that occurs during acceleration and deceleration is found to cause buckling-like deflection in the middle of the FLM.

In this paper we focus on quantitatively showing how a pair of cables attached to the tip of an FLM increases the fundamental natural frequency and thus stiffens the FLM. This is done through numerical simulations in ANSYS software as well as experimental data. Further, the cause of buckling-like mid-link deflection when one pair of cables are used for stiffening is investigated analytically. It will be shown that adding a second pair of cables to the same rigid hub will further increase the fundamental frequency making the FLM stiffer. A systematic way of deciding the location of cable attachment on the FLM is also discussed.

2. Mathematical Model and Simulation

The flexible link manipulator studied in the paper is modeled using finite segment method (also called rigid finite element method). This method is known to give good predictions even in the presence of contact forces, large translations, and rotations. The mathematical model of the FLM is similar to the one presented in Dixit and Prasanth Kumar [27]. However, since the current paper also deals with two pairs of cables, the following description is briefly repeated here for reference. The flexible link is discretized into a finite number of rigid links interconnected by revolute joints with torsional springs and dampers. The damping coefficients are usually obtained from experiments. Torsional spring constants are determined from flexural rigidity as follows:

$$K^{i(i-1)} = \frac{EI}{L} \quad i = 2, \dots, N,$$

$$K^{1(0)} = \frac{2EI}{L} \tag{1}$$

for the first torsional spring at the hub,

where L is the length of each of N segments.

Hub is considered as the body-0 and payload as body-$(N + 1)$, with N rigid segments connecting the hub and payload. Therefore, there are $N + 2$ bodies in total for N segment discretization of the flexible link. The outer end of Nth rigid segment is connected to the payload through a rigid joint. Since the cables from hub support are also connected at the same point, spring forces due to stiffness of the cables act at that point. Equations of motion of the planar multibody system can be written as follows:

$$\mathbf{M}\ddot{\mathbf{q}} = \mathbf{h} + {}^{(c)}\mathbf{h}, \tag{2}$$

where mass matrix $\mathbf{M} = \text{diag}([\mathbf{M}_0 \ \mathbf{M}_1 \ \cdots \ \mathbf{M}_N \ \mathbf{M}_{N+1}])$, array of body coordinates $\mathbf{q} = \begin{bmatrix} \mathbf{q}_0^T & \mathbf{q}_1^T & \cdots & \mathbf{q}_N^T & \mathbf{q}_{N+1}^T \end{bmatrix}^T$, array of applied forces $\mathbf{h} = \begin{bmatrix} \mathbf{h}_0^T & \mathbf{h}_1^T & \cdots & \mathbf{h}_N^T & \mathbf{h}_{N+1}^T \end{bmatrix}^T$, and array of constraint forces ${}^{(c)}\mathbf{h} = \begin{bmatrix} {}^{(c)}\mathbf{h}_0^T & {}^{(c)}\mathbf{h}_1^T & \cdots & {}^{(c)}\mathbf{h}_N^T & {}^{(c)}\mathbf{h}_{N+1}^T \end{bmatrix}^T$.

The applied force vector $\mathbf{h}$ consists of all zeros, except for $\mathbf{h}_N$ and $\mathbf{h}_s$ which depend on cable stiffness and end deflection. While $\mathbf{h}_N$ is the applied force on the tip of the link due to first pair of cables, $\mathbf{h}_s$ is the force due to second pair of cables on body-s. At any point of time, either one of the cables is in tension applying transverse and longitudinal forces on the tip. Constraint forces are determined from Jacobian matrix and Lagrange multipliers. Apart from rigid joint constraints and revolute joint constraints on the bodies, an additional driver constraint on the hub specifies desired angular displacement trajectory as a 3-4-5 polynomial as follows:

$$\phi_0(t) = a_0 + a_1 t + a_2 t^2 + a_3 t^3 + a_4 t^4 + a_5 t^5. \tag{3}$$

TABLE 1: Simulation parameters of FLM.

Parameter	Value
Flexible link size	$500 \times 30 \times 1.5$
Mass of link	0.175 kg
Mass of clamp	0.150 kg
Distance of cable attachment from shaft axis	70 mm
Mass of payload	0.18 kg, 0.12 kg, 0.06 kg
Density of link	7800 kg/m^3
Diameter of cable	0.8 mm
Number of cables	1 pair and 2 pairs
Cable material density	7800 kg/m^3
Target rotation	45 deg.
Time duration	1 sec

Coefficients of this polynomial are determined from the following conditions:

$$\phi_0(0) = 0,$$

$$\dot{\phi}_0(0) = 0,$$

$$\ddot{\phi}_0(0) = 0,$$

$$\phi_0(1) = \frac{\pi}{4}, \tag{4}$$

$$\dot{\phi}_0(1) = 0,$$

$$\ddot{\phi}_0(1) = 0.$$

Table 1 lists the parameter and values used for simulation. A damping coefficient of 0.01 is found to closely match the vibration decay observed in experiments on flexible link. The results of simulations are shown in Figures 3 and 4. Without cables, the tip deflection is quite high. Using one pair of cables attached between tip and hub, deflections are brought down to less than a millimeter. Although this may seem reasonably small deflection, it will be shown in the next section that this could lead to high mid-link deflection resulting in foreshortening.

3. Deflection Analysis and Cable Attachment Location

Analysis of mode shapes of a flexible link gives valuable information on the location of attachment of cables on the FLM to minimize tip vibration. FLMs in clamped-free condition will have maximum deflection at the free tip for its first normal mode. This is similar to a cantilever beam which has maximum deflection at the free end. Figure 5(a) shows the first and second normal modes of a cantilever beam. For the given smooth trajectory input, first normal mode is more prominent than the second normal mode. Therefore, the location of attachment of the first pair of cables will be at the tip. With this attachment at the tip resisting the tip

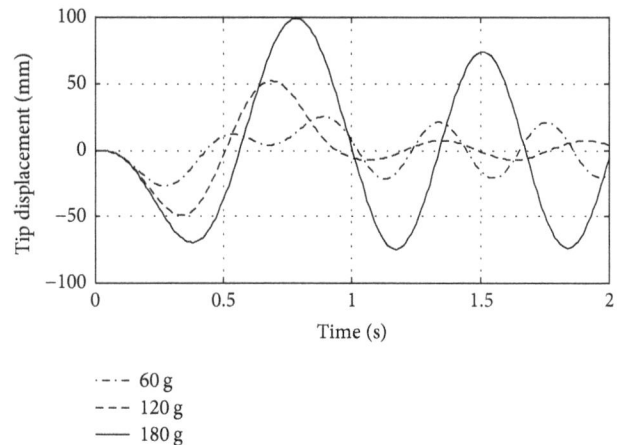

FIGURE 3: FLM tip deflection without cables for various payloads.

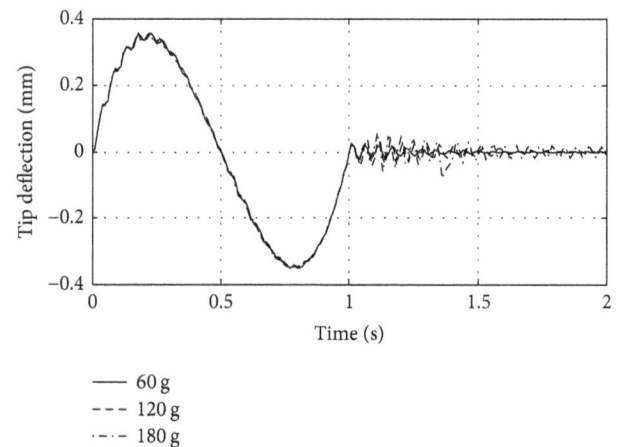

FIGURE 4: FLM tip deflection with one pair of cables for various payloads.

deflection, the flexible link will behave similar to a clamped-pinned beam rather than a clamped-free beam. Figure 5(b) shows the normal mode shapes of a cantilever beam with the new clamped-pinned boundary condition.

It should be noted that the behavior of FLM with the first pair of cables will be similar to that of a cantilever beam with the end pinned but will not be identical. It is because a pinned joint does not allow translation in any direction whereas cable attachment does allow translation consistent with the constraints imposed by both cables. For this reason, we also try to validate the results of maximum deflection with multibody dynamics simulation and mode analysis through ANSYS simulation which are closer to the actual physical model.

After attachment of the first pair of cables, multibody dynamics analysis as described in previous section for an FLM with one pair of cables at the tip has been carried out for the one second trajectory in (3). For each time step, deflection of the FLM in local frame (attached at the hub of the FLM) with respect to the coordinate along its length has been obtained from simulation. At each time step there will be a deflection curve with maximum deflection at some point

(a) Cantilever mode shapes for clamped-free boundary condition

(b) Cantilever mode shapes for clamped-pinned boundary condition

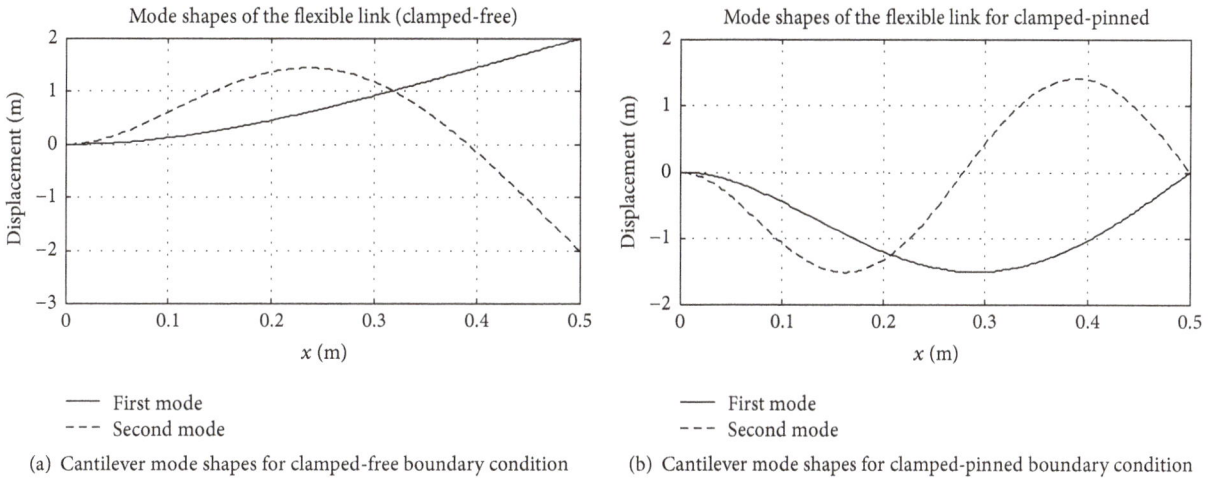

FIGURE 5: Mode shapes of a cantilever beam with clamped-free and clamped-pinned boundary conditions.

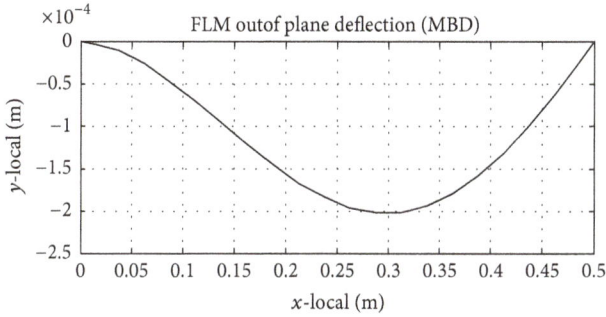

FIGURE 6: FLM maximum deflection curve with one pair of cables at the tip with 180 gram payload.

FIGURE 7: Forces acting on the payload at the tip of FLM.

along the x-axis of the local frame. A deflection curve whose maximum deflection is larger than the maximum deflection at all other time steps is chosen. The location of maximum deflection would be the ideal location for the attachment of second pair of cables to minimize the maximum deflection. It can be seen from Figure 6 that the optimal location for second pair attachment can be at 0.3 m for an FLM of 0.5 m length. It is interesting to note that the FLM maximum deflection pattern resembles the first normal mode shape of a cantilever beam with one end fixed and the other end pinned plotted in Figure 5(b). It can be seen that the location of maximum deflection also matches very closely the multibody dynamics analysis. Therefore, it can be concluded that the optimal location for the attachment of the second pair of stiffening cables will be at $x = 0.3$ m.

3.1. Buckling-Like Mid-Link Deflection. One of the important advantages of stiffened FLM is its fast speed of operation. Upper limit of acceleration of the FLM is limited by maximum bending moment on the link. The higher the acceleration or deceleration is, the more the FLM is prone to mid-link deflection (deflection between hub and tip). Figure 6 shows that the mid-link deflection is as high as 2 mm. In the following analysis, we assume the link is in dynamic

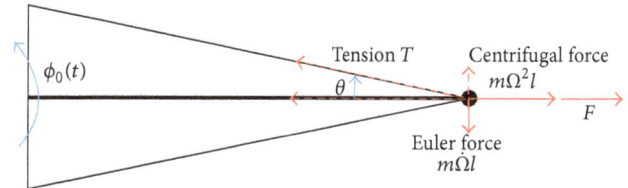

equilibrium under constant angular acceleration, and the FLM mass is negligible compared to payload mass. Further, the radial buckling load can be assumed to be negligible since the FLM is thin and long. Force balance as shown in Figure 7 provides us the following set of equations:

$$m\Omega^2 l - T \cos\theta = F,$$
$$m\dot{\Omega}l = T \sin\theta, \quad (5)$$

where T is tension in the cable, θ is the angle made by the cable with the FLM, F is the net horizontal force on the payload in the local frame, and $\Omega = \dot{\phi}_0(t)$ is the angular velocity of the hub. The payload is subjected to centrifugal force along the radial direction away from the center and Euler force perpendicular to the radial direction and opposite to the direction of acceleration. Since the payload is assumed to be stationary in the local frame, there is no Coriolis force.

If F is negative, the radial load on the link is compressive which is to be avoided. For no compressive load ($F \geq 0$ or $m\Omega^2 l \geq T \cos\theta$),

$$\dot{\Omega} \leq \Omega^2 \tan\theta. \quad (6)$$

Equation (6) provides the condition for compressive tip load on the manipulator link. It can be observed that the above-mentioned condition cannot be satisfied for any trajectory starting from rest and hence there will always be a compressive load on the link that causes mid-link deflection. If accelerations are allowed for short duration at the beginning and end of trajectory such that (6) is satisfied for most of the time of trajectory's duration, then these accelerations have to be higher than normal. In the extreme case, these sudden accelerations act as impact on the cables causing permanent deformation or even complete failure due to snapping. Permanent deformation increases the cable length, introduces slackness in the cable, and thus increases the tip deflection.

Ω^2 term on the right-hand side of (6) is scaled by $\tan\theta$. The angle θ depends on the distance from motor axis to the point where cable is attached to the hub. For the same trajectory, the larger the value of θ is, the sooner the condition is satisfied. However, due to practical limitations, θ cannot be increased beyond a certain limit. For example, $\theta = 45°$ requires a hub as big as the length of the link itself on either side of the motor axis.

Hence, the simplest way to resist mid-link deflection due to acceleration and deceleration is to use a second pair of cables attached between the rigid hub and the ideal location discussed earlier. This solves the problem with only a negligible increase in the rotating inertia.

3.2. Buckling of FLM. It has been shown in Section 3.1 that after attachment of single pair of strings at the tip of FLM, the tip boundary condition can be taken as pinned; that is, displacement is very close to zero and the slope needs not be zero at the tip. Therefore, considering fixed-pinned boundary condition for the hub and tip, respectively, critical buckling load can be given by

$$P_{cr} = \frac{2\pi^2 EI}{l^2}. \tag{7}$$

When a second pair of strings is attached at $0.6l$ as discussed in Section 3, the attachment point can be assumed to act like a node. Figure 8 shows the configuration of FLM with two pairs of strings.

With reference to Figure 8, after attachment of second pair of string at $0.6l$, the FLM is divided into two segments. Boundary conditions for segment-1 of length l_1 will be fixed-pinned and for segment-2 of length l_2, it will be pinned-pinned. We write the equation for critical buckling load at the tip for full FLM and at the mid-link for segment-1 as follows:

$$P_{cr1} = \frac{2\pi^2 EI}{l_1^2},$$
$$P_{cr2} = \frac{\pi^2 EI}{l_2^2}, \tag{8}$$

where l_1 and l_2 are the lengths of segment-1 and segment-2, respectively.

In order to have higher critical buckling load that allows higher accelerations to be used after the attachment of second pair of strings, the following conditions have to be satisfied:

$$P_{cr1} > P_{cr},$$
$$P_{cr2} > P_{cr}. \tag{9}$$

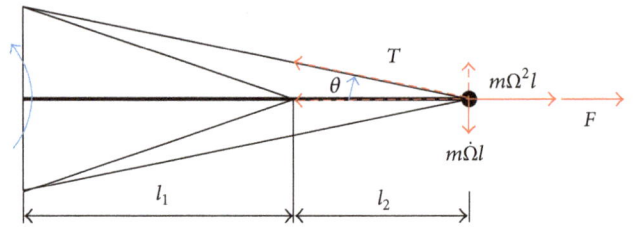

FIGURE 8: FLM with two pairs of strings.

For segment-1,

$$\frac{2\pi^2 EI}{l_1^2} > \frac{2\pi^2 EI}{l^2}. \tag{10}$$

Therefore, $l_1 < l$, which is true as the second pair attachment location lies between hub and the tip.

Similarly, for the second segment,

$$\frac{\pi^2 EI}{l_2^2} > \frac{2\pi^2 EI}{l^2}. \tag{11}$$

Therefore, $l_2 < 0.707l$, which is also true as $l_2 = 0.4l$ as presented in Section 3.1.

Therefore, it can be concluded that the attachment of second pair of string enhances the buckling load thereby helping to achieve higher accelerations than achievable with a single pair of strings.

Another important conclusion derived from this analysis is that if the location of second pair attachment is at $0.293l$ from the hub, the critical buckling load due to attachment of second pair will not be higher than that of the critical buckling with only one pair of strings. Therefore, in order to have higher critical buckling load, second pair of strings should be attached beyond $0.293l$ from the hub. The location of attachment of second pair of strings proposed in this paper is at $0.6l$.

3.3. Effect of Stiffening. In order to know the effect of cable attachment quantitatively, fundamental natural frequencies and mode shapes were obtained for three different cases of FLM with no cables, one pair of cables, and two pairs of cables in ANSYS and plotted in Figure 9. The natural frequencies obtained were 1.191 Hz, 6.244 Hz, and 12.466 Hz, respectively, for these three cases. With the addition of second pair of cables at the decided location, the fundamental natural frequency almost doubles. This provides evidence that the FLM will be significantly stiffened by adding a second pair of cables leading to lower deflections.

4. Experimental Setup and Validation

Two experimental setups were constructed: one for finding the natural frequencies of FLM and another for acquiring deflection trajectories at the tip and mid-link. The first experimental setup is shown in Figure 10 where the FLM is rigidly held in a vice. After exciting with a hammer, deflection data at a point on the link is acquired. FFT of the data gives

1
Displacement
STEP = 1
SUB = 1
FREQ = 1.191
DMX = 5.313   Y

Z X

(a) First normal mode of FLM without cables

1
Displacement
STEP = 1
SUB = 1
FREQ = 6.244
DMX = 3.968   Y

Z X

(b) First normal mode of FLM with one pair of cables

1
Displacement
STEP = 1
SUB = 1   Y
FREQ = 12.466
DMX = 4.615   Z X

(c) First normal mode of FLM with two pairs of cables

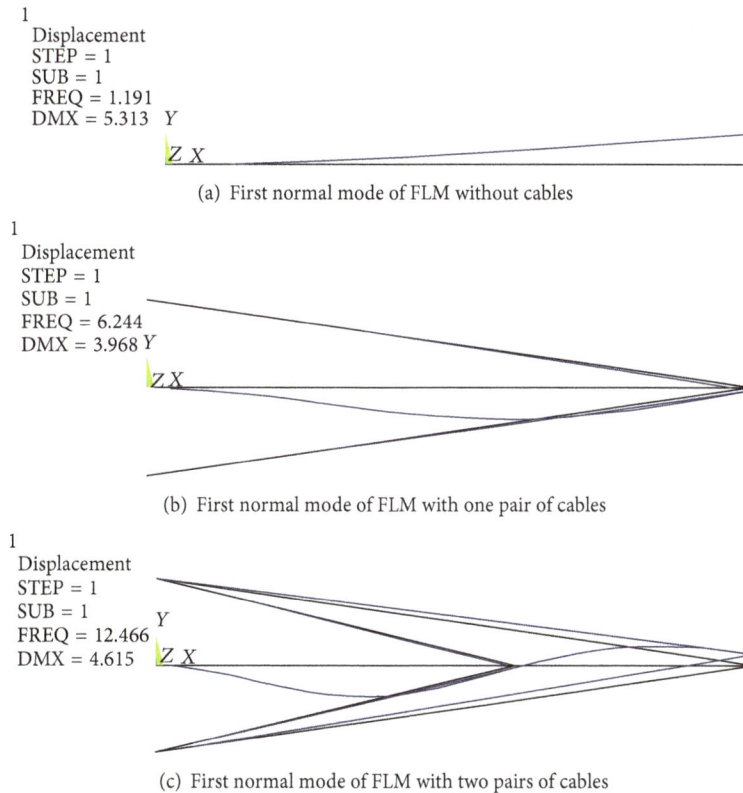

FIGURE 9: Modal analysis in ANSYS.

FIGURE 10: Experimental setup for finding natural frequencies for one pair of cables.

the natural frequency. The experiment was repeated for three cases: FLM without cables, FLM with one pair of cables, and FLM with two pairs of cables. The purpose of this experiment is to qualitatively see how much one and two pairs of cables stiffen the FLM.

Figure 11 shows the FFT of FLM deflection data from experiments for three cases. The fundamental natural frequencies obtained are 1.356 Hz, 6.66 Hz, and 12.31 Hz, respectively. These fundamental natural frequencies shown by FFT plots closely match those obtained from ANSYS.

The second experimental setup consists of a DC servo motor, power supply, FLM, cables, two displacement sensors, an attachment for mounting displacement sensors, a data acquisition system, and a computer. The DC servo motor is a 100 watt Dynamixel Pro servo H54-100-S500-R which can accept position commands over RS485 communication link from the computer. FLM is mounted on the motor

shaft by means of a rigid hub. The attachment for mounting displacement sensors is a rigid manipulator arm attached to the same hub and parallel to the FLM in undeflected condition.

Unlike the setup discussed in Dixit and Prasanth Kumar [27] where deflection is measured only at the end of the trajectory, deflection in local frame can be measured with the current setup throughout the trajectory. The two displacement sensors are noncontact laser based sensors opto-NCDT 1402 with a range of 50 mm to 150 mm and a resolution of 10 microns. The sensor outputs are analog which are acquired by a data acquisition system at 1 kHz from which deflection data is obtained. The DC servo motor has digital output pins, one of which is connected to data acquisition system. This output pin, which is usually held low, is triggered to high at the beginning of trajectory command received from computer and held low again from the end of trajectory. The input and output data are time synced using this trigger. The schematic of experimental setup and photograph of cable stiffened FLM are shown in Figures 12(a) and 12(b), respectively. Note that due to difficulty in measuring deflection exactly at the tip where the payload is attached, deflection close to the tip is measured as tip deflection. When comparing with simulation results, deflection of the corresponding point in simulation model is considered tip deflection.

Experiments were conducted with 180 gram payload and deflection at tip and mid-link (where second pair of cables are attached) was measured. Figures 13 and 14 show the simulation and experimental results of FLM deflections with

(a) FFT of FLM without cables

(b) FFT of FLM with one pair of cables

(c) FFT of FLM with two pairs of cables

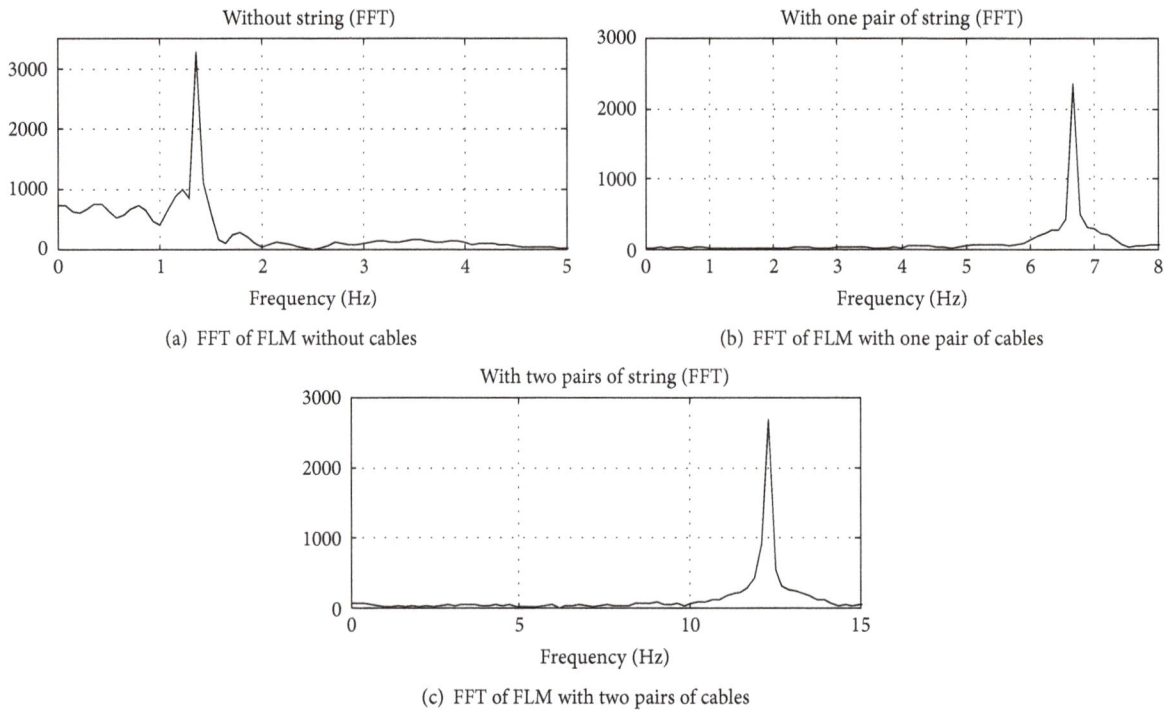

FIGURE 11: FFT of FLM deflection data for three cases.

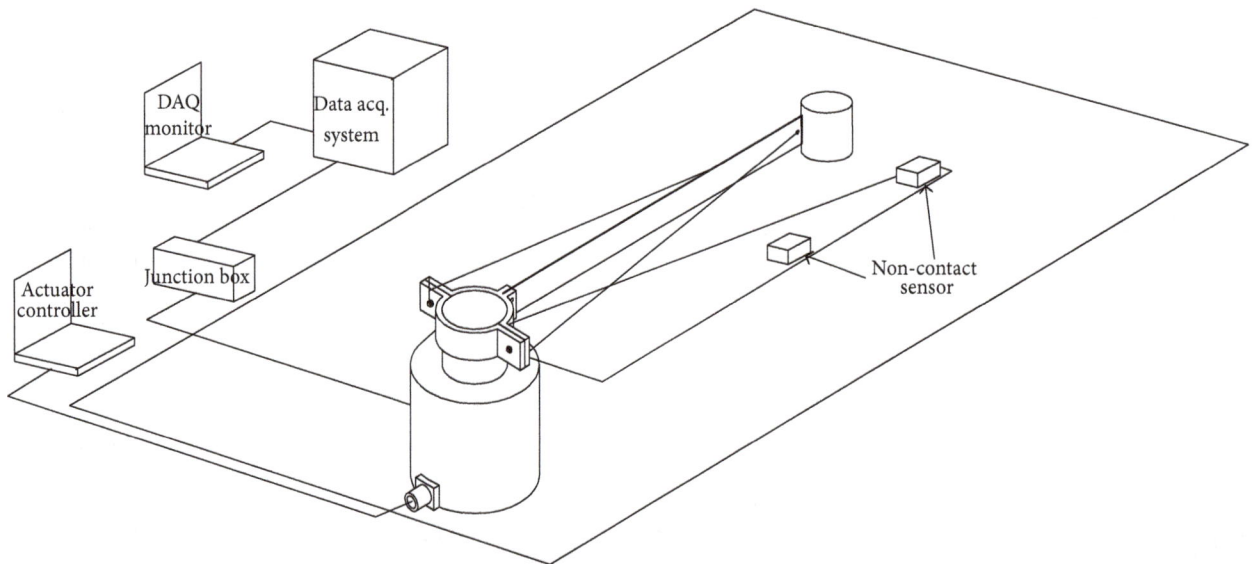

(a) Schematic of experimental setup of cable stiffened FLM

(b) Experimental setup for one pair of cables

FIGURE 12: Experimental setup of cable stiffened FLM.

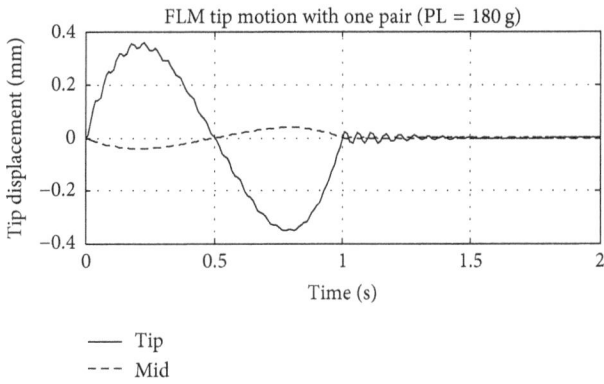

FIGURE 13: Simulation results of tip and mid-link deflections for one pair of cables.

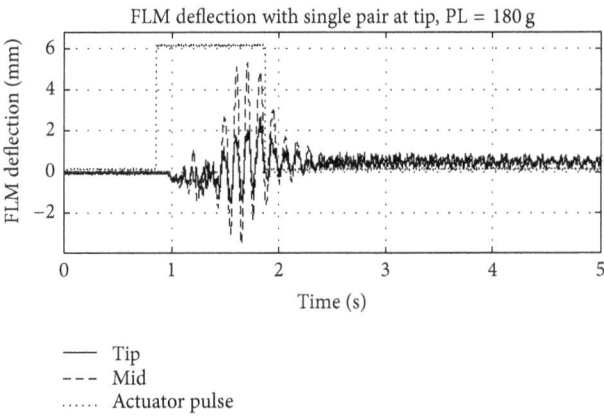

FIGURE 14: Experimental results of tip and mid-link deflections for one pair of cables.

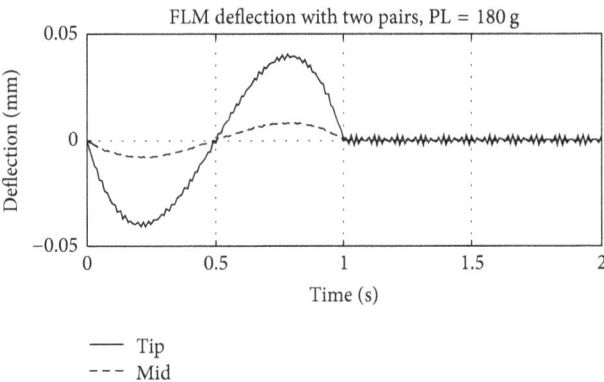

FIGURE 15: Simulation results of tip and mid-link deflections for two pairs of cables.

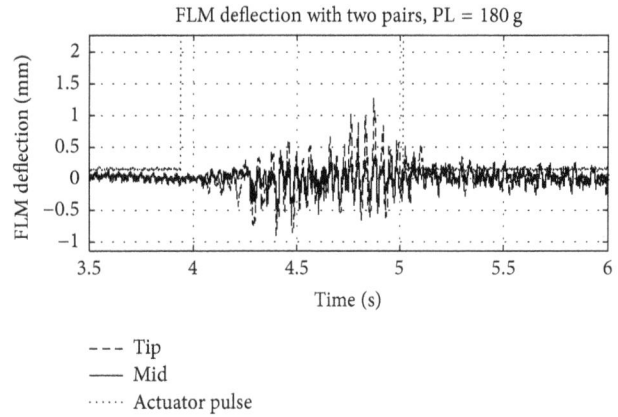

FIGURE 16: Experimental results of tip and mid-link deflections for two pairs of cables.

one pair of cables attached at the tip, respectively. Similarly, Figures 15 and 16 show the simulation and experimental results of FLM deflections with two pairs of cables, respectively. The actuator pulse shown in experimental results indicate the duration of trajectory command.

Comparing the simulation and experimental results, it is clear that addition of second pair of cables in fact reduces the deflection at both the tip and the mid-link. However, deflections occurring during experiments are about one order of magnitude higher than those of simulation results and higher frequency components also appear. As a first guess, we believed that the discrepancy was due to the presence of slackness in the cables allowing the deflections to increase. Even when we tried to reduce the slackness by introducing a small pretension in the cables, results were similar. Another possibility is the presence of small bends in the steel cables connecting the link to the rigid hub. These bends act as curved beams with low bending stiffness compared to the axial stiffness of the cable. We theorize that when the payload is subjected to inertia forces during the link motion, low stiffness, till the bend opens up completely, causes connected points on the link to deflect significantly. This is a major limitation of cable stiffening method.

5. Conclusions

It is known that any flexible body has infinity degrees of freedom. Therefore, ideally a single input at the hub will not be able to track arbitrary trajectories of FLMs making them underactuated systems. As the interest is to restrict oscillations of the tip alone, the methodology presented in this paper attempts to restrict DOFs of the tip by attachment of cables. A systematic way is presented for the attachment of cables to minimize link deflections. Foreshortening effect due to mid-link deflection gets minimized by attachment of additional pairs of cables at the specified locations. Two sets of experimental setups were constructed. It is concluded that the simulation results and experimental results have very good agreement as far as the fundamental natural frequencies are concerned.

Trajectories obtained from second experimental setup show one order of magnitude discrepancy with simulation results, although the trend of reduced deflection with increase in the number of pairs of cables is present. The major limitation of large deflections found in experiments may be overcome by more accurate fabrication and assembly of cables on the FLM.

Competing Interests

The authors declare that there is no conflict of interests regarding the publication of this paper.

Acknowledgments

This work was supported by the project "Cable Stiffened Flexible Link Manipulator for Pick-and-Place Tasks" (YSS/2014/000010) sponsored by Department of Science and Technology, Government of India.

References

[1] R. H. Cannon Jr. and E. Schmitz, "Initial experiments on the end-point control of a flexible one-link robot," *The International Journal of Robotics Research*, vol. 3, no. 3, pp. 62–75, 1984.

[2] W. D. Zhu and C. D. Mote Jr., "Dynamic modeling and optimal control of rotating Euler-Bernoulli beams," *Journal of Dynamic Systems, Measurement and Control*, vol. 119, no. 4, pp. 802–808, 1997.

[3] M. H. Korayem, H. N. Rahimi, and A. Nikoobin, "Mathematical modeling and trajectory planning of mobile manipulators with flexible links and joints," *Applied Mathematical Modelling*, vol. 36, no. 7, pp. 3229–3244, 2012.

[4] W. Chen, "Dynamic modeling of multi-link flexible robotic manipulators," *Computers & Structures*, vol. 79, no. 2, pp. 183–195, 2001.

[5] A. Fenili and J. M. Balthazar, "The rigid-flexible nonlinear robotic manipulator: modeling and control," *Communications in Nonlinear Science and Numerical Simulation*, vol. 16, no. 5, pp. 2332–2341, 2011.

[6] M. Khairudin, "Dynamic modelling of a flexible link manipulator robot using AMM," *TELKOMNIKA Indonesian Journal of Electrical Engineering*, vol. 6, no. 3, pp. 185–190, 2008.

[7] S. Bošnjak and N. Zrnić, "On the dynamic modelling of exible manipulators," *FME Transactions*, vol. 34, pp. 231–237, 2006.

[8] A. A. Ata, E. H. Haraz, A. E. A. Rizk, and S. N. Hanna, "Kinematic analysis of a single link flexible manipulator," in *Proceedings of the IEEE International Conference on Industrial Technology (ICIT '12)*, pp. 852–857, Athens, Greece, March 2012.

[9] J.-C. Piedboeuf, "Six methods to model a flexible beam rotating in the vertical plane," in *Proceedings of the IEEE International Conference on Robotics and Automation (ICRA '01)*, vol. 3, pp. 2832–2839, IEEE, Seoul, Republic of Korea, May 2001.

[10] H. Sugiyama, N. Kobayashi, and Y. Komaki, "Modeling and experimental methods for dynamic analysis of the spaghetti problem," *Journal of Vibration and Acoustics*, vol. 127, no. 1, pp. 44–51, 2005.

[11] G. S. Kiran, A. Kumar, P. M. Pathak, and N. Sukavanam, "Trajectory control of flexible space robot," in *Proceedings of the IEEE International Conference on Mechatronics and Automation (ICMA '08)*, pp. 738–743, IEEE, Takamatsu, Japan, August 2008.

[12] I. A. Mahmood, S. O. R. Moheimani, and B. Bhikkaji, "Precise tip positioning of a flexible manipulator using resonant control," *IEEE/ASME Transactions on Mechatronics*, vol. 13, no. 2, pp. 180–186, 2008.

[13] V. Etxebarria, A. Sanz, and I. Lizarraga, "Control of a lightweight flexible robotic arm using sliding modes," *International Journal of Advanced Robotics*, vol. 2, no. 2, pp. 103–110, 2005.

[14] G. Mamani, J. M. A.-D. Silva, and V. Feliu-Batlle, "Least squares state estimator based sliding mode control of a very lightweight single-link flexible robot arm," in *Proceedings of the IEEE 2009 International Conference on Mechatronics (ICM '09)*, pp. 1–6, IEEE, Malaga, Spain, April 2009.

[15] R. Fareh, M. Saad, and M. Saad, "Adaptive control for a single flexible link manipulator using sliding mode technique," in *Proceedings of the 6th International Multi-Conference on Systems, Signals and Devices (SSD '09)*, pp. 1–6, IEEE, Djerba, Tunisia, March 2009.

[16] S. Kurode and P. Dixit, "Output feedback control of flexible link manipulator using sliding modes," in *Proceedings of the 2012 7th International Conference on Electrical and Computer Engineering (ICECE '12)*, pp. 949–952, IEEE, Dhaka, Bangladesh, December 2012.

[17] S. S. Ge, T. H. Lee, and G. Zhu, "A nonlinear feedback controller for a single-link flexible manipulator based on a finite element model," *Journal of Robotic Systems*, vol. 14, no. 3, pp. 165–178, 1997.

[18] R. Morales, V. Feliu, and V. Jaramillo, "Position control of very lightweight single-link flexible arms with large payload variations by using disturbance observers," *Robotics and Autonomous Systems*, vol. 60, no. 4, pp. 532–547, 2012.

[19] M. Baroudi, M. Saad, and W. Ghie, "State-feedback and linear quadratic regulator applied to a single-link flexible manipulator," in *Proceedings of the 2009 IEEE International Conference on Robotics and Biomimetics (ROBIO '09)*, pp. 1381–1386, IEEE, Guilin, China, December 2009.

[20] M. A. Ahmad and Z. Mohamed, "Techniques of vibration and end-point trajectory control of flexible manipulator," in *Proceedings of the 6th International Symposium on Mechatronics and its Applications (ISMA '09)*, pp. 1–6, IEEE, Sharjah, United Arab Emirates, March 2009.

[21] H. Yang, M. H. Ang Jr., and H. Krishnan, "Control of a tip-loaded flexible-link robot using shaped input command," in *Proceedings of the American Control Conference (ACC '98)*, vol. 5, pp. 3075–3076, IEEE, Philadelphia, Pa, USA, June 1998.

[22] W. Chatlatanagulchai, V. M. Beazel, and P. H. Meckl, "Command shaping applied to a flexible robot with configuration-dependent resonance," in *Proceedings of the American Control Conference*, IEEE, Minneapolis, Minn, USA, June 2006.

[23] K.-P. Liu and Y.-C. Li, "Vibration suppression for a class of flexible manipulator control with input shaping technique," in *Proceedings of the International Conference on Machine Learning and Cybernetics*, pp. 835–839, August 2006.

[24] K.-P. Liu, "Experimental evaluation of preshaped inputs to reduce vibration for flexible manipulator," in *Proceedings of the 6th International Conference on Machine Learning and Cybernetics (ICMLC '07)*, vol. 4, pp. 2411–2415, IEEE, Hong Kong, August 2007.

[25] M. Romano, B. N. Agrawal, and F. Bernelli-Zazzera, "Experiments on command shaping control of a manipulator with flexible links," *Journal of Guidance, Control, and Dynamics*, vol. 25, no. 2, pp. 232–239, 2002.

[26] N. Seth and K. S. Rattan, "Vibration control of flexible manipulators," in *Proceedings of the IEEE National Aerospace and Electronics Conference (NAECON '92)*, pp. 876–882, 1992.

[27] R. Dixit and R. Prasanth Kumar, "Cable stiffened flexible link manipulator," in *Proceedings of the IEEE/RSJ International Conference on Intelligent Robots and Systems (IROS '14)*, pp. 871–876, Chicago, Ill, USA, September 2014.

4

A Discrete Model for Nonlinear Vibration of Beams Resting on Various Types of Elastic Foundations

A. Khnaijar and R. Benamar

Equipe des Etudes et Recherches en Simulation, Instrumentation et Mesures (ERSIM), Université Mohammed V,
Ecole Mohammadia des Ingénieurs, Avenue Ibn Sina, BP 765, Rabat, Morocco

Correspondence should be addressed to A. Khnaijar; khnaijar@gmail.com

Academic Editor: Kim M. Liew

This paper presents a discrete physical model to approach the problem of nonlinear vibrations of beams resting on elastic foundations. The model consists of a beam made of several small bars, evenly spaced. The bending stiffness is modeled by spiral springs, and the Winkler soil stiffness is modeled using linear vertical springs. Concentrated masses, presenting the inertia of the beam, are located at the bar ends. Finally, the nonlinear effect is presented by the axial forces in the bars, assumed to behave as longitudinal springs, due to the change in their length induced by the Pythagorean Theorem. This model has the advantage of simplifying parametric studies, because of its discrete nature, allowing any modification in the mass matrix, the stiffness matrix, and the nonlinearity tensor to be made separately. Therefore, once the model is established, various practical applications may be performed without the need of going through all the formulation again. The study of the nonlinear behavior makes the solution of the movement equation rise in complexity. By considering this discrete model and using the linearization method, one can achieve an idealized approach to this nonlinear problem and obtain quite easily approximate solutions.

1. Introduction

The analysis of the vibration of beams resting on elastic foundations wears a practical and theoretical interest in many fields such as civil, mechanical, and transportation engineering. Analytical and numerical methods applied to the modeling of such a problem are extensively addressed in the literature. However, the combination between the emergence of high-performance computers and problems complexity made the discrete methods more appealing.

A discrete method such as the Finite Element Method (FEM) is the first to address the problem numerically [1–3]. The Differential Quadrature Method (DQM) is also employed for the solution to similar engineering problems involving beam vibrations and foundations [4–6]. Therefore, Malekzadeh and Karami [7] gather the advantages of both previous methods (DQM and FEM) to perform the free vibration and buckling analysis of thick beams on two-parameter elastic foundations.

On the other hand, soil behavior wears a great complexity, because of its nonlinear nature. In that order, literature has investigated multiple soil models starting from the idealized elastic theories modeled by Winkler to the more complex viscoelastic theories of Pasternak and Hetinyi. Similarly, Kerr [8] carries a study showing the evolution of soil theories with their extensions.

The aim of this work is the development of a flexible general discrete model for linear and nonlinear vibrations of beams resting on elastic foundations, via the adaptation and the extension of the approach presented in [9]. Accordingly, the process of discretization carried out is intended to allow efficient variation of beam and soil geometrical and mechanical characteristics, in order to make it easy to perform multiple parametric studies.

For nonlinear vibrations, the multimode analysis leads to a coupled nonlinear differential system involving the contribution to various modes. Using the harmonic balance method, Benamar [18] and Benamar et al. [19] reduce the nonlinear free vibration problem to a set of nonlinear algebraic equations. Consequently, the objective here is the investigation into linear and nonlinear vibrations, using the discrete physical model developed for various types of soil stiffness, and beam end conditions.

This study establishes a simplification to the nonlinear problem induced by large deflections. However, the nonlinearity tensor obtained becomes heavy in terms of calculation for high values of the discretization parameter N (presenting the number of masses in the physical model in Figure 2). In fact, we addressed this complexity and proposed a simplification of the calculation, leading to an efficient algorithm.

The structure of this paper is as follows: Section 2.1 presents the general formulation, where the continuous and discrete models are detailed, in order to show their theoretical similarities. After the establishment of the discrete model, the following Section 2.2 is concerned with determination of the discrete model parameters and the nondimensional formulation. Finally, Section 2.3 introduces a solution to the multimodal nonlinear problem using a linearization procedure.

The physical discrete model validation is done in Section 3, devoted to details and comments on the results obtained in the cases considered. An examination is made of how far the results deviate from the linear and the nonlinear continuous theories, and a discussion is given about their range of validity under the different assumptions adopted.

The main purpose of this work is achieved in Section 4, in which straight applications of the theory are developed and validated via comparisons with previous references. The first application deals with a partially supported beam on elastic foundations, where the springs presenting the soil reaction are applied to only a limited portion of the beam span. In the second application, the beam examined is resting on a variable elastic foundation, where the distribution of soil spring stiffness is linear or parabolic.

2. General Theory

A nonlinear discrete model of a beam, made of extensional bars, concentrated masses and rotational springs located at the bar ends, is introduced in [9] to approach the continuous theory using different assumptions. In the present work, a beam resting on an elastic soil is undergoing the same discretization process by extending and completing the previous model.

2.1. General Formulation. This section presents the discrete and continuous models for nonlinear vibrations of beams resting on elastic foundations. Both models are based on the Lagrangian principle, leading to a formally identical description. The linear behavior of the mechanical system examined is determined by the classical mass and rigidity matrices m_{ij} and k_{ij}. On the other hand, the nonlinear behavior, presented by nonlinearity tensor b_{ijkl} due to the axial forces induced by the large vibration amplitudes in the continuous case, is established by the application of the Pythagorean Theorem in the discrete physical model.

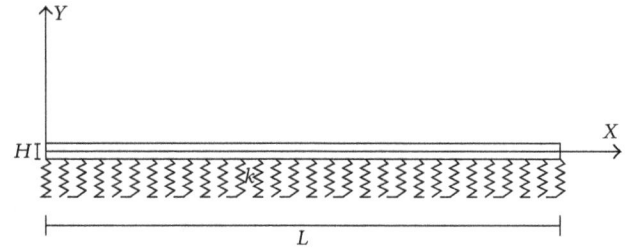

FIGURE 1: A continuous beam supported on Winkler elastic foundation.

2.1.1. Continuous Model. Before introducing the extended formulation of the discrete model, a quick review of the theory behind the continuous model [18] is introduced. Let us consider transverse vibrations of a beam, shown in Figure 1, resting on elastic foundations, having the characteristics (S, L, H, I, E, ρ, and k) defined in Nomenclature. The total beam strain energy can be written as the sum of the strain energy due to bending denoted as V_l^s and the strain energy induced by Winkler springs V_l^l plus the axial strain energy V_{nl} due to the axial load induced by large deflections. Thus V_l^s, V_l^l, V_{nl}, and the kinetic energy T are as follows [20]:

$$V_l^s = \frac{EI}{2} \int_0^L \left(\frac{\partial^2 W}{\partial x^2} \right)^2 dx,$$

$$V_l^l = \frac{EI}{2} \int_0^L (W)^2 dx,$$

$$V_{nl} = \frac{ES}{8L} \left(\int_0^L \left(\frac{\partial W}{\partial x} \right)^2 dx \right)^2, \tag{1}$$

$$T = \frac{\rho S}{2} \int_0^L \left(\frac{\partial W}{\partial t} \right)^2 dx,$$

where W is the beam transverse displacement. Using a generalized parameterization and the usual summation convention used in [18], the transverse displacement can be described as [20]

$$W(x,t) = q_i(t) w_i(x) \quad \text{for } i, j = 1, \dots, N. \tag{2}$$

By replacing W (2) in its new form into the energy expressions V_l^s, V_l^l, V_{nl}, and T (1), one gets

$$V_l^s = \frac{\overline{k^s_{ij}}}{2} q_i q_j, \tag{3}$$

$$V_l^l = \frac{\overline{k^l_{ij}}}{2} q_i q_j, \tag{4}$$

$$V_{nl} = \frac{\overline{b_{ijkl}}}{2} q_i q_j q_k q_l, \tag{5}$$

$$T = \frac{\overline{m_{ij}}}{2} \dot{q}_i \dot{q}_j \tag{6}$$

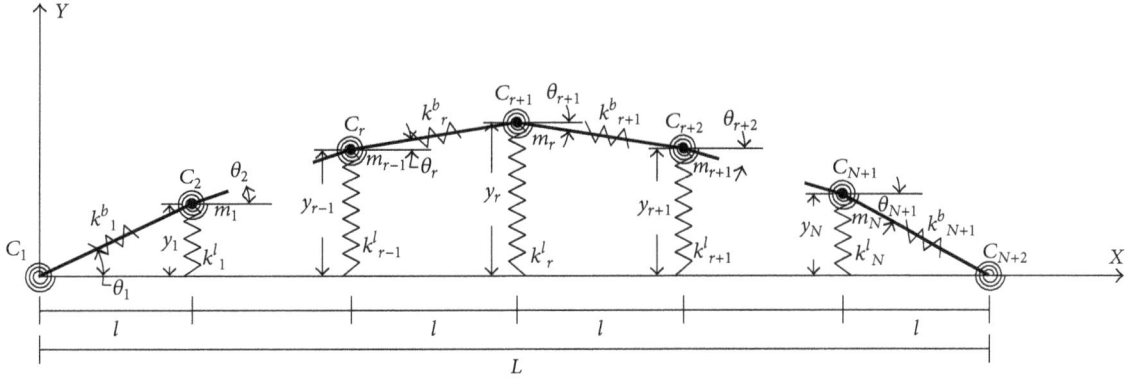

FIGURE 2: A multi-degree-of-freedom discrete model made of N concentrated masses, connected by longitudinal and spiral springs (bar characteristics) and longitudinal springs (soil stiffness).

in which $\overline{m_{ij}}$, $\overline{k^s_{ij}}$, $\overline{k^l_{ij}}$, and $\overline{b_{ijkl}}$ are defined as follows in the modal basis (MB):

$$\overline{m_{ij}} = \rho S \int_0^L w_i w_j dx \quad \text{for } i, j = 1, \ldots, N,$$

$$\overline{k^s_{ij}} = EI \int_0^L \frac{\partial^2 w_i}{\partial x^2} \frac{\partial^2 w_j}{\partial x^2} dx \quad \text{for } i, j = 1, \ldots, N,$$

$$\overline{k^l_{ij}} = k \int_0^L w_i w_j dx \quad \text{for } i, j = 1, \ldots, N, \tag{7}$$

$$\overline{b_{ijkl}} = \frac{ES}{4L} \int_0^L \frac{\partial w_i}{\partial x} \frac{\partial w_j}{\partial x} dx \int_0^L \frac{\partial w_k}{\partial x} \frac{\partial w_l}{\partial x} dx$$

$$\text{for } i, j, k, l = 1, \ldots, N.$$

For a conservative system, the dynamic behavior of the structure may be obtained by Lagrange's equations [20]:

$$-\frac{\partial}{\partial t}\left(\frac{\partial T}{\partial \dot{q}_r}\right) + \frac{\partial T}{\partial q_r} - \frac{\partial V}{\partial q_r} = 0 \quad \text{for } r = 1, \ldots, N, \tag{8}$$

where $V = V^s_l + V^l_l + V_{\text{nl}}$. By introducing the energy expressions, that is, (3) to (6), into (8), one gets

$$m_{ir}\ddot{q}_i + k^s_{ir}q_i + k^l_{ir}q_i + 2q_i q_j q_k b_{ijkr} = 0$$

$$\text{for } r = 1, \ldots, N. \tag{9}$$

This nonlinear differential system (9) may be expressed in a matrix form as

$$[\mathbf{M}]\{\ddot{\mathbf{q}}\} + [\mathbf{K}^s]\{\mathbf{q}\} + [\mathbf{K}^l]\{\mathbf{q}\} + 2[\mathbf{B}(\mathbf{q})]\{\mathbf{q}\} = 0, \tag{10}$$

where $[\mathbf{M}]$, $[\mathbf{K}^s]$, $[\mathbf{K}^l]$, and $[\mathbf{B}(\mathbf{q})]$ are, respectively, the mass matrix, the bending rigidity matrix, the Winkler springs rigidity matrix, and the nonlinear rigidity matrix.

According to the conclusion of [9, 18, 21, 22], a harmonic distortion of the nonlinear response of a harmonically exited structure happens at large transverse vibration amplitudes. However, it is shown both experimentally and theoretically that the higher harmonic components remain very small,

compared to the response first harmonic component. Consequently, the nonlinear free response of the beam is assumed in the present paper, as in many previous ones, to be drawn as [18, 20]

$$q_i = a_i \cos\left(\omega^{\text{nl}} t\right), \tag{11}$$

where a_i's are the contribution coefficients to the beam nonlinear mode shapes. By replacing (11) in (10) and applying the harmonic balance method, one has

$$\left([\mathbf{K}^s] + [\mathbf{K}^l]\right)\{\mathbf{A}\} - (\omega)^2 [\mathbf{M}]\{\mathbf{A}\} + \frac{3}{2}[\mathbf{B}(\mathbf{A})]\{\mathbf{A}\}$$
$$= 0. \tag{12}$$

At this point, the vibration of a continuous Euler-Bernoulli beam resting on elastic foundations is put under a matrix form. It is interesting to note that the model presented in this section and summarized in (12) can be considered as a multidimensional form of the Duffing equation, very often encountered in nonlinear vibration analyses of structures with cubic nonlinearities. The solution to this nonlinear equation is discussed in Section 2.3.

2.1.2. Discrete Model. In order to adapt the discrete model [9] to the case of a beam resting on Winkler elastic foundations and to compare it to the continuous model established in Section 2.1.1, let us consider the N-degree-of-freedom (DOF) system shown in Figure 2, made of N concentrated masses $m_1, \ldots, m_N$, $N + 1$ bars presented by longitudinal linear springs, $N + 2$ torsional springs, and N longitudinal linear soil springs. C_r is the stiffness coefficient of the rth spiral spring, for $r = 1$ to $N + 2$. The moment $\mathcal{M}$ in the spiral spring is given by $\mathcal{M} = -C_r \Delta\theta$; $\Delta\theta = \theta_r - \theta_{r-1}$ being the angle between the bars adjacent to the node r, considered as longitudinal springs of length l_r and stiffness k^b_r, $r = 1$, to N. The Winkler foundations are modeled using the longitudinal vertical spring's distribution, with k^l_r presenting the stiffness coefficient of the rth linear spring, for $r = 1$ to N.

Using vector notation, $\{\mathbf{d}_1\}; \{\mathbf{d}_2\}; \ldots; \{\mathbf{d}_i\}; \ldots; \{\mathbf{d}_N\}$ is the displacement basis, denoted as DB, and y_1 to y_N are, respectively, the transverse displacements of the masses m_1 to m_N from the horizontal equilibrium positions ($y = 0$) of the system corresponding to the nondeformed positions of the springs (spiral and linear). The vectors $\{\mathbf{d}_i\}$ are defined by

$$\{\mathbf{d}_1\}^T = \begin{bmatrix} 1 & 0 & 0 & 0 & \cdots & 0 \end{bmatrix};$$

$$\{\mathbf{d}_2\}^T = \begin{bmatrix} 0 & 1 & 0 & 0 & \cdots & 0 \end{bmatrix};$$

$$\vdots$$

$$\{\mathbf{d}_i\}^T = \begin{bmatrix} 0 & 0 & \cdots & 1 & \cdots & 0 \end{bmatrix}; \tag{13}$$

$$\vdots$$

$$\{\mathbf{d}_N\}^T = \begin{bmatrix} 0 & 0 & 0 & 0 & \cdots & 1 \end{bmatrix},$$

with $\{\mathbf{d}_i\}^T$ being the unit displacement of the ith mass. The displacement vector can be written in DB as

$$\{\mathbf{y}\} = y_1 \{\mathbf{d}_1\} + y_2 \{\mathbf{d}_2\} + \cdots + y_i \{\mathbf{d}_i\} + \cdots + y_N \{\mathbf{d}_N\}; \tag{14}$$

by considering the modal basis $\{\{\mathbf{\Phi}_1\}, \{\mathbf{\Phi}_2\}, \ldots, \{\mathbf{\Phi}_i\}, \ldots, \{\mathbf{\Phi}_N\}\}$, obtained by solution of the linear eigenvalue problem and denoted as MB, the displacement vector can be expressed as

$$\{\mathbf{y}\} = \overline{y_1} \{\mathbf{\Phi}_1\} + \overline{y_2} \{\mathbf{\Phi}_2\} + \cdots + \overline{y_i} \{\mathbf{\Phi}_i\} + \cdots + \overline{y_N} \{\mathbf{\Phi}_N\}, \tag{15}$$

where $\{\mathbf{\Phi}_i\}$ is the ith linear mode shape of the system and $\overline{y_1}, \overline{y_2}, \ldots, \overline{y_i}, \ldots, \overline{y_N}$ are the components of the mass displacements expressed in MB. $\{\mathbf{\Phi}_i\}$ is denoted in DB as

$$\{\mathbf{\Phi}_i\}^T = \begin{bmatrix} \varphi_{1i} & \varphi_{2i} & \cdots & \varphi_{1N} \end{bmatrix} \quad \text{for } i = 1, \ldots, N. \tag{16}$$

Using the transition matrix $[\mathbf{\Phi}]$ from DB to MB, the two sets of coordinates in the two bases are related by

$$\begin{bmatrix} \overline{y_1} & \overline{y_2} & \cdots & \overline{y_N} \end{bmatrix} = [\mathbf{\Phi}] \begin{bmatrix} y_1 & y_2 & \cdots & y_N \end{bmatrix}^T. \tag{17}$$

So, the displacement vector can be written by analogy with the continuous displacement vector q_i, as [9]

$$y_i = A_i^{\text{discr}} \cos\left(\omega^{\text{nl}}_{\text{discr}} t\right) = a_j \varphi_{ij} \cos\left(\omega^{\text{nl}}_{\text{discr}} t\right) \tag{18}$$

$$\text{for } i, j = 1, \ldots, N,$$

where A_i^{discr} is the modulus of the displacement y_i expressed in DB (or the contribution of the normalized vector $\{\mathbf{d}_i\}$ of DB) and a_j is the modulus of the displacement y_i expressed in MB (or the contribution of the normalized vector $\{\mathbf{\Phi}_i\}$ of MB). $\omega^{\text{nl}}_{\text{discr}}$ is the nonlinear frequency of the discrete system associated with the amplitude A_i^{discr}. Using the usual summation convention for the repeated indices i, j, k, and l, the kinetic, linear, and nonlinear strain energy expressions,

obtained by replacement of the expression (18) of y_i in the energy expressions (3) to (6), are

$$T = \frac{1}{2} a_i a_j \left(\omega^{\text{nl}}_{\text{discr}}\right)^2 \overline{m_{ij}} \sin^2\left(\omega^{\text{nl}}_{\text{discr}} t\right) \tag{19}$$

$$\text{for } i, j = 1, \ldots, N,$$

$$V_l^s = \frac{1}{2} a_i a_j \overline{k^s_{ij}} \cos^2\left(\omega^{\text{nl}}_{\text{discr}} t\right) \quad \text{for } i, j = 1, \ldots, N, \tag{20}$$

$$V_l^l = \frac{1}{2} a_i a_j \overline{k^l_{ij}} \cos^2\left(\omega^{\text{nl}}_{\text{discr}} t\right) \quad \text{for } i, j = 1, \ldots, N, \tag{21}$$

$$V_{\text{nl}} = \frac{1}{2} a_i a_j a_k a_l \overline{b_{ijkl}} \cos^4\left(\omega^{\text{nl}}_{\text{discr}} t\right) \tag{22}$$

$$\text{for } i, j, k, l = 1, \ldots, N,$$

where $\overline{m_{ij}}$, $\overline{k^s_{ij}}$, $\overline{k^l_{ij}}$, and $\overline{b_{ijkl}}$ are, respectively, the general terms of the mass tensor, the linear rigidity tensors corresponding to the spiral and longitudinal vertical springs, and the nonlinear rigidity tensor in MB. The relationships between the expressions for these tensors in DB and MB can be obtained using the transition matrix $[\mathbf{\Phi}]$ as follows [19, 23]:

$$\overline{m_{ij}} = \varphi_{si} \varphi_{tj} m_{st},$$

$$\overline{k^s_{ij}} = \varphi_{si} \varphi_{tj} k^s_{st},$$

$$\overline{k^l_{ij}} = \varphi_{si} \varphi_{tj} k^l_{st}, \tag{23}$$

$$\overline{b_{ijkl}} = \varphi_{si} \varphi_{tj} \varphi_{pk} \varphi_{ql} b_{stpq}.$$

Replacing in (8) the expressions for T, V_l^l, V_l^s, and V_{nl} given in (19)–(22) and applying the harmonic balance method, in MB, lead to [24, 25]

$$\left(\left[\overline{\mathbf{K}}^s\right] + \left[\overline{\mathbf{K}}^l\right]\right) \{\mathbf{A}\} - \left(\omega^{\text{nl}}_{\text{discr}}\right)^2 \left[\overline{\mathbf{M}}\right] \{\mathbf{A}\}$$

$$+ \frac{3}{2} \left[\overline{\mathbf{B}}(\mathbf{A})\right] \{\mathbf{A}\} = 0. \tag{24}$$

The linearized approximate methods of solution of nonlinear algebraic systems of the type (24) are proposed in [20]. In the present paper, to improve the accuracy of the numerical solutions in the range of vibration amplitudes taken into consideration, the so-called second formulation in [20] is combined with the Newton-Raphson method in Section 2.3.

2.2. The Expressions for the General Terms of the Tensors m_{ij}, k^s_{ij}, k^l_{ij}, and b_{ijkl}. One of the major advantages of this model is its ability to explicitly express each tensor and give a clear insight of the parameters chosen, before going through the solution of the dynamic problem. However, in a discrete problem, the calculation time depends on the discretization number N. For the linear parameters m_{ij} and k_{ij}, increasing the number of masses in the physical discrete model does not heavily affect the calculation time, because filling the matrices and performing the change of bases from the DB to the MB do not require more than a two loops' process.

Nevertheless, for the nonlinear parameter b_{ijkl} (4D tensor), it requires only one loop in the filling process but does need eight nested loops for the inversion, in order to transform the tensor from DB to MB. In fact, this limitation made the solution of the nonlinear problem for high values of the discretization parameter N a quite lengthy process. Accordingly, by analyzing the form of the nonlinearity tensor, we noticed a certain number of patterns. Then, we carried a meticulous analysis leading to a simplified procedure as detailed in Appendix A, where the eight nested loops are reduced to only three.

Appendix B gives a summary of the calculation, carried out in [9], for the expression of the parameters m_{ij}, k^s_{ij}, and b_{ijkl}. Identically, Sections 2.2.1(1) and 2.2.1(2) give detailed calculations of the new parameter k^l_{ij}.

2.2.1. The Expressions for the General Terms of the Tensors m_{ij}, k^s_{ij}, k^l_{ij}, and b_{ijkl}.

The general expression of m_{ij}, k^s_{ij}, k^l_{ij}, and b_{ijkl} for the discrete model are as follows:

(i) The Mass tensor $[\mathbf{M}]/m_{ij}$ terms are expressed as (Appendix B)

$$m_{ij} = \frac{\rho S L}{(N+1)} \delta_{ij} \quad \text{for } i, j = 1, \ldots, N. \quad (25)$$

(ii) The spiral springs tensor $[\mathbf{K}^s]/k^s_{ij}$ terms are expressed as (Appendix B)

$$k^s_{(r-2)r} = \frac{(N+1)^3}{L^3} EI \quad \text{for } r = 3, \ldots, N,$$

$$k^s_{(r-1)r} = -4 \frac{(N+1)^3}{L^3} EI \quad \text{for } r = 2, \ldots, N, \quad (26)$$

$$k^s_{rr} = 6 \frac{(N+1)^3}{L^3} EI \quad \text{for } r = 1, \ldots, N$$

and the other values of k^s_{ij} are obtained by symmetry relations or are equal to zero. The boundaries conditions of the system determine the stiffness k^s_{ii} for $i = 1$ and $N + 1$. For simply supported ends, the corresponding torsional springs C_1 and C_{N+1} stiffness are equal to zero, since the rotation is free; then $k^s_{ii} = 5(N+1)^3 EI/L^3$ for $i = 1$ and $N + 1$. For clamped ends, the corresponding torsional spring stiffness is infinite, since the rotation is prevented, so $k^s_{ii} = +\infty$ for $i = 1$ and $N + 1$.

(iii) The Winkler springs tensor $[\mathbf{K}^l]/k^l_{ij}$ terms are expressed as (details given in Sections 2.2.1(1) and 2.2.1(2))

$$k^l_{ij} = k_i l \delta_{ij} = \frac{k_i L}{(N+1)} \delta_{ij} \quad \text{for } i, j = 1, \ldots, N. \quad (27)$$

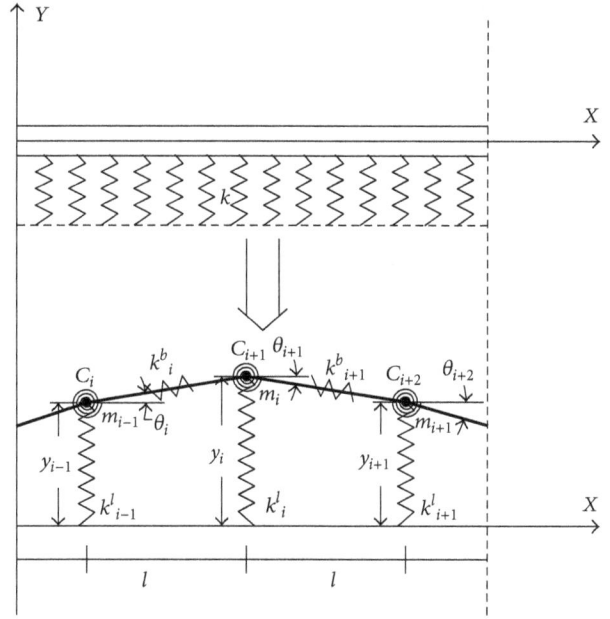

FIGURE 3: Discretization of the linear vertical springs (soil stiffness).

(iv) The nonlinear rigidity tensor $[\mathbf{B}]/b_{ijkl}$ terms are expressed as (Appendix B)

$$b_{iiii} \cong 2 \frac{ES}{8l^3} = 2(N+1)^3 \frac{ES}{8L^3}$$

$$\text{for } i = 1, \ldots, N,$$

$$b_{i(i-1)(i-1)(i-1)} = b_{(i-1)i(i-1)(i-1)} = b_{(i-1)(i-1)i(i-1)}$$

$$= b_{(i-1)(i-1)(i-1)i} \cong -(N+1)^3 \frac{ES}{8L^3}$$

$$\text{for } i = 2, \ldots, N,$$

$$b_{ii(i-1)(i-1)} = b_{(i-1)ii(i-1)} = b_{(i-1)(i-1)ii} = b_{i(i-1)(i-1)i}$$

$$= b_{i(i-1)i(i-1)} = b_{(i-1)i(i-1)i}$$

$$\cong (N+1)^3 \frac{ES}{8L^3} \quad \text{for } i = 2, \ldots, N,$$

$$b_{iii(i-1)} = b_{(i-1)iii} = b_{i(i-1)ii} = b_{ii(i-1)i}$$

$$\cong -(N+1)^3 \frac{ES}{8L^3} \quad \text{for } i = 2, \ldots, N$$

$$(28)$$

and other values of b_{ijkl} are obtained by symmetry relations or are equal to zero.

(1) Calculation of the Equivalent Stiffness for the Winkler Linear Springs.
The linear potential energy stored in the Winkler vertical spring k^l_i, shown in the discrete model of Figure 3, can be written as [9]

$$V^l_{li} = \frac{1}{2} k^l_i (y_i)^2 \quad \text{with } y_0 = y_{N+1} = 0. \quad (29)$$

On the other hand, the elementary potential energy corresponding to an elementary section of length dx in the continuous beam resting on the elastic foundations is given by

$$dV_l^l = \frac{1}{2}k\left(y\left(x\right)\right)^2 dx \qquad (30)$$

with k corresponding to the soil stiffness per length unit. By approximation, one can write

$$\left(y\left(x\right)\right)_i = y_i,$$
$$\left(k\right)_i = k_i, \qquad (31)$$
$$dx = \frac{l_i + l_{i+1}}{2}.$$

By substituting (31) in (30), one obtains

$$dV_l^l = \frac{1}{2}k_i\frac{l_i + l_{i+1}}{2}\left(y_i\right)^2 \quad \text{for } i = 1,\ldots,N. \qquad (32)$$

Since the bars have the same length (30) can be reduced to

$$dV_l^l = \frac{1}{2}k_i l\left(y_i\right)^2 \quad \text{for } i = 1,\ldots,N. \qquad (33)$$

The identification between (29) and (33) gives

$$k_i^l = k_i l = \frac{k_i L}{(N+1)} \quad \text{for } i = 1,\ldots,N. \qquad (34)$$

(2) Expressions for the General Terms of the Tensors k_{ij}^l. The relationship existing between the longitudinal elongations Δy_i of the ith linear spring and the transverse displacements y_i is

$$\Delta y_i = y_i. \qquad (35)$$

The potential energy stored in the ith linear spring, of stiffness k_i^l, is given directly in terms of the stiffness k_i^l and the spring length variation y_i, by the very well-known relationship, that is, $1/2\,k_i^l(y_i)^2$, which leads, after summation, to

$$V_l^l = \frac{1}{2}\sum_{i=1}^{N}k_i^l\left(y_i\right)^2 \quad \text{for } i = 1 \text{ to } N. \qquad (36)$$

The identification between (4) and (30) leads to the following expression for the rigidity tensor associated with the Winkler foundations in DB:

$$k_{ij}^l = \delta_{ij}k_i^l = \delta_{ij}\frac{k_i L}{(N+1)} \quad \text{for } i,j = 1,\ldots,N, \qquad (37)$$

where δ_{ij} is Kronecker's symbol. Equation (37) can be written in a matrix form as

$$\left[\mathbf{K}^l\right] = k_i^l\left[\mathbf{I}\right] = \frac{k_i L}{(N+1)}\left[\mathbf{I}\right] \qquad (38)$$

with $[\mathbf{I}]$ being the identity matrix.

2.2.2. Nondimensionalization. In order to establish the necessary comparisons with previous publications without further manipulations, the nondimensional formulation of the problem presented is essential. To define the nondimensional parameters, let us put

$$\frac{m_{ij}}{m^*_{ij}} = \frac{\overline{m}_{ij}}{m^*_{ij}} = \frac{\rho S L}{N+1} \quad \text{for } i,j = 1,\ldots,N, \qquad (39)$$

$$\frac{k_{ij}^l}{k_{ij}^{l*}} = \frac{\overline{k}_{ij}^l}{k_{ij}^{l*}} = \frac{EI\left(N+1\right)^3}{L^3} \quad \text{for } i,j = 1,\ldots,N, \qquad (40)$$

$$\frac{k_{ij}^s}{k_{ij}^{s*}} = \frac{\overline{k}_{ij}^s}{k_{ij}^{s*}} = \frac{EI\left(N+1\right)^3}{L^3} \quad \text{for } i,j = 1,\ldots,N, \qquad (41)$$

$$\frac{b_{ijkl}}{b^*_{ijkl}} = \frac{\overline{b}_{ijkl}}{b^*_{ijkl}} = \frac{ES\left(N+1\right)^3}{8L^3} \qquad (42)$$

$$\text{for } i,j,k,l = 1,\ldots,N,$$

$$\frac{\omega^*}{\omega} = \sqrt{\frac{EI}{\rho S L^4}}. \qquad (43)$$

The Winkler stiffness coefficient is given by

$$k_{ij}^l = k l \delta_{ij} = \alpha_i\frac{EI\left(N+1\right)^3}{L^3}\delta_{ij} \quad \text{for } i,j = 1,\ldots,N \qquad (44)$$

α_i and λ_i being the nondimensional parameter:

$$\alpha_i = \frac{k_i L^4}{EI\left(N+1\right)^4} = \frac{\lambda_i}{(N+1)^4} \quad \text{with } \lambda_i = \frac{k_i L^4}{EI}. \qquad (45)$$

In the case of a constant soil distribution $\alpha_i = \alpha$ and $\lambda_i = \lambda$.

The nondimensional amplitude A^* is expressed as

$$A^* = \frac{A}{R} \quad \text{with } R = \sqrt{\frac{I}{S}}, \qquad (46)$$

where R is the radius of gyration.

The expression for the nonlinear frequency [19] is

$$\omega_{\text{nl}}^2 = \frac{a_i a_j\left(\overline{k}_{ij}^s + \overline{k}_{ij}^l\right) + \dfrac{3}{2}a_i a_j a_k a_l\overline{b}_{ijkl}}{\overline{m}_{ij}a_i a_j} \qquad (47)$$

$$\text{for } i,j,l,k = 1,\ldots,N.$$

As the nondimensional formulation is established, the general expressions for m^*_{ij}, k^{*s}_{ij}, k^{*l}_{ij}, and b^*_{ijkl} for the discrete model become as follows:

(i) The nondimensional mass tensor $[\mathbf{M}^*]/m^*_{ij}$ is obtained by replacing (25) in (39):

$$m^*_{ij} = \delta_{ij} \quad \text{for } i,j = 1,\ldots,N. \qquad (48)$$

(ii) The nondimensional spiral springs tensor $[\mathbf{K}^{*s}]/k^{*s}{}_{ij}$ is obtained by replacing (26) in (40):

$$k^{*s}{}_{(r-2)r} = 1 \quad \text{for } r = 3, \dots, N,$$

$$k^{*s}{}_{(r-1)r} = -4 \quad \text{for } r = 2, \dots, N, \qquad (49)$$

$$k^{*s}{}_{rr} = 6 \quad \text{for } r = 1, \dots, N$$

and the other values of $k^{*s}{}_{ij}$ are obtained by symmetry relations or are equal to zero.

(iii) The nondimensional Winkler springs tensor $[\mathbf{K}^{*l}]/k^{*l}{}_{ij}$ is obtained by replacing (27) in (41):

$$k^{*l}{}_{ij} = \alpha\delta_{ij} \quad \text{for } i, j = 1, \dots, N. \qquad (50)$$

(iv) The nondimensional nonlinear rigidity tensor $[\mathbf{B}^*]/b^*{}_{ijkl}$ is obtained by replacing (28) in (42):

$$b^*{}_{iiii} = 2 \quad \text{for } i = 1, \dots, N,$$

$$b^*{}_{i(i-1)(i-1)(i-1)} = b^*{}_{(i-1)i(i-1)(i-1)} = b^*{}_{(i-1)(i-1)i(i-1)}$$

$$= b^*{}_{(i-1)(i-1)(i-1)i} = -1$$

$$\text{for } i = 2, \dots, N,$$

$$b^*{}_{ii(i-1)(i-1)} = b^*{}_{(i-1)ii(i-1)} = b^*{}_{(i-1)(i-1)ii} \qquad (51)$$

$$= b^*{}_{i(i-1)(i-1)i} = b^*{}_{i(i-1)i(i-1)}$$

$$= b^*{}_{(i-1)i(i-1)i} = 1 \quad \text{for } i = 2, \dots, N,$$

$$b^*{}_{iii(i-1)} = b^*{}_{(i-1)iii} = b^*{}_{i(i-1)ii} = b^*{}_{ii(i-1)i} = -1$$

$$\text{for } i = 2, \dots, N$$

and other values of $b^*{}_{ijkl}$ are equal to zero.

2.3. Solution of the Nonlinear Algebraic System: Linearization. In order to solve the nonlinear amplitude equation and establish a good approximation for large vibration amplitudes of the beam examined, a linearization of the nonlinear algebraic problem is performed, based on the so-called second formulation introduced in [20]. For this linearization to be accomplished, only the second-order terms of the type $\varepsilon_i\varepsilon_j a_1 b_{ijkr}$ are neglected when considering the first nonlinear mode, in (24), rewritten in summation form as

$$\left(\overline{k^l_{ir}} + \overline{k^s_{ir}} - \omega^2\overline{m_{ir}}\right)a_i + \frac{3}{2}a_ia_ja_k\overline{b_{ijkr}} = 0 \qquad (52)$$

$$\text{for } r = 1, \dots, N.$$

The nonlinear expression $a_ia_ja_k\overline{b^*{}_{ijkr}}$ is split to terms proportional to $a_1{}^3$, terms proportional to $a_1{}^2\varepsilon_i$, and terms proportional to $a_1{}^2\varepsilon_i\varepsilon_j$ which are neglected, leading to

$$a_ia_ja_k\overline{b^*{}_{ijkr}} = a_1{}^3\overline{b^*{}_{111r}} + a_1{}^2\varepsilon_i\overline{b^*{}_{11ir}} \qquad (53)$$

$$\text{for } r = 1, \dots, N.$$

The 1D ratio of the nonlinear nondimensional frequency established from (47) is given by

$$\omega^*{}_{\text{nl}}{}^2 = (N+1)^4 \frac{\left(\overline{k^{s*}{}_{11}} + \overline{k^{l*}{}_{11}}\right)}{\overline{m^*{}_{11}}} \left[1 \right.$$

$$\left. + \frac{3}{16} \frac{\overline{b^*{}_{1111}}}{\left(\overline{k^{s*}{}_{11}} + \overline{k^{l*}{}_{11}}\right)} \left(\frac{a_1}{R}\right)^2 \right]. \qquad (54)$$

Then (52) can be written in a matrix form as

$$\left(\left[\overline{\mathbf{K}^{*l}{}_{RI}}\right] + \left[\overline{\mathbf{K}^{*s}{}_{RI}}\right] - \omega_{\text{nl}}{}^{*2}\left[\overline{\mathbf{M}^*{}_{RI}}\right]\right)\{\mathbf{A}^*{}_{RI}\}$$

$$+ \frac{3}{16}\left[\boldsymbol{\alpha}^*_I\right]\{\mathbf{A}^*{}_{RI}\} = \left\{-\frac{3}{16}\left(\frac{a_1}{R}\right)^3\overline{b^*{}_{111r}}\right\} \qquad (55)$$

with

$$\left[\boldsymbol{\alpha}^*_I\right] = \left(\frac{a_1}{R}\right)^2\overline{b^*{}_{11ir}}, \qquad (56)$$

where $\left[\overline{\mathbf{K}^{*s}{}_{RI}}\right] = \overline{k^{*s}_{ij}}$, $\left[\overline{\mathbf{K}^{*l}{}_{RI}}\right] = \overline{k^{*l}_{ij}}$, and $\left[\overline{\mathbf{M}^*{}_{RI}}\right] = \overline{m^*_{ij}}$ are, respectively, the reduced beam rigidity matrix, the Winkler foundation rigidity matrix, and the mass matrix associated with the first nonlinear mode, obtained by varying i and j in the set $(3, 5, \dots, 11)$. $[\boldsymbol{\alpha}^*_I]$ is a 5×5 square matrix, depending on a_1, whose general term $\alpha_{ij}{}^*$ is equal to $\overline{b_{11ir}}(a_1/R)^2$, and $\{-3/16(a_1/R)^3\overline{b_{111r}}\}$ is a column vector representing the right side of the linear system (55) in which the reduced unknown vector is $\{\mathbf{A}^*{}_{RI}\}^T = [\varepsilon_3/R, \dots, \varepsilon_i/R, \dots, \varepsilon_N/R]$. The modal contributions $\varepsilon_3, \dots, \varepsilon_i, \dots, \varepsilon_N$ can be calculated simply by solving the linear system (55) of five equations and five unknowns [20].

This formulation is proven [20] to give good results when applied to various nonlinear vibration problems. However, to get an even better approximation, it is combined with a Newton-Raphson algorithm, which requires a good initial estimate of the solution (given by the actual linearization method). Accurate results (up to 10^{-14}) are obtained within few numbers of iteration.

3. Results Validation

In Section 2, a general theory is developed for the physically discrete model, with demonstrations carried out for calculating the new model parameters. In order to confirm the extended theory, this section is devoted to the validation of the results in the linear and nonlinear cases for simply supported and clamped beams.

3.1. Simply Supported Beams Resting on Elastic Foundations. In Figure 2, a discrete physical model, made of masses, bars, and springs, is described. In the following section, a validation of the results obtained for a simply supported beam resting on Winkler elastic foundation is discussed. Namely, calculations are performed in the linear and nonlinear cases using Matlab's software, based on the procedure described in Section 2.

TABLE 1: Comparison of the nondimensional frequencies corresponding to the first 3 modes and the nondimensional Winkler soil stiffness for a S-S beam.

| | S-S beam (linear) | | | | | | | | | |
λ	Present study $(N = 10)$	Present study $(N = 50)$	Present study $(N = 100)$	Ref. [10]	Ref. [11]	Exact Ref. [12]	Average	Error % $(N = 10)$	Error % $(N = 50)$	Error % $(N = 100)$
	3,131	3,141	3,141	3,150			3,144	0,427	0,104	0,092
0	6,198	6,279	6,2822	6,280	3,141	3,141	6,280	1,304	0,013	−0,035
	9,139	9,411	9,421	9,420			9,420	2,982	0,092	−0,014
	3,742	3,748	3,7483	3,750			3,749	0,181	0,022	0,016
100	6,300	6,378	6,3807	6,380	3,748	3,7484	6,380	1,245	0,034	−0,010
	9,172	9,441	9,4511	9,450			9,450	2,945	0,093	−0,012
	10,024	10,024	10,024	10,020			10,023	−0,013	−0,016	−0,016
10000	10,350	10,368	10,368	10,360	10,024	10,024	10,360	0,095	−0,076	−0,082
	11,415	11,558	11,563	11,570			11,570	1,343	0,104	0,057
	31,623	31,623	31,623	31,620			31,622	−0,004	−0,004	−0,004
1000000	31,634	31,635	31,635	31,630	31,623	31,623	31,630	−0,016	−0,014	−0,014
	31,678	31,685	31,685	31,720			31,720	0,133	0,112	0,111

3.1.1. Linear Case. One has now to make the form of the matrices introduced in Section 2 explicit, knowing that the calculations are developed in the MB. The nondimensional mass matrix is reduced to the unity matrix $[\mathbf{I}]$, and the stiffness matrix $[\mathbf{K}_N^{SS}]$, combining the effects of the extensional and spiral springs, obtained by addition of the Winkler soil stiffness matrix and the spiral spring matrix, for the simply supported case, can be presented as follows [9]:

$$[\mathbf{K}_N^{SS}] = \frac{(N + 1)^3 EI}{L^3}$$

$$\cdot \begin{bmatrix} 5+\alpha & -4 & 1 & 0 & \cdot & \cdot & 0 \\ -4 & 6+\alpha & -4 & 1 & 0 & \cdot & \cdot \\ 1 & -4 & \cdot & \cdot & \cdot & \cdot & \cdot \\ 0 & 1 & \cdot & \cdot & \cdot & \cdot & 0 \\ \cdot & 0 & \cdot & \cdot & \cdot & -4 & 1 \\ 0 & \cdot & 0 & 1 & -4 & 6+\alpha & -4 \\ 0 & \cdot & \cdot & 0 & 1 & -4 & 5+\alpha \end{bmatrix}, \quad (57)$$

$$[\mathbf{K}_N^{SS}] = \frac{(N + 1)^3 EI}{L^3} [\mathbf{K}_N^{SS*}].$$

With this new expression for the nondimensional rigidity matrix $[\mathbf{K}_N^{SS*}]$ (24), corresponding to linear vibrations, can be written as

$$[\mathbf{K}_N^{SS*}]\{\mathbf{y}\} + (\omega_{\text{discr}})^2 \frac{\rho SL^4}{EI (N + 1)^4} \{\mathbf{y}\} = 0 \quad (58)$$

with $\{\mathbf{y}\}$ being the vector of mass displacements in the MB. Solving (58) gives the eigenvalues $\{\beta\}$, which allows the corresponding frequencies to be written as

$$\omega_l^{SS}{}_{\text{discr}} = (N + 1)^2 \sqrt{\beta_i} \sqrt{\frac{EI}{\rho SL^4}}. \quad (59)$$

Table 1 gives, for the first 3 modes of the simply supported beam, the nondimensional frequencies, for increasing number of masses ($N = 10, 50, 100$) and increasing values of the Winkler foundation stiffness. Accordingly, these nondimensional frequencies are calculated from (59), by extracting the $\sqrt{EI/\rho SL^4}$ term, where the eigenvalues β_i are those of the nondimensional stiffness matrix $[\mathbf{K}_N^{SS*}]$. So, they can be presented as

$$\omega_l^{SS}{}_{\text{discr}}{}^* = (N + 1)^2 \sqrt{\beta_i}. \quad (60)$$

In order to validate the model introduced in Section 2, for the simply supported beam, a comparison of the first 3 natural frequencies is carried out for increasing values of N (the number of masses) and λ (the soil stiffness), and results are drawn in Table 1.

The natural frequencies, Table 1, converge faster by increasing the N or λ. A good approximation is already obtained for ($N = 10$, $\lambda = 0$) where the overall error ((average − present study (N))/average %) is below three percent limit. On the other hand, by increasing the soil stiffness to the case ($N = 10$, $\lambda = 1000000$) the error drops below 0.15%. However, once the discretization number N is higher than 50 the overall error is under the 0.12%. Thus, increasing one of the parameters λ and N, or both, makes the discrete model converge to the reference models [10–12].

3.1.2. Nonlinear Case

(1) Mode Shapes. In [9], the discussions covered only the validation of nonlinear frequencies. In the present work, the validation is extended to the beam mode shapes. Likewise, Table 2 gives a comparison between the mode shapes calculated using the present discrete physical model with previous results [13].

The results are identical to the exact theoretical mode shape in the linear case. However, in the nonlinear case

TABLE 2: Comparison of the normalized first mode shape at different positions of the beam in the S-S beam case for $N = 100$.

$\xi = x/L$	Exact Ref. [13] $A^* = 0$	Present $A^* = 0$	FEM Ref. [13] $A^* = 0$	Present $A^* = 1$	FEM Ref. [13] $A^* = 1$	Present $A^* = 2$	FEM Ref. [13] $A^* = 2$	Present $A^* = 3$	FEM Ref. [13] $A^* = 3$
0,00	0,0000	0,0000	0,0000	0,0000	0,0000	0,0000	0,0000	0,0000	0,0000
0,05	0,1564	0,1564	0,1564	0,1544	0,1564	0,1498	0,1564	0,1449	0,1564
0,10	0,3090	0,3090	0,3090	0,3054	0,3090	0,2969	0,3090	0,2876	0,3090
0,15	0,4540	0,4540	0,4540	0,4493	0,4540	0,4381	0,4540	0,4258	0,4540
0,20	0,5878	0,5878	0,5880	0,5827	0,5880	0,5707	0,5880	0,5570	0,5880
0,25	0,7071	0,7071	0,7071	0,7024	0,7071	0,6911	0,7071	0,6780	0,7071
0,30	0,8091	0,8090	0,8090	0,8053	0,8090	0,7962	0,8090	0,7853	0,8090
0,35	0,8910	0,8910	0,8910	0,8886	0,8910	0,8825	0,8910	0,8750	0,8910
0,40	0,9511	0,9511	0,9511	0,9498	0,9511	0,9468	0,9511	0,9430	0,9511
0,45	0,9877	0,9877	0,9877	0,9874	0,9877	0,9865	0,9877	0,9855	0,9877
0,50	1,0000	1,0000	1,0000	1,0000	1,0000	1,0000	1,0000	1,0000	1,0000

TABLE 3: Comparison of the first three modes' nondimensional nonlinear frequency $\omega^{ss*}_{nl\ discr}/\omega^{ss*}_{l\ discr}$, in S-S beam, for various values of the nondimensional Winkler soil stiffness.

		SIMPLY SUPPORTED BEAM							
λ		$0,1\pi^4$			$1\pi^4$			$10\pi^4$	
A^*	$N = 100$	Ref. [14]	Error %	$N = 100$	Ref. [14]	Error %	$N = 100$	Ref. [14]	Error %
0,2	1,0026	1,0034	0,0843	1,0014	1,0019	0,0494	1,0003	1,0003	0,0044
0,4	1,0102	1,0135	0,3278	1,0056	1,0075	0,1876	1,0010	1,0014	0,0377
0,6	1,0228	1,0301	0,7128	1,0126	1,0167	0,4053	1,0023	1,0031	0,0799
0,8	1,0401	1,0529	1,2144	1,0223	1,0295	0,7036	1,0041	1,0054	0,1310
1	1,0620	1,0814	1,7928	1,0346	1,0457	1,0649	1,0064	1,0085	0,2110
2	1,2294	1,2924	4,8729	1,1319	1,1708	3,3188	1,0253	1,0334	0,7885
3	1,4666	1,5795	7,1497	1,2779	1,3519	5,4771	1,0560	1,0736	1,6424

(high amplitudes), in spite of the result often accepted in the literature, according to which the mode shape in the simply supported case is amplitude independent [18], this discrete physical model leads to mode shapes which seem to be amplitude dependent. As a result, the amplitude dependency may be due to the hypothesis (small triangle angle) taken for the nonlinearity effect which is sensitive to high amplitudes.

(2) Comparison of Nondimension Frequency $\omega^{ss*}_{nl\ discr}/\omega^{ss*}_{l\ discr}$. For validating the results in the nonlinear case, comparing the ratio of the nonlinear frequency parameter $(\omega^{ss*}_{nl\ discr}/\omega^{ss*}_{l\ discr})$ (54) and (60) may be one of the best indicators of the nonlinearity type and acuity. The number of masses considered in the calculation is $(N = 100)$, because of the good approximation obtained for this degree of discretization in the linear case (error below 0.1%).

Table 3 gives the nonlinear frequencies for increasing values of the vibration amplitude and the soil parameter λ. These numerical results give a comparison of the beam nonlinear behavior by the mean of frequency ratio $(\omega^{ss*}_{nl\ discr}/\omega^{ss*}_{l\ discr})$.

For small values of the nondimensional vibration amplitude (<1), the results are close to those of [14] (error $\approx$ 1%). However, once the nondimensional amplitude rises to values higher than one, the difference becomes more noticeable making the error exceed seven percent. This late conclusion may be due to the hypothesis taken for the nonlinearity effect. Another aspect that one may extract from this comparison is that the soil stiffness affects the convergence between the discrete and the continuous models [14], since the error decreases by increasing the soil stiffness coefficient.

3.2. Clamped Beam Resting on Elastic Foundations. This section is concerned with the validation of the results obtained for the clamped beam case. Also, calculations of the frequencies and mode shapes are performed for the linear and nonlinear cases using Matlab's software, based on the procedure described in Section 2.

3.2.1. Linear Case. As it is stated in [9], to establish the clamped beam stiffness matrix, an infinite coefficient should

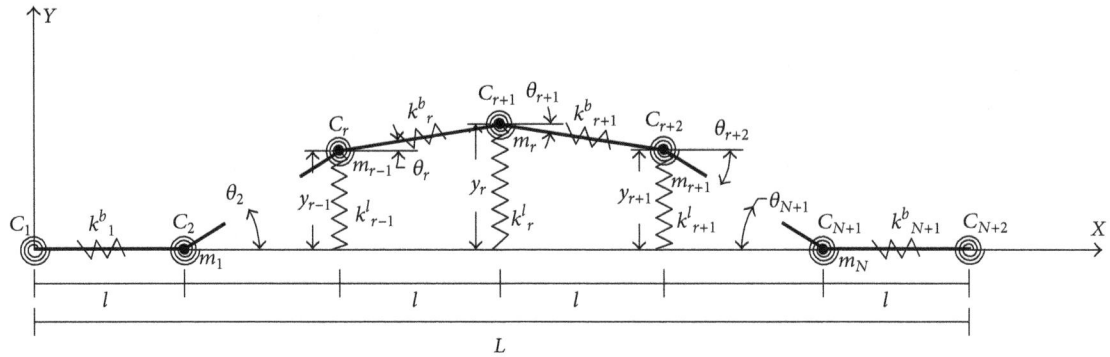

FIGURE 4: The model considered for the discrete C-C beam by blocking the rotation of the first and the last bars.

be taken for the first and last nodes. In that order the following matrix is built:

$$
\left[\mathbf{K}_N^{\text{CC}}\right] = \frac{(N+1)^3 \, EI}{L^3}
$$

$$
\cdot
\begin{bmatrix}
+\infty & -4 & 1 & 0 & \cdot & \cdot & 0 \\
-4 & 6+\alpha & -4 & 1 & 0 & \cdot & \cdot \\
1 & -4 & \cdot & \cdot & \cdot & \cdot & \cdot \\
0 & 1 & \cdot & \cdot & \cdot & \cdot & 0 \\
\cdot & 0 & \cdot & \cdot & \cdot & -4 & 1 \\
0 & \cdot & 0 & 1 & -4 & 6+\alpha & -4 \\
0 & \cdot & \cdot & 0 & 1 & -4 & +\infty
\end{bmatrix}
\qquad (61)
$$

However, once introduced in the calculation, a mathematical error about the conditioning of the matrix arises (ill-conditioned matrix).

Since the present physical model considers that the bars are rigid transversally, it may be concluded that, by blocking (nonrotation) the first bar, the second node located at its end would be blocked too at $y = 0$ as shown in Figure 4. Using this method for taking into account the clamped ends conditions, the N-DOF clamped beam problem is reduced to a $(N-2)$-DOF problem by removing the first and last row and column of the stiffness matrix; then (61) becomes

$$
\left[\mathbf{K}_{N-2}^{\text{CC}}\right] = \frac{(N+1)^3 \, EI}{L^3}
$$

$$
\cdot
\begin{bmatrix}
6+\alpha & -4 & 1 & 0 & \cdot & \cdot & 0 \\
-4 & 6+\alpha & -4 & 1 & 0 & \cdot & \cdot \\
1 & -4 & \cdot & \cdot & \cdot & \cdot & \cdot \\
0 & 1 & \cdot & \cdot & \cdot & \cdot & 0 \\
\cdot & 0 & \cdot & \cdot & \cdot & -4 & 1 \\
0 & \cdot & 0 & 1 & -4 & 6+\alpha & -4 \\
0 & \cdot & \cdot & 0 & 1 & -4 & 6+\alpha
\end{bmatrix},
\qquad (62)
$$

$$
\left[\mathbf{K}_{N-2}^{\text{CC}}\right] = \frac{(N+1)^3 \, EI}{L^3} \left[\mathbf{K}_{N-2}^{\text{CC}}{}^*\right].
$$

With this new expression for the nondimensional rigidity matrix $[\mathbf{K}_{N-2}^{\text{CC}}{}^*]$, the linear vibration equation (24) can be described by

$$
\left[\mathbf{K}_{N-2}^{\text{CC}}{}^*\right]\{\mathbf{y}\} + \left(\omega_l^{\text{CC}}{}_{\text{discr}}\right)^2 \frac{\rho S L^4}{EI\,(N+1)^4}\{\mathbf{y}\} = 0 \qquad (63)
$$

with $\{\mathbf{y}\}$ being the vector of mass displacements in MB. Solving (63) gives the eigenvalues $\{\boldsymbol{\beta}\}$ so one can write the frequencies as

$$
\omega_l^{\text{CC}}{}_{\text{discr}} = (N+1)^2 \sqrt{\beta_i}\sqrt{\frac{EI}{\rho S L^4}}. \qquad (64)
$$

Table 4 gives the first three modes of the clamped beam nondimensional frequencies for increasing number of masses ($N = 10, 50, 100$) and increasing values of the Winkler foundation stiffness. These nondimensional frequencies are calculated from (64) by extracting the $\sqrt{EI/\rho S L^4}$ term, where the eigenvalues β_i are those of the nondimensional stiffness matrix $[\mathbf{K}_{N-2}^{\text{CC}}{}^*]$, so they can be presented as

$$
\omega_l^{\text{CC}}{}_{\text{discr}}^* = (N+1)^2 \sqrt{\beta_i}. \qquad (65)
$$

In order to validate the discrete model, for clamped beams, Table 4 shows a comparison between the results obtained here for the first three modes of a clamped beam, for various values of the soil stiffness coefficients and increasing number of masses $N = (10/50/100)$, and the theoretical values of the natural frequencies of continuous beams obtained in [12].

In conclusion, the natural frequencies drawn in Table 4 converge slower to the continuous model [12] than the simply supported case. The error percentage ((results [12] − present study (N))/results [12]%), for the case of no soil stiffness applied $\lambda = 0$, do not get below the one percent until $N = 100$. However, once the soil stiffness λ is increased, a good approximation is already obtained for the case of $N = 10$. This convergence rate is due to the approximations taken by contraction of the matrices in the clamped beam case.

TABLE 4: Comparison of the first three modes' nondimension frequency and the nondimensional Winkler soil stiffness for a C-C beam.

λ	Present study $(N = 10)$	Present study $(N = 50)$	Present study $(N = 100)$	Ref. [11]	Exact Ref. [12]	Error % $(N = 10)$	Error % $(N = 50)$	Error % $(N = 100)$
				C-C beam (linear)				
0	5,2153	4,8251	4,7775	4,7300	4,7300	−10,261	−2,011	−1,003
	8,5727	8,0083	7,9313	7,8500	7,8540	−9,150	−1,965	−0,984
	11,7658	11,2052	11,1030	11,0000	10,9960	−7,001	−1,902	−0,973
100	5,3833	5,0338	4,9919	4,9500	4,9500	−8,753	−1,692	−0,846
	8,6121	8,0565	7,9809	7,9000	7,9040	−8,959	−1,930	−0,973
	11,7811	11,2229	11,1212	11,0100	11,0140	−6,965	−1,897	−0,974
10000	10,1800	10,1328	10,1278	10,1200	10,1230	−0,563	−0,097	−0,047
	11,1400	10,8995	10,8692	10,8400	10,8390	−2,777	−0,558	−0,279
	13,0681	12,6693	12,5991	12,5300	12,5260	−4,328	−1,144	−0,583
1000000	31,6286	31,6271	31,6269	31,6400	31,6260	−0,008	−0,003	−0,003
	31,6654	31,6552	31,6540	31,6700	31,6530	−0,039	−0,007	−0,003
	31,7732	31,7467	31,7422	31,7500	31,7380	−0,111	−0,027	−0,013

TABLE 5: Comparison of the first mode shape at different positions of the beam in the C-C beam case.

$\xi = x/L$	Present $N = 100$	ASM Ref. [15]	FEM Ref. [13]	Present $N = 100$	GFEM Ref. [16]	FEM Ref. [13]	Present $N = 100$	GFEM Ref. [16]	FEM Ref. [13]
		C-C beam (nonlinear)							
		$A^* = 0$			$A^* = 2$			$A^* = 5$	
0,00	0,0000	0,0000	0,0000	0,0000	0,0000	0,0000	0,0000	0,0000	0,0000
0,05	0,0324	0,0325	0,0329	0,0341	0,0339	0,0341	0,0464	0,0411	0,0413
0,10	0,1196	0,1119	0,1203	0,1242	0,1234	0,1237	0,1555	0,1429	0,1432
0,15	0,2444	0,2435	0,2454	0,2508	0,2505	0,2506	0,2927	0,2795	0,2796
0,20	0,3906	0,3900	0,3922	0,3969	0,3981	0,3982	0,4363	0,4307	0,4308
0,25	0,5435	0,5435	0,5455	0,5480	0,5511	0,5513	0,5757	0,5815	0,5817
0,30	0,6895	0,6901	0,6918	0,6917	0,6960	0,6964	0,7058	0,7200	0,7204
0,35	0,8168	0,8178	0,8189	0,8174	0,8214	0,8219	0,8217	0,8372	0,8376
0,40	0,9156	0,9164	0,9170	0,9155	0,9182	0,9185	0,9159	0,9259	0,9262
0,45	0,9783	0,9787	0,9789	0,9782	0,9792	0,9293	0,9783	0,9812	0,9813
0,50	1,0000	1,0000	1,0000	1,0000	1,0000	1,0000	1,0000	1,0000	1,0000

3.2.2. Nonlinear Case

(1) Mode Shapes. In [9], the discussions covered only the validation of frequencies. In order to have a more complete validation, Table 5 compares the present work mode shapes with previous works.

For clamped beams, the mode shapes are shown both theoretically and experimentally to be amplitude dependent at large vibration amplitudes. Table 5 shows that the discrete model leads to amplitude dependent mode shapes. By comparing these mode shapes to the ones obtained in [13, 15, 16], it can be seen that these results complies with the results obtained by the two methods FEM [13] and GFEM [16].

(2) Comparison of Nondimension Frequency $\omega^{cc*}_{nl\ discr}/$ $\omega^{cc*}_{l\ discr}$. As it is explained in the linear case Section 3.2.1, the

same procedure for implementing the beam end conditions is applied for the nonlinear rigidity tensor, which is also subjected to dimension reduction by removing the first and last column and row of the 4D tensor. Once this manipulation done, the results of $(\omega^{cc*}_{nl\ discr}/\omega^{cc*}_{l\ discr})$ given by (54) and (65) for various values of the vibration amplitude and soil stiffness are summarized in Table 6. The number of masses taken is $(N = 100)$, for which a good convergence is reached in the linear case (error below 1%).

The nonlinear behavior of a clamped beam resting on elastic foundations is presented in Table 6 by means of the amplitude dependence of the parameter $(\omega^{cc*}_{nl\ discr}/\omega^{cc*}_{l\ discr})$. For small values of nondimension amplitude (<1), the results are very close to [14] (error <1%). However, once the nondimensional amplitude rises (>1), the difference becomes more significant. The last conclusion may be due as it is stated

TABLE 6: Comparison of first modes nondimensional nonlinear frequency $\omega_{nl\ discr}^{cc*}/\omega_{l\ discr}^{cc*}$, for various values of the nondimensional Winkler soil stiffness for a C-C beam.

	Clamped-clamped beam								
λ	$0{,}1\pi^4$			$1\pi^4$			$10\pi^4$		
A^*	$N = 100$	Ref. [14]	Error %	$N = 100$	Ref. [14]	Error %	$N = 100$	Ref. [14]	Error %
0,2	1,0013	1,0009	−0,0397	1,0011	1,0008	−0,0314	1,0005	1,0003	−0,0161
0,4	1,0052	1,0037	−0,1475	1,0044	1,0032	−0,1244	1,0018	1,0013	−0,0541
0,6	1,0116	1,0082	−0,3391	1,0100	1,0071	−0,2859	1,0041	1,0029	−0,1236
0,8	1,0206	1,0146	−0,5879	1,0177	1,0125	−0,5110	1,0073	1,0052	−0,2137
1	1,0320	1,0227	−0,9046	1,0275	1,0195	−0,7829	1,0115	1,0081	−0,3330
2	1,1224	1,0876	−3,1963	1,1058	1,0756	−2,8113	1,0451	1,0320	−1,2670
3	1,2587	1,1870	−6,0403	1,2253	1,1625	−5,4062	1,0988	1,0705	−2,6458

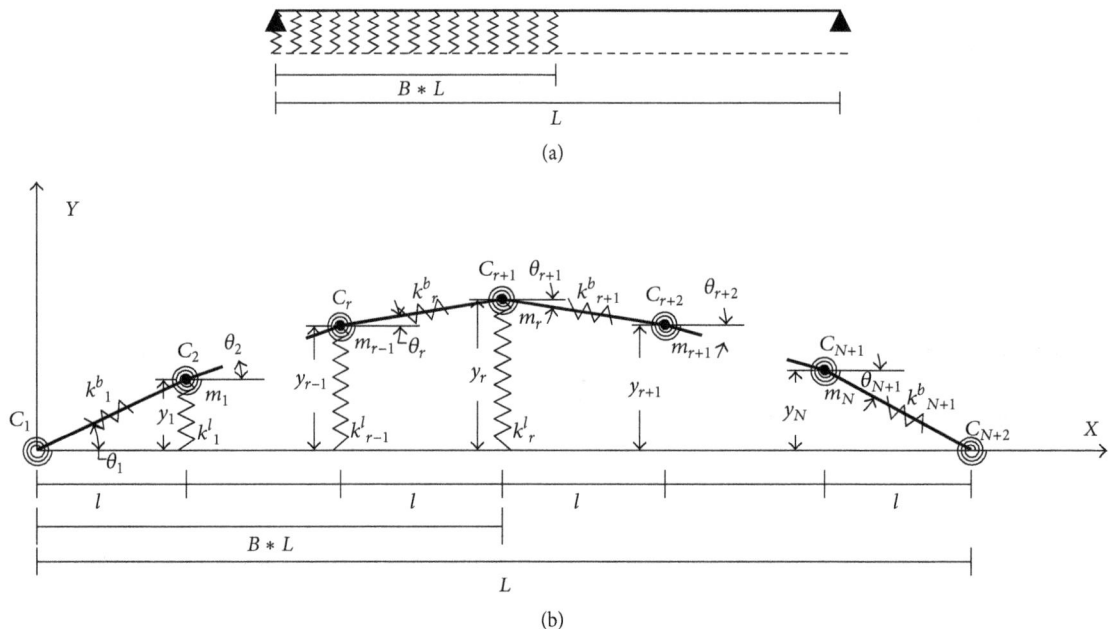

FIGURE 5: A partially supported continuous S-S beam.

for the simply supported case to the model taken for the nonlinearity effect. It may also be noticed that the effect of the soil stiffness on the nonlinear frequency affects the error between the discrete and the continuous model where the error decreases for the same nondimensional amplitude for high values of the soil stiffness parameter λ.

4. Applications

One of the main purposes in developing the discrete physical model presented here is to allow easy variation of the different parameters, compared with the continuous models, especially in the nonlinear case for which analytical solutions may be very laborious and, quite often, impossible to obtain. Since the key factor in the present work is the soil stiffness, two applications of an important practical interest are examined. The first application is concerned with a beam supported partially on elastic foundations. The second is concerned with a beam supported on a Winkler elastic foundation with a variable stiffness. In both cases, simply supported and clamped beams are examined.

4.1. A Partially Supported Beam. In order to get a good approximation in the nonlinear case, the number N of masses used in the discretization process is taken equal to 100, taking the same algorithm developed according to the steps discussed in Section 2 and varying the soil stiffness parameter λ values and the repartition inside the algorithm between the supported span ($\lambda_i \neq 0$) and the nonsupported span ($\lambda_i = 0$).

4.1.1. A Simply Supported Beam (Linear Case). In this section, a simply supported beam is assumed to be subjected to the effect of partial intermediate supports and comparison is made between the results of the continuous model of Figure 5(a) and the discrete model of Figure 5(b).

For a practical convenience, a semilogarithmic curve is drawn, showing the relative stiffness parameter λ on the abscissa and the nondimensional frequency parameter

$(\sqrt{\omega_l^{SS*}}_{discr})$ on the ordinate. The ratio of the supported span to the total span B [26] chosen covers the cases where the beam is supported along 25, 50, and 75 percent of its span; the extreme cases are also taken where $B = 0$, no-intermediate supports, and $B = 1$, fully supported. The results correspond to the case $N = 100$, which also enabled a simplification of B ratios (25% corresponds to 25 nodes supported and so on).

In order to establish a comparison with the results given in [26], the first five linear modes NDF(λ) curves are drawn. Since the results extracted from [26] consisted only of curves, a graphical comparison is conducted showing a very good approximation. The conclusions drawn from Figure 6 are that the first modes are more sensitive to the soil stiffness variation. For the first mode the influence on the frequency is noticeable at an early stage ($\lambda = 10$). However, the 5th mode seems to be less sensitive to the soil stiffness variation since the frequency jump is noticeable only for $\lambda = 1000$.

4.1.2. A Simply Supported Beam (Nonlinear Case). After validating the physical model in the linear regime, one may investigate the nonlinearity effect on the NDF(λ)($\sqrt{\omega_l^{SS*}}_{discr}$).

It may be concluded in Figure 7(b), corresponding to a high nondimensional amplitude ($A^* = 3$), that the difference between the curves is more accentuated than in Figure 7(a) ($A^* = 1$), corresponding to a lower nondimensional amplitude and, consequently, a less accentuated nonlinear effect. By deepening the analysis, the gaps between the curves seem to take different paths: In the case of $B = 1$, the gap is very small, but by taking a zoom, the gap is more pronounced for low soil stiffness and tends to become narrower as the soil stiffness rises. In the case of $B = 0.75$, contrary to the case $B = 1$, the gap between the curves seems to widen by increasing the soil stiffness. In the case of $B = 0.5$, the same conclusion may be made as in the case $B = 0.75$ but the gap seems to remain constant. For $B = 0.25$, the same behavior is obtained as for $B = 0.5$. In the case of $B = 0$, the two curves are parallel.

4.1.3. A Clamped Beam (Linear Case). In this section, the case of partially supported beam with clamped ends is investigated by comparing the results of the continuous model of Figure 8(a) to the discrete model of Figure 8(b).

The study of partially supported beams in [26] covered only the simply supported and free ends cases. In [27] and based on the work of [26], the clamped beam case is investigated.

In [27], a study is conducted using continuous beam models and results are given for the 4 first linear modes shapes. By graphically comparing the curves NDF(λ)($\sqrt{\omega_l^{CC*}}_{discr}$) of Figure 9 to [27], a very good approximation is reached by the present physically discrete model. The same conclusion is obtained for the simply supported beam in terms of frequency variation.

4.1.4. A Clamped Beam (Nonlinear Case). After establishing and validating the physical model in the linear regime in Section 4.1.3, a comparison is carried out here for the

nonlinear case. To show the impact of nonlinearity, the NDF(λ)($\sqrt{\omega_l^{CC*}}_{discr}$) curves in the linear case ($A^* = 0$) and in the nonlinear case ($A^* \neq 0$) are drawn in the same graph Figure 10.

In Figure 10(b), which corresponds to a high value of the nondimensional amplitude ($A^* = 3$), the difference between the curves is more expressed in Figure 10(a) ($A^* = 1$), corresponding to a lower nondimensional amplitude and, consequently, a less accentuated nonlinear effect. By deepening the analysis, the gaps between the curves seem to take different paths: in the case of $B = 1$, the gap is very small, but by taking a zoom, the gap is more pronounced for low soil stiffness and tends to become narrower as the soil stiffness rises. In the case of $B = 0.75$, contrary to the case $B = 1$, the gap between the curves seems to widen by increasing the soil stiffness. In the case of $B = 0.5$, the same conclusion may be made as in the case $B = 0.75$ but the gap seems to remain constant. For $B = 0.25$, the same behavior is obtained as for $B = 0.5$. In the case of $B = 0$, the two curves are parallel. However, the gaps between the curves are narrower than in the simply supported case.

4.2. Variable Elastic Foundation. Foundations are often not as homogenous as one may expect, especially for long beams such as pipes and piers, because soil is made of multiple layers with a variable stiffness. In order to approach this reality, the foundation may be sometimes modeled with a distribution of stiffness with linear (66) or parabolic (67) variations. In [17], this problem is solved using the Differential Transform Method (DTM) and in the following Sections 4.2.1 and 4.2.2, comparison of frequencies is made between the discrete model introduced in this paper and the DTM.

$$k_l(x) = \lambda(1 - \pi x), \tag{66}$$

$$k_p(x) = \lambda\left(1 - \tau x^2\right). \tag{67}$$

Equations (66) and (67) are introduced in the calculation algorithm and results are obtained.

4.2.1. A Simply Supported Beam (Linear Case). Simplification of variable implementation is the aim of the present discrete model. Figure 11 shows a simply supported beam resting on a variable elastic foundation and Table 7 gives the variation of the frequency parameter Ω_i associated with the ith mode shape, for $i = 1$ to 8, in accordance with the soil variation for the simply supported beam.

The results obtained for the simply supported beam case, drawn in Table 7 for a linear soil distribution and in Table 8 for a parabolic distribution, for the first eight modes (frequency parameter Ω_i), show that the soil distribution has a low impact on the convergence speed. On the other hand, the discrete model converges well to the results of [17] for a low value of the number of masses since the error is already below one percent for $N = 10$.

4.2.2. A Simply Supported Beam (Nonlinear Case). Figure 12 shows the backbone curve of a simply supported beam

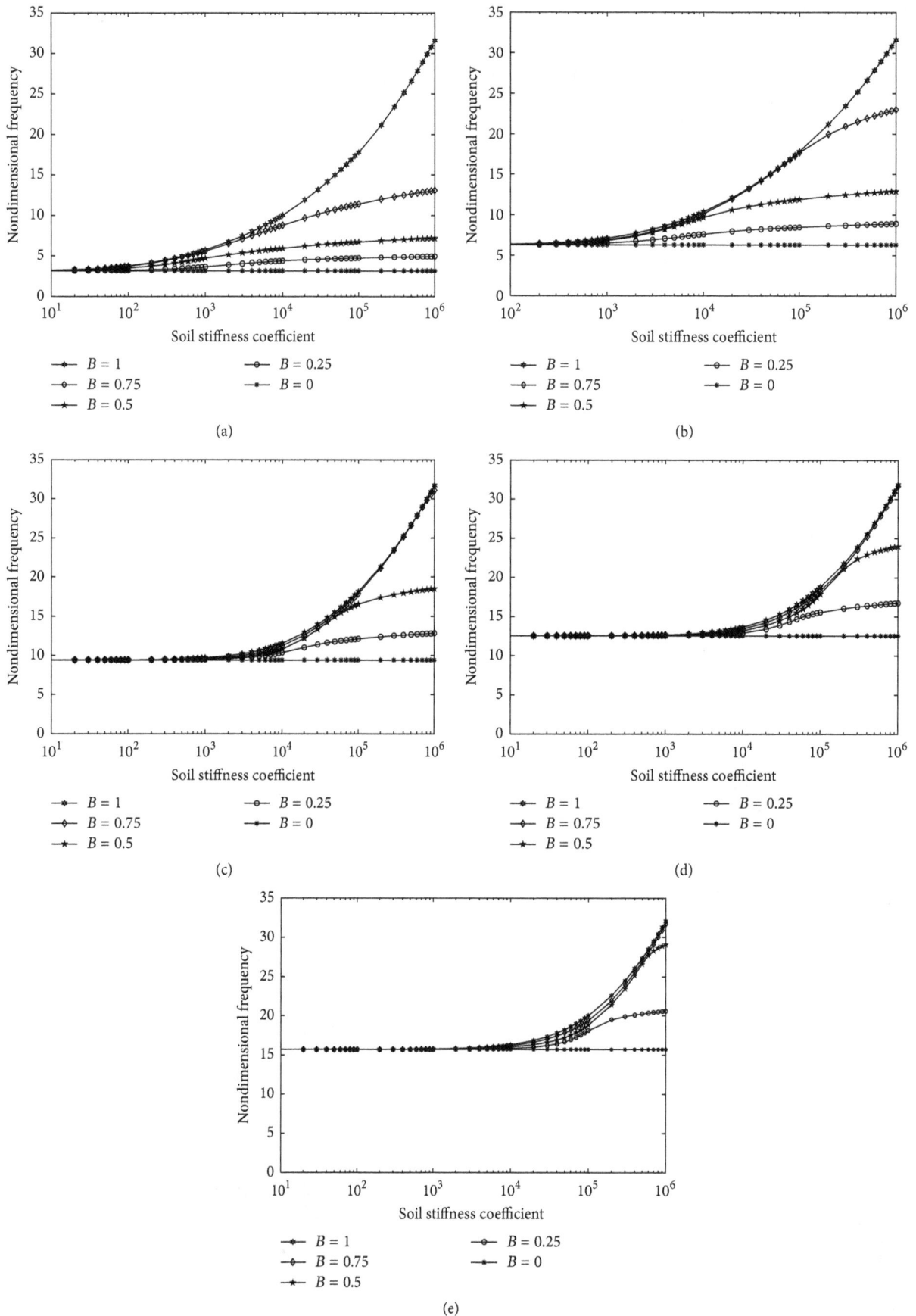

FIGURE 6: Nondimensional frequency $\sqrt{\omega_i^{SS*}{}_{discr}}(\lambda)$ for a simply supported beam: (a) first mode; (b) second mode; (c) third mode; (d) forth mode; (e) fifth mode.

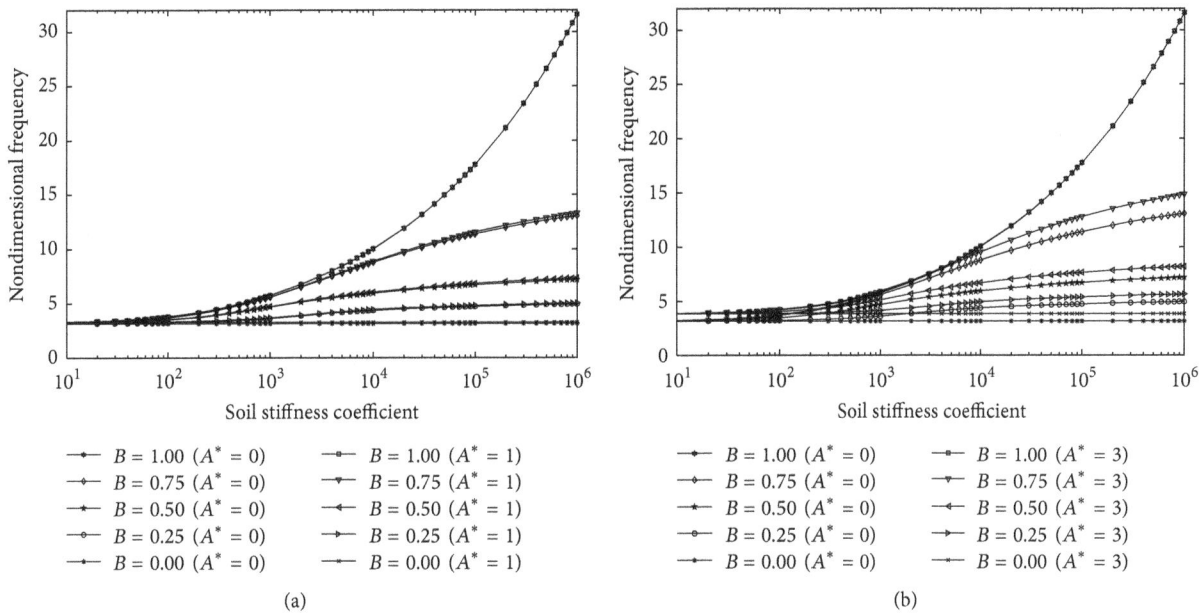

FIGURE 7: Nondimensional frequency $\sqrt{\omega_l^{SS*}}_{\text{discr}}(\lambda)$ for a simply supported beam: (a) first mode ($A^* = 0/A^* = 1$); (b) first mode ($A^* = 0/A^* = 3$).

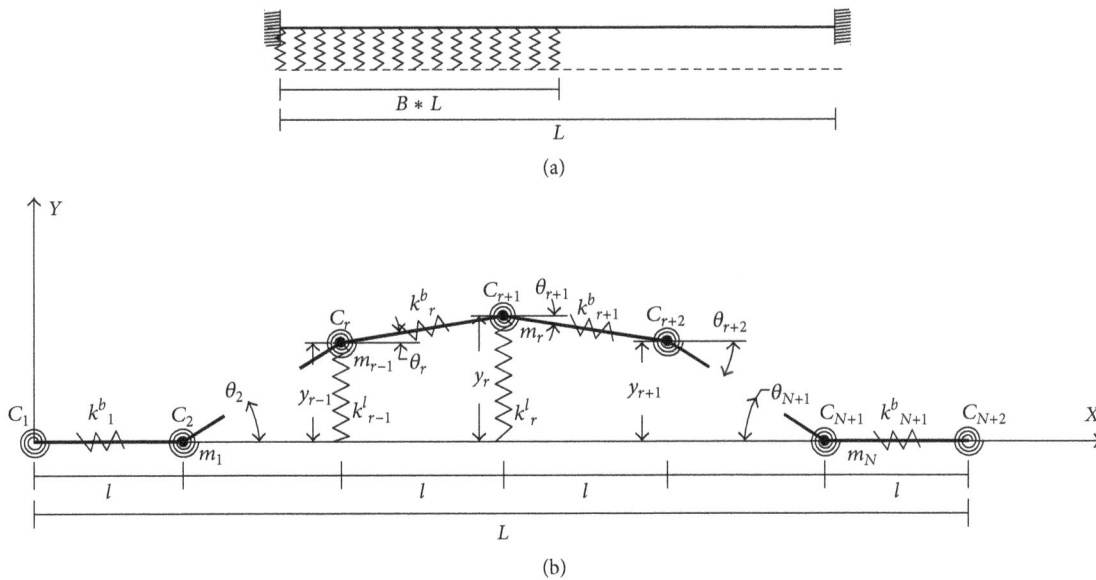

FIGURE 8: A partially supported C-C beam: (a) continuous model; (b) discrete model.

resting on a variable elastic foundation. In Figure 12(a), corresponding to a low soil stiffness ($\lambda = 10$), the nonlinear behavior is more expressed since the nonlinear frequency is higher than the one of the high soil stiffness Figure 12(b) ($\lambda = 1000$). However, for this last case the curves are more distinct.

4.2.3. A Clamped Beam (Linear Case). Figure 13 shows a clamped beam resting on a variable elastic foundation.

Table 9 indicates the variation of the frequency parameter Ω_i associated with the ith mode shape, with $i = 1$ to 3, for the linear and parabolic soil stiffness variation, by increasing both the number of masses from $N = 10$ to $N = 100$ and the soil stiffness from 1 to 1000. It is obvious form the results obtained that the soil distribution (linear and parabolic) has no effect on the convergence. However, a low number of masses ($N = 10$) is not sufficient to get a good approximation (error about

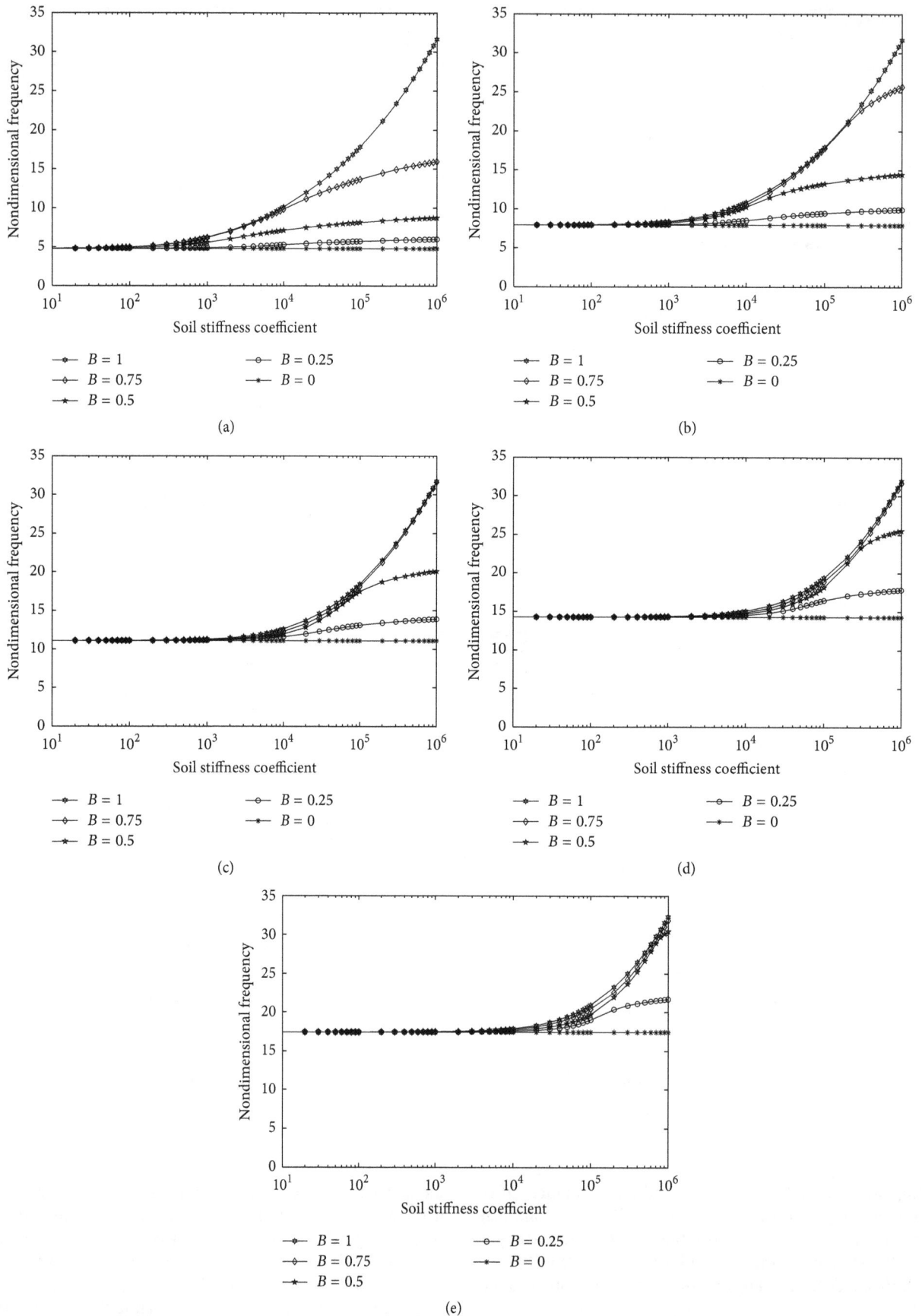

FIGURE 9: Nondimensional frequency $\sqrt{\omega_l^{CC*}}_{discr}(\lambda)$ for a clamped beam: (a) first mode; (b) second mode; (c) third mode; (d) forth mode; (e) fifth mode.

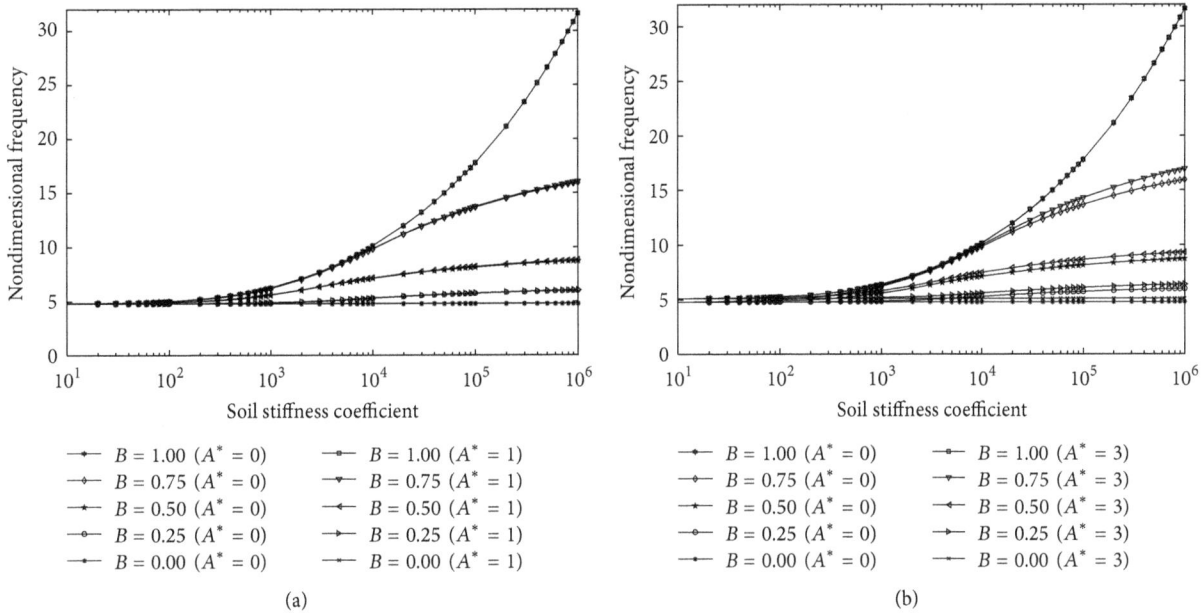

FIGURE 10: Nondimensional frequency $\sqrt{\omega_l^{CC*}{}_{discr}}(\lambda)$ for a clamped beam: (a) first mode ($A^* = 0/A^* = 1$); (b) first mode ($A^* = 0/A^* = 3$).

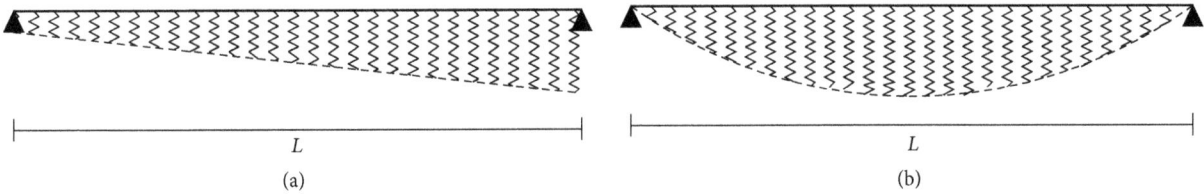

FIGURE 11: Model of S-S beam supported on variable soil. (a) Linear distribution; (b) parabolic distribution.

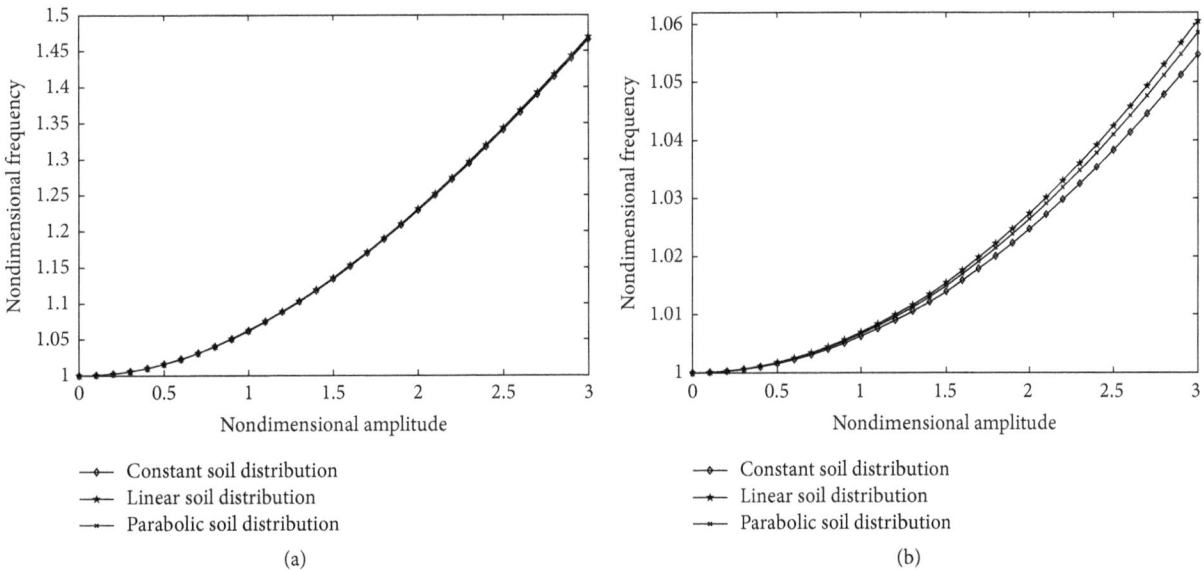

FIGURE 12: Backbone curve nondimensional frequency $\omega_{nl}^{SS*}{}_{discr}/\omega_l^{SS*}{}_{discr}(A^*)$: (a) ($\tau = 0.2$, $\pi = 0.2$, and $\lambda = 10$); (b) ($\tau = 0.2$, $\pi = 0.2$, and $\lambda = 1000$).

TABLE 7: The first eight modes' frequencies for the linear and parabolic soil distribution in S-S beam case.

λ			Linear ($\pi = 0.2$)							
			Ω_1	Ω_2	Ω_3	Ω_4	Ω_5	Ω_6	Ω_7	Ω_8
10	Ref. [17]		3,2118	6,2922	9,4275	12,5675	15,7085	18,8499	21,9914	25,1329
	Present work	$N = 10$	3,2010	6,2074	9,1420	11,8954	14,4077	16,6270	18,5079	20,0122
		$N = 100$	3,2116	6,2912	9,4240	12,5594	15,6927	18,8225	21,9479	25,0681
100	Ref. [17]		3,6999	6,3720	9,4515	12,5777	15,7138	18,8529	21,9933	25,1342
	Present work	$N = 10$	3,6884	6,2895	9,1681	11,9073	14,4144	16,6313	18,5111	20,0147
		$N = 100$	3,6993	6,3709	9,4481	12,5696	15,6980	18,8256	21,9499	25,0694
1000	Ref. [17]		5,6185	7,0420	9,6828	12,6783	15,7657	18,8831	22,0123	25,1469
	Present work	$N = 10$	5,6021	6,9748	9,4177	12,0242	14,4808	16,6747	18,5426	20,0396
		$N = 100$	5,6171	7,0406	9,6794	12,6702	15,7500	18,8558	21,9690	25,0822

TABLE 8: The first eight modes' frequencies for the linear and parabolic soil distribution in S-S beam case.

λ			Parabolic ($\tau = 0.2$)							
			Ω_1	Ω_2	Ω_3	Ω_4	Ω_5	Ω_6	Ω_7	Ω_8
10	Ref. [17]		3,2150	6,2926	9,4276	12,5675	15,7086	18,8499	21,9914	25,1329
	Present work	$N = 10$	3,2042	6,2078	9,1421	11,8955	14,4077	16,6270	18,5079	20,0122
		$N = 100$	3,2148	6,2916	9,4241	12,5594	15,6927	18,8226	21,9480	25,0681
100	Ref. [17]		3,7212	6,3755	9,4526	12,5781	15,7140	18,8530	21,9933	25,1342
	Present work	$N = 10$	3,7090	6,2928	9,1691	11,9077	14,4146	16,6315	18,5112	20,0148
		$N = 100$	3,7206	6,3744	9,4491	12,5700	15,6982	18,8257	21,9499	25,0694
1000	Ref. [17]		5,6788	7,0676	9,6923	12,6824	15,7679	18,8843	22,0131	25,1474
	Present work	$N = 10$	5,6600	6,9987	9,4269	12,0285	14,4832	16,6763	18,5438	20,0406
		$N = 100$	5,6772	7,0660	9,6888	12,6744	15,7521	18,8571	21,9697	25,0827

TABLE 9: The 3 first modes' frequencies for the linear and parabolic soil distribution in C-C beam case.

λ			Linear ($\pi = 0.2$)			Parabolic ($\tau = 0.2$)		
			Ω_1	Ω_2	Ω_3	Ω_1	Ω_2	Ω_3
1	Ref. [17]		4,7322	7,8537	10,9958	4,7323	7,8537	10,9958
	Present work	$N = 10$	4,2684	7,0143	9,6267	4,2685	7,0143	9,6267
		$N = 100$	4,6849	7,7746	10,8833	4,6850	7,7747	10,8833
10	Ref. [17]		4,7512	7,8579	10,9973	4,7522	7,8581	10,9974
	Present work	$N = 10$	4,2799	7,0169	9,6277	4,2805	7,0170	9,6278
		$N = 100$	4,7029	7,7786	10,8848	4,7039	7,7788	10,8848
100	Ref. [17]		4,9297	7,8993	11,0125	4,9391	7,9013	11,0132
	Present work	$N = 10$	4,3897	7,0427	9,6378	4,3953	7,0437	9,6381
		$N = 100$	4,8730	7,8180	10,8992	4,8820	7,8199	10,8999
1000	Ref. [17]		6,1172	8,2815	11,1611	6,1665	8,2988	11,1677
	Present work	$N = 10$	5,1964	7,2860	9,7362	5,2301	7,2957	9,7398
		$N = 100$	6,0167	8,1825	11,0407	6,0643	8,1989	11,0469

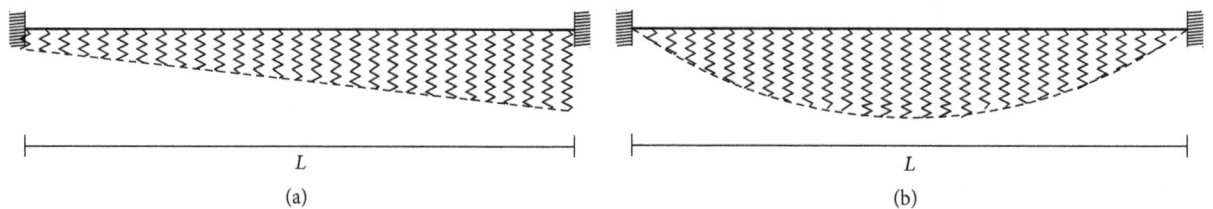

FIGURE 13: Model of C-C beam supported by a soil with a variable stiffness: (a) linear distribution; (b) parabolic distribution.

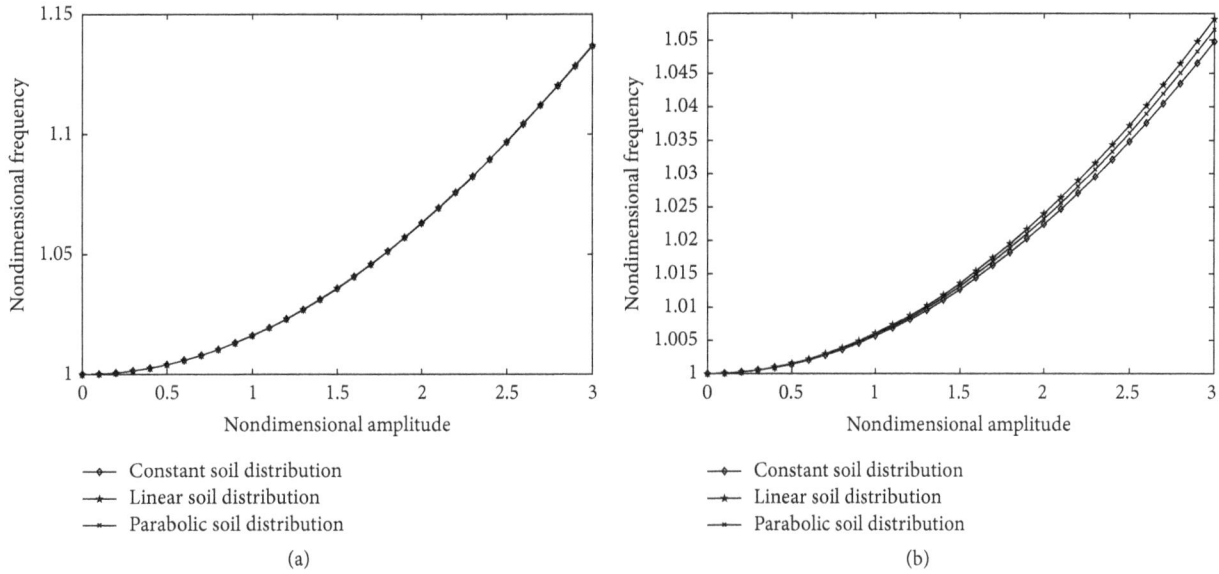

FIGURE 14: Backbone curve nondimensional frequency $\omega_{nl\;discr}^{CC*}/\omega_{l\;discr}^{CC*}(A^*)$: (a) ($\tau = 0.2$, $\pi = 0.2$, and $\lambda = 10$); (b) ($\tau = 0.2$, $\pi = 0.2$, and $\lambda = 1000$).

10%). For $N = 100$, the results become accurate with an error around one percent.

In conclusion, by applying the present model to the problem of vibrations of beams resting on variable elastic foundations, the simply supported beam converges faster than the clamped beam. This convergence may be expected because of the hypothesis taken to solve the clamped beam problem, based on the reduced matrix, and form the behavior of the clamped beam, which is more restrained than the simply supported one.

4.2.4. A Clamped Beam (Nonlinear Case). The backbone curve presenting the nondimensional frequency is drawn for the case of a clamped beam in Figure 14. In Figure 14(a), with a low soil stiffness ($\lambda = 10$), the nonlinear behavior is more expressed since the nonlinear frequency is higher than the one of the high soil stiffness Figure 14(b) ($\lambda = 1000$). On the other hand, the nonlinear frequency seems to be lower than the simply supported case. This frequency difference is less distinguishable for high soil stiffness $\lambda = 1000$.

5. Conclusion

A discrete model of a beam made of multiple bars, masses, and spiral springs is introduced in [9]. Accordingly, the present paper extends the previous model to the case of beams resting on elastic foundations, and validations are carried out for both the linear and nonlinear cases with various soil stiffness's and beam's end conditions.

Before validation of the results obtained via the extended theory and to complete the results of the previous paper [9], a mode shape analysis using the present model is investigated in the linear and nonlinear cases for simply supported and clamped beams. For the clamped beam, a good approximation is established, since amplitude dependent mode shapes are obtained, in agreement with previous theoretical and experimental works. However, in the simply supported case, the mode shape seemed to be slightly deviating from the original shape for high amplitudes, which is not conformed to the result mentioned in [9, 18], according to which it is amplitude independent, namely, because of the hypothesis taken in the discrete model for the nonlinearity effect, which is sensitive to the amplitude variation (small triangle angle approximation). On the other hand, the nonlinear/linear frequency ratios are in a good agreement with the result given in [13], based on the FEM method. For the case of a simply supported beam, the linear theory is validated with an error less than three percent for $N = 10$ to an error less than 0.1% for $N = 100$ in all the cases considered. In the nonlinear case, the error increased for high amplitudes to seven percent but decreased to one percent by increasing the soil stiffness.

For the clamped beam, the convergence is relatively slower, which is expected because of the hypothesis taken. In the linear case, the error is about ten percent for $N = 10$ and dropped to one percent for $N = 100$. In the nonlinear case, the error increased for high amplitudes to six percent but it decreased to about three percent by increasing the soil stiffness.

The main objective of this paper is to construct a general discrete physical model enabling easy solutions in various configurations of a practical interest. Two applications were conducted. The first application deals with beams partially supported on elastic foundations, and the results obtained complies graphically with [26, 27] in the linear case. For the nonlinear case, similar curves corresponding to the first nonlinear mode are plotted for various amplitudes ($A^* = 1$

and $A^* = 3$), to show the shifting from the linear to the nonlinear curves. The second application is concerned with beams resting on elastic foundations with a variable stiffness (linear and parabolic). The comparison of the results obtained with previous results [17] shows the convergence of the present method for different modes, soil stiffness, and beam ends conditions. For this application, the simply supported case is proven faster to converge on the linear case. For the nonlinear case, the backbone curves are drawn with various soil stiffness coefficients ($\lambda = 10$, $\lambda = 1000$), showing the deviation introduced by the nonlinearity effect.

Appendix

A. Simplified Calculation $\overline{b_{ijkl}}$

As stated in Section 2.1, the nonlinearity tensor $\overline{b_{ijkl}}$ may be written as

$$\overline{b_{ijkl}} = \varphi_{si}\varphi_{tj}\varphi_{pk}\varphi_{ql}b_{stpq}. \tag{A.1}$$

The summation (A.1) may be rewritten, that is, Matlab program, as

$$
\begin{aligned}
&\text{for } i = 1:N; \\
&\quad \text{for } j = 1:N; \\
&\quad\quad \text{for } k = 1:N; \\
&\quad\quad\quad \text{for } l = 1:N; \\
&\quad\quad\quad\quad \text{for } s = 1:N; \\
&\quad\quad\quad\quad\quad \text{for } t = 1:N; \\
&\quad\quad\quad\quad\quad\quad \text{for } p = 1:N; \\
&\quad\quad\quad\quad\quad\quad\quad \text{for } q = 1:N; \\
&\overline{b}(i,j,k,l) = b(i,j,k,l) + \varphi(s,i)*\varphi(t,j)*\varphi(p,k)*\varphi(q,l)*b(s,t,p,q); \\
&\quad\quad\quad\quad\quad\quad\quad\quad \text{end}\ldots
\end{aligned}
\tag{A.2}
$$

This means the number of calculations to establish the nonlinearity tensor $\overline{b_{ijkl}}$ terms is N^8.

In Section 2.3 a method to solve the complex nonlinear problem is proposed. In this method the $\overline{b_{ijkl}}$ term required for solving the equation is $\overline{b_{11kl}}$, so the earlier (A.2) summation becomes

$$
\begin{aligned}
&\text{for } k = 1:N; \\
&\quad \text{for } l = 1:N; \\
&\quad\quad \text{for } s = 1:N; \\
&\quad\quad\quad \text{for } t = 1:N; \\
&\quad\quad\quad\quad \text{for } p = 1:N; \\
&\quad\quad\quad\quad\quad \text{for } q = 1:N; \\
&\overline{b}(1,1,k,l) = b(1,1,k,l) + \varphi(s,1)*\varphi(t,1)*\varphi(p,k)*\varphi(q,l)*b(s,t,p,q); \\
&\quad\quad\quad\quad\quad\quad \text{end}\ldots
\end{aligned}
\tag{A.3}
$$

Now the number of calculation dropped to N^6.

By analyzing the form and symmetry patterns of the nonlinear tensor, a simplification is established reducing the six earlier nested loops to only three, as for $p = 1 : N;$

   for $q = 1 : N;$

     $\bar{b}(1,1,p,q) = 2 * \varphi(1,s) * \varphi(1,t) * \varphi(1,p) * \varphi(1,q);$

     for $i = 2 : N;$

     $M0 = 2 * \varphi(i,s) * \varphi(i,t) * \varphi(i,p) * \varphi(i,q);$

     $M1 = -1 * \varphi(i,s) * \varphi(i-1,t) * \varphi(i-1,p) * \varphi(i-1,q);$

     $M2 = -1 * \varphi(i-1,s) * \varphi(i,t) * \varphi(i-1,p) * \varphi(i-1,q);$

     $M3 = -1 * \varphi(i-1,s) * \varphi(i-1,t) * \varphi(i,p) * \varphi(i-1,q);$

     $M4 = -1 * \varphi(i-1,s) * \varphi(i-1,t) * \varphi(i-1,p) * \varphi(i,q);$

     $M5 = -1 * \varphi(i-1,s) * \varphi(i,t) * \varphi(i,p) * \varphi(i,q);$

     $M6 = -1 * \varphi(i,s) * \varphi(i-1,t) * \varphi(i,p) * \varphi(i,q);$ (A.4)

     $M7 = -1 * \varphi(i,s) * \varphi(i,t) * \varphi(i-1,p) * \varphi(i,q);$

     $M8 = -1 * \varphi(i,s) * \varphi(i,t) * \varphi(i,p) * \varphi(i-1,q);$

     $M9 = 1 * \varphi(i,s) * \varphi(i-1,t) * \varphi(i-1,p) * \varphi(i,q);$

     $M10 = 1 * \varphi(i-1,s) * \varphi(i,t) * \varphi(i-1,p) * \varphi(i,q);$

     $M11 = 1 * \varphi(i-1,s) * \varphi(i-1,t) * \varphi(i,p) * \varphi(i,q);$

     $M12 = 1 * \varphi(i-1,s) * \varphi(i,t) * \varphi(i,p) * \varphi(i-1,q);$

     $M13 = 1 * \varphi(i,s) * \varphi(i-1,t) * \varphi(i,p) * \varphi(i-1,q);$

     $M14 = 1 * \varphi(i,s) * \varphi(i,t) * \varphi(i-1,p) * \varphi(i-1,q);$

     $\bar{b}(1,1,p,q) = \bar{b}(1,1,p,q) + M0 + M1 + M2 + M3 + M4 + M5 + M6 + M7 + M8 + M9 + M10 + M11 + M12 + M13 + M14;$

end...

B. Parameters Calculation Summary

The following appendix summarized the demonstrations already carried out in [9].

B.1. Expressions for the General Terms of the Tensors $\mathbf{m_{ij}}$. The kinetic energy of the N-DOF system is the sum of the kinetic energies of the masses, considered as particles, concentrated at the nodes:

$$T = \frac{1}{2}\dot{y}_i^{\,2} m_i. \qquad (B.1)$$

The value of the concentrated mass m_i for a beam divided into $(N + 1)$ bars is equal to $m_i = \rho SL/(N + 1)$, $i = 1,\dots,N$.

Identifying the terms of (6) and (B.1) leads to the following expression for the mass tensor in DB:

$$m_{ij} = \frac{\rho SL}{(N+1)}\delta_{ij} \quad \text{for } i, j = 1,\dots,N, \qquad (B.2)$$

where δ_{ij} is Kronecker's symbol.

B.2. Expressions for the General Terms of the Tensors $\mathbf{k^s_{ij}}$. By considering that transverse displacements are relatively small, compared to the length of the bars, of the discrete

system shown in Figure 15, the following approximation can be established:

$$\sin \theta_i = \frac{y_i - y_{i-1}}{l} \approx \theta_i \qquad (B.3)$$

for $i = 1,\dots,N + 1$ with $y_{N+1} = y_0 = 0$.

The linear potential energy V_l^s of the N-DOF system, resulting from the linear counterpart of the stretching forces in the $N + 2$ spiral springs, is given by

$$V_l^s = \frac{1}{2}\sum_{i=1}^{N+2} C_i (\Delta\theta_i)^2 = \frac{1}{2}\sum_{i=1}^{N+2} C_i (\theta_i - \theta_{i-1})^2 \qquad (B.4)$$

with $\theta_0 = \theta_{N+2} = 0.$

By substituting (B.4), $\theta_i - \theta_{i-1}$, with its expression in (B.3), V_l^s, one can write

$$V_l^s = \frac{1}{2}\sum_{i=1}^{N+2} C_i (\Delta\theta_i)^2 = \frac{1}{2l^2}\sum_{i=1}^{N+2} C_i (y_i - 2y_{i-1} + y_{i-2})^2 \qquad (B.5)$$

with $y_{N+2} = y_{N+1} = y_0 = y_{-1} = 0.$

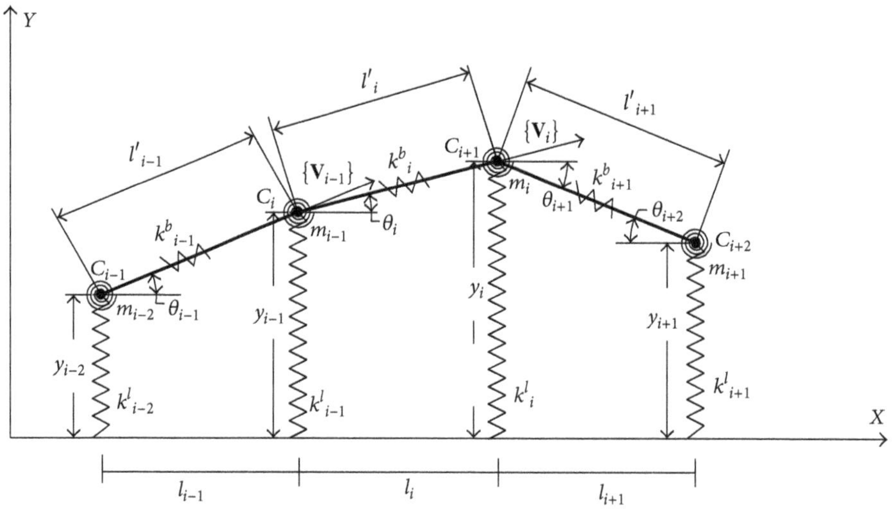

FIGURE 15: Present beam model with a zoom in the neighborhood of ith bar.

The mass m_r is subjected from the three spirals springs r, $r+1$, and $r+2$ to the resulting elastic force given by

$$F^r_{\ l} = -\frac{1}{l^2}\left[C_r\left(y_r - 2y_{r-1} + y_{r-2}\right)\right.$$
$$\left. - 2C_{r+1}\left(y_{r+1} - 2y_r + y_{r-1}\right)\right. \tag{B.6}$$
$$\left. + C_{r+2}\left(y_{r+2} - 2y_{r+1} + y_r\right)\right].$$

The expression for the linear potential energy V^s_l of the system, associated with the spiral springs, is given in (20). The linear spring force applied to the mass r can be derived from V^s_l as follows:

$$\frac{\partial V^s_l}{\partial y_r} = -\frac{1}{2}\left(y_i k^s_{\ ir} + y_j k^s_{\ rj}\right) = -y_i k^s_{\ ir} \tag{B.7}$$

$$\text{for } i, j, r = 1, \ldots, N$$

in which the classical symmetry relation, that is, $k^s_{\ ij} = k^s_{\ ji}$, is adopted. Equations (B.6) and (B.7) lead to the following expressions for the rigidity matrix general terms:

$$k^s_{\ (r-2)r} = \frac{C_r}{l^2} \quad \text{for } r = 3, \ldots, N,$$

$$k^s_{\ (r-1)r} = -\frac{2}{l^2}\left(C_r + C_{r+1}\right) \quad \text{for } r = 2, \ldots, N, \tag{B.8}$$

$$k^s_{\ rr} = \frac{1}{l^2}\left(C_r + 4C_{r+1} + C_{r+2}\right) \quad \text{for } r = 1, \ldots, N.$$

The others values of $k^s_{\ ij}$ are obtained by symmetry relations from (B.8) or are equal to zero.

The linear potential energy stored in the spiral spring C_i, subjected to the rotations shown in Figure 15, can be written as

$$V_{Cl} = \frac{1}{2}C_i\left(\Delta\theta_i\right)^2 = \frac{1}{2}C_i\left(\theta_i - \theta_{i-1}\right)^2 \tag{B.9}$$

$$\text{for } \theta_0 = \theta_{N+1} = 0.$$

On the other hand, [9] gives the elementary bending potential energy corresponding to elementary sections of length dx in a continuous Euler-Bernoulli beam:

$$dV_b = \frac{1}{2}EI\left(\frac{d^2y}{dx^2}\right)^2 dx = \frac{1}{2}EI\left(\theta_i - \theta_{i-1}\right)^2 \tag{B.10}$$

$$\text{with } \theta_i = \left(\frac{dy}{dx}\right)_i.$$

I is the quadratic moment relative to the neutral fibre of this section.

By identification between (B.9) and (B.10), we get

$$C_i = \frac{EI}{l} = (N+1)\frac{EI}{L} \quad \text{with } 2 \leq i \leq N+1. \tag{B.11}$$

B.3. Expression for the General Term of the Nonlinear Rigidity Tensor $\mathbf{b_{ijkl}}$. The stretching of the axial spring having the equivalent stiffness $k^b_{\ i} = E_i S_i / l_i$ [9], in which E_i (N·m^{-2}) is the Young modulus of the material, S_i (m^2) is the area of cross section and l_i (m) is the length of the bar i, and Δl_i is the longitudinal displacement following the vector $\{v_i\}$, Figure 15.

Assuming that the masses $m_1, \ldots, m_N$ are transversally displaced by $y_1, \ldots, y_N$. By applying the Pythagorean Theorem to the triangle formed by $l_i / (y_i - y_{i-1}) / l'_i$ and developing the square root to the first order, the following nonlinear energy expression is obtained [9]:

$$V_{nl} = \frac{1}{2}\sum_{i=1}^{N+1}\frac{k^b_{\ i}}{4l_i^2}\left(y_i^4 - 4y_{i-1}^3 y_i + 6y_{i-1}^2 y_i^2\right.$$
$$\left. - 4y_i^3 y_{i-1} + y_{i-1}^4\right) y_0 = y_{N+1} = 0. \tag{B.12}$$

One obvious conclusion coming to sight is that the last expression for the potential energy stored in the $(N+1)$ linear longitudinal springs contains the terms y_i^4, y_{i-1}^4, $y_i^3 y_{i-1}$, $y_{i-1}^2 y_i^2$, and y_i^4. On the other hand, the symmetry relations usually encountered in the previous cases examined by the present method, for example, [20, 23], are adopted here as follows:

$$b_{ijkl} = b_{ijlk},$$

$$b_{ijkl} = b_{klij},$$

$$b_{ijkl} = b_{jikl},$$

$$b_{ijkl} = b_{klij},$$

$$b_{ijkl} = b_{ikjl}$$

$$\text{(B.13)}$$

$$\text{for } i, j, k, l = 1, \dots, N.$$

With bar having the following characteristics, E, Young's modulus, S, area of the section, and l, length of the bar may then be assimilated to a longitudinal spring whose stiffness is [9]

$$k^b_i = \frac{ES}{l}. \quad \text{(B.14)}$$

Adopting the symmetry relations (B.13) in the relation (5), one has

$$V_{\text{nl}} = \sum_{i=1}^{N+1} \left(b_{iiii} y_i^4 + 4 b_{(i-1)(i-1)(i-1)} y_{i-1}^3 y_i \right.$$

$$+ 6 b_{(i-1)(i-1)ii} y_{i-1}^2 y_i^2 + 4 b_{iii(i-1)} y_i^3 y_{i-1}$$

$$\left. + b_{(i-1)(i-1)(i-1)(i-1)} y_{i-1}^4 \right) \quad \text{with } y_0 = y_{N+1} = 0.$$

$$\text{(B.15)}$$

The identification between (B.12) and (B.15) leads to the following expressions for the general terms of the nonlinear rigidity tensor b_{ijkl}:

$$b_{iiii} \cong 2 \frac{ES}{8l^3} = 2 (N + 1)^3 \frac{ES}{8L^3}$$

$$\text{for } i = 1, \dots, N,$$

$$b_{i(i-1)(i-1)(i-1)} = b_{(i-1)i(i-1)(i-1)} = b_{(i-1)(i-1)i(i-1)}$$

$$= b_{(i-1)(i-1)(i-1)i} \cong -\frac{ES}{8l^3}$$

$$= -(N + 1)^3 \frac{ES}{8L^3} \quad \text{for } i = 2, \dots, N,$$

$$b_{ii(i-1)(i-1)} = b_{(i-1)ii(i-1)} = b_{(i-1)(i-1)ii} = b_{i(i-1)(i-1)i}$$

$$= b_{(i-1)i(i-1)} = b_{(i-1)i(i-1)i} \cong \frac{ES}{8l^3}$$

$$= (N + 1)^3 \frac{ES}{8L^3} \quad \text{for } i = 2, \dots, N,$$

$$b_{iii(i-1)} = b_{(i-1)iii} = b_{i(i-1)ii} = b_{ii(i-1)i} \cong -\frac{ES}{8l^3}$$

$$= -(N + 1)^3 \frac{ES}{8L^3} \quad \text{for } i = 2, \dots, N.$$

$$\text{(B.16)}$$

The rest of b_{ijkl} are equal to zero.

Nomenclature

$*$:	Nondimensional parameter symbol
a_j:	Contribution coefficient of the jth linear mode shape in MB for the discrete system
A_i:	The discrete modulus of the displacement y_i expressed in DB
$\{A\}$:	Displacement amplitudes of the masses $m_1; \dots; m_i; \dots; m_N$ in DB
b_{ijkl}:	The general term of the nonlinear rigidity tensor in DB
$\overline{b}_{ijkl}$:	The general term of the nonlinear rigidity tensor in MB
B:	The ratio of the supported span to the total span
$[B]$:	The nonlinear rigidity matrix in DB
$[\overline{B}]$:	The nonlinear rigidity matrix in MB
C_r:	The stiffness coefficient of the rth spiral spring
DB:	Displacement basis
Δl_i:	The longitudinal displacement following the vector
δ_{ij}:	Kronecker's symbol
$\{d_i\}$:	The ith displacement vector in DB
E:	The Young modulus of the bar's material
E_i:	The Young modulus of the ith bar
H:	The thickness of the bar in m
I:	The quadratic moment relative to the neutral fibre for the cross section in m^4
$[I]$:	The identity matrix
k:	The soil stiffness spring stiffness
k^s_{ij}:	The general term of the linear rigidity tensor in DB (spiral spring)
$\overline{k}^s_{ij}$:	The general term of the linear rigidity tensor in MB (spiral spring)
$[K^s]$:	Matrix of linear rigidity in DB (spiral spring)
$[\overline{K^s}]$:	Matrix of linear rigidity in MB (spiral spring)
k^l_{ij}:	The general term of the linear rigidity tensor in DB (Winkler soil spring)
$\overline{k}^l_{ij}$:	The general term of the linear rigidity tensor in MB (Winkler soil spring)
k^b_i:	The general term of the axial rigidity tensor
$k_l(x)$:	Linear soil distribution function
$k_p(x)$:	Parabolic soil distribution function
$[K^l]$:	Matrix of linear rigidity in DB (Winkler soil spring)
$[\overline{K^l}]$:	Matrix of linear rigidity in MB (Winkler soil spring)
$[K_N^{SS*}]$:	Linear rigidity matrix of the N-DOF discrete system presenting a simply supported beam
$[K_N^{CC}]$:	Linear rigidity matrix of the N-DOF discrete system presenting a clamped-clamped beam

$[\mathbf{K}_{N-2}^{CC}]$: Linear rigidity matrix of the N-DOF discrete system presenting a clamped-clamped beam

l_i: Length of the bar i in m

l_i': The length of the bar i after deformation in m

L: Total length of the bars

m_{ij}: The general term of the mass tensor in DB

$\overline{m_{ij}}$: The general term of the mass tensor in MB

MB: Modal basis

Ω_i: Frequency parameter of the ith mode

$[\mathbf{M}]$: Matrix of masses in DB

$[\overline{\mathbf{M}}]$: Matrix of masses in MB

$\mathcal{M}$: Moment in the spiral spring

N: Number of degrees of freedom (number of masses)

q_i: The time component of the transverse displacement of the ith mass

$\dot{q}_i$: The first derivative with respect to time of q_i

$\ddot{q}_i$: The second derivative with respect to time of q_i

R: The radius of gyration

S_i: The area of cross section of the ith bar in m^2

S: The area of cross section for uniform beam in m^2

T: The kinetic energy

$\{v_i\}$: Vector following the ith bar

V: The total potential energy

$V_{l\,i}^l$: The linear potential energy stored in the ith Winkler spring

V_l: The linear potential energy

V_{nl}: The total nonlinear strain energy stored in the $(N+1)$ longitudinal springs

V_l^s: The linear potential energy stored in the $(N+2)$ spiral springs

V_l^l: The total linear potential energy stored in the (N) Winkler springs

W: Beam transverse displacement

x: The distance on the beam

y_i: Transverse displacement of the ith mass in DB

$\overline{y}_i$: Transverse displacement of the ith mass in MB

α: Soil stiffness parameter

$\{\beta\}$: The eigenvalue of nondimensional rigidity vector

β_i: The ith eigenvalue of nondimensional rigidity

ε_i: The small contribution of the ith mode

θ_r: The angular displacement of the bar r in rad

λ: Soil stiffness parameter

λ_i: The ith soil stiffness parameter

ξ: Relative length x/l

ρ: Mass per unit volume of the bar (kg m^{-3})

φ_r: The rth component of $\{\mathbf{\Phi}_i\}$

$\{\mathbf{\Phi}_i\}^T$: Transpose of the ith linear mode shape of the system (when considering a N-DOF model)

$\{\mathbf{\Phi}_i\}$: The ith linear mode shape of the system in MB (when considering a N-DOF model)

$[\mathbf{\Phi}]$: Transition matrix

ω^{nl}_{discr}: The nonlinear frequency parameter of the discrete system

$\omega^{SS}_{l\,discr}$: The first linear frequency of the discrete system in the case of simply supported beam

$\omega^{SS}_{nl\,discr}$: The first nonlinear frequency of the discrete system in the case of simply supported beam

$\omega^{CC}_{l\,discr}$: The first linear frequency of the discrete system in the case of clamped-clamped beam

$\omega^{CC}_{nl\,discr}$: The first nonlinear frequency of the discrete system in the case of clamped-clamped beam.

References

[1] T. Yokoyama, "Vibration analysis of Timoshenko beam-columns on two-parameter elastic foundations," *Computers and Structures*, vol. 61, no. 6, pp. 995–1007, 1996.

[2] Y. C. Hou, C. H. Tseng, and S. F. Ling, "A new high-order non-uniform Timoshenko beam finite element on variable two-parameter foundations for vibration analysis," *Journal of Sound and Vibration*, vol. 191, no. 1, pp. 91–106, 1996.

[3] E. Ergüven and A. Gedikli, "A mixed finite element formulation for Timoshenko beam on Winkler foundation," *Computational Mechanics*, vol. 31, no. 3-4, pp. 229–237, 2003.

[4] C.-N. Chen, "Vibration of prismatic beam on an elastic foundation by the differential quadrature element method," *Computers and Structures*, vol. 77, no. 1, pp. 1–9, 2000.

[5] W. Q. Chen, C. F. Lv, and Z. G. Bian, "Free vibration analysis of generally laminated beams via state-space-based differential quadrature," *Composite Structures*, vol. 63, no. 3-4, pp. 417–425, 2004.

[6] P. Malekzadeh and A. R. Vosoughi, "DQM large amplitude vibration of composite beams on nonlinear elastic foundations with restrained edges," *Communications in Nonlinear Science and Numerical Simulation*, vol. 14, no. 3, pp. 906–915, 2009.

[7] P. Malekzadeh and G. Karami, "A mixed differential quadrature and finite element free vibration and buckling analysis of thick beams on two-parameter elastic foundations," *Applied Mathematical Modelling*, vol. 32, no. 7, pp. 1381–1394, 2008.

[8] A. D. Kerr, "Elastic and Viscoelastic Foundation Models," *Journal of Applied Mechanics*, vol. 31, no. 3, pp. 491–498, 1964.

[9] A. Rahmouni, Z. Beidouri, and R. Benamar, "A discrete model for geometrically nonlinear transverse free constrained vibrations of beams with various end conditions," *Journal of Sound and Vibration*, vol. 332, no. 20, pp. 5115–5134, 2013.

[10] A. J. Valsangkar and R. Pradhanang, "Vibrations of beam-columns on two-parameter elastic foundations," *Earthquake Engineering & Structural Dynamics*, vol. 16, no. 2, pp. 217–225, 1988.

[11] C. Franciosi and A. Masi, "Free vibrations of foundation beams on two-parameter elastic soil," *Computers and Structures*, vol. 47, no. 3, pp. 419–426, 1993.

[12] M. A. De Rosa and M. J. Maurizi, "The influence of concentrated masses and pasternak soil on the free vibrations of euler beams—exact solution," *Journal of Sound and Vibration*, vol. 212, no. 4, pp. 573–581, 1998.

[13] B. S. Sarma and T. K. Varadan, "Lagrange-type formulation for finite element analysis of non-linear beam vibrations," *Journal of Sound and Vibration*, vol. 86, no. 1, pp. 61–70, 1983.

[14] G. V. Rao, "Large amplitude vibrations of slender, uniform beams on elastic foundation," *Indian Journal of Engineering & Materials Sciences*, vol. 10, no. 1, pp. 87–91, 2003.

[15] D. A. Evensen, "Nonlinear vibrations of beams with various boundary conditions," *AIAA Journal*, vol. 6, no. 2, pp. 370–372, 1968.

[16] G. R. Bhashyam and G. Prathap, "Galerkin finite element method for non-linear beam vibrations," *Journal of Sound and Vibration*, vol. 72, no. 2, pp. 191–203, 1980.

[17] A. Kacar, H. T. Tan, and M. O. Kaya, "Free vibration analysis of beams on variable winkler elastic foundation by using the differential transform method," *Mathematical and Computational Applications*, vol. 16, no. 3, pp. 773–783, 2011.

[18] R. Benamar, *Nonlinear dynamic behavior of fully clamped beams and rectangular isotropic and laminated plates [Ph.D. thesis]*, University of Southampton, Southampton, UK, 1990.

[19] R. Benamar, M. M. K. Bennouna, and R. G. White, "The effects of large vibration amplitudes on the mode shapes and natural frequencies of thin elastic structures, part I: simply supported and clamped-clamped beams," *Journal of Sound and Vibration*, vol. 149, no. 2, pp. 179–195, 1991.

[20] M. El Kadiri, R. Benamar, and R. G. White, "Improvement of the semi-analytical method, for determining the geometrically non-linear response of thin straight structures. Part I: application to clamped-clamped and simply supported-clamped beams," *Journal of Sound and Vibration*, vol. 249, no. 2, pp. 263–305, 2002.

[21] M. M. Bennouna and R. G. White, "The effects of large vibration amplitudes on the fundamental mode shape of a clamped-clamped uniform beam," *Journal of Sound and Vibration*, vol. 96, no. 3, pp. 309–331, 1984.

[22] H. F. Wolfe, *An experimental investigation of nonlinear behaviour of beams and plates excited to high levels of dynamic response [Ph.D. thesis]*, University of Southampton, 1995.

[23] A. Eddanguir, Z. Beidouri, and R. Benamar, "Geometrically nonlinear transverse vibrations of discrete multi-degrees of freedom systems with a localised non-linearity," *International Journal of Mathematics and Statistics*, vol. 4, no. S09, pp. 73–87, 2009.

[24] R. E. Mickens, "Comments on the method of harmonic balance," *Journal of Sound and Vibration*, vol. 94, no. 3, pp. 456–460, 1984.

[25] M. N. Hamdan and T. D. Burton, "On the steady state response and stability of nonlinear oscillators using harmonic balance," *Journal of Sound and Vibration*, vol. 166, no. 2, pp. 255–266, 1993.

[26] P. F. Doyle and M. N. Pavlovic, "Vibration of beams on partial elastic foundations," *Earthquake Engineering & Structural Dynamics*, vol. 10, no. 5, pp. 663–674, 1982.

[27] M. Eisenberger, D. Z. Yankelevsky, and M. A. Adin, "Vibrations of beams fully or partially supported on elastic foundations," *Earthquake Engineering & Structural Dynamics*, vol. 13, no. 5, pp. 651–660, 1985.

5

Development of an Experimental Model for a Magnetorheological Damper Using Artificial Neural Networks (Levenberg-Marquardt Algorithm)

Ayush Raizada, Pravin Singru, Vishnuvardhan Krishnakumar, and Varun Raj

Birla Institute of Technology and Science-Pilani, K.K. Birla Goa Campus, Goa 403726, India

Correspondence should be addressed to Pravin Singru; pmsingru@goa.bits-pilani.ac.in

Academic Editor: Toru Otsuru

This paper is based on the experimental study for design and control of vibrations in automotive vehicles. The objective of this paper is to develop a model for the highly nonlinear magnetorheological (MR) damper to maximize passenger comfort in an automotive vehicle. The behavior of the MR damper is studied under different loading conditions and current values in the system. The input and output parameters of the system are used as a training data to develop a suitable model using Artificial Neural Networks. To generate the training data, a test rig similar to a quarter car model was fabricated to load the MR damper with a mechanical shaker to excite it externally. With the help of the test rig the input and output parameter data points are acquired by measuring the acceleration and force of the system at different points with the help of an impedance head and accelerometers. The model is validated by measuring the error for the testing and validation data points. The output of the model is the optimum current that is supplied to the MR damper, using a controller, to increase the passenger comfort by minimizing the amplitude of vibrations transmitted to the passenger. Besides using this model for cars, bikes, and other automotive vehicles it can also be modified by retraining the algorithm and used for civil structures to make them earthquake resistant.

1. Introduction

Isolation of the forces transmitted by external application is the most important function of a suspension system. The suspension system comprises a spring element and a dissipative element, which when placed between the object to be protected and the excitation reduces the vibration transmitted to the object.

Suspension systems range from active to passive suspensions. MR damper lies in between this range and behaves like a semiactive suspension system. The damping of a passive suspension system is a property of the system and cannot be varied, whereas in an active suspension system the damping of the system can be altered by using an actuator to give it an external force. This external force helps in improving the ride quality. The shortcoming of this model is that it requires considerable amount of external power source; it is costly and is difficult to incorporate in the system due to the added mass. A variation of the active suspension system is the semiactive or

the adaptive suspension system. In these systems the damping of the system is varied by controlling the current thereby changing the viscous properties of the damping elements in the suspension system. PID neural network controller is one such controller used to develop a model to predict the displacement and velocity behavior of the MR damper [1]. In comparison to active suspension semiactive suspension systems' power consumption is considerably less.

The magnetorheological (MR) dampers and electrorheological dampers are the most common examples of semiactive dampers. In the past few years, research on MR damper has improved its capabilities and reduced the gap between adaptive suspension system and truly active suspension systems. Structurally, the MR damper is similar to a simple fluid damper, except that the viscosity of the fluid in the MR damper can be changed by altering the current in the system that induces a magnetic field. The MR fluid is a non-Newtonian fluid composed of mineral oil with suspended iron nanoparticles. When there is no magnetic flux (zero

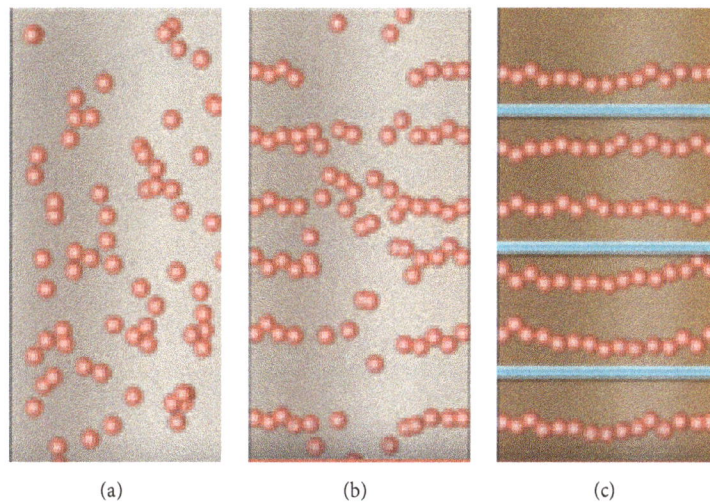

FIGURE 1: Alignment of iron particles at (a) zero current; (b) low current; (c) high current.

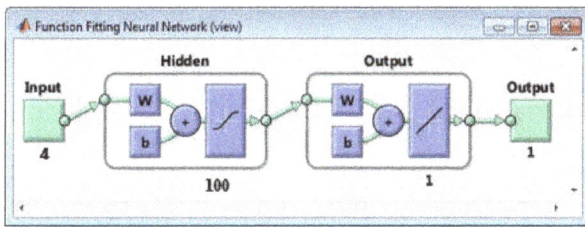

FIGURE 2: Neural Network Architecture.

FIGURE 3: Experimental setup of MR damper.

current), the MR damper behaves like a normal fluid damper in which the iron nanoparticles are randomly oriented as seen in Figure 1(a).

When a magnetic field is applied to the MR damper, the iron nanoparticles align themselves along the magnetic flux, as seen in Figures 1(b) and 1(c), and form chains which makes the fluid partly semisolid. These particles help in reinforcing the damping by forming chains in the oil that obstructs the movement of the oil as the magnetic field developed is in the direction perpendicular to the movement of the oil. Hence increasing the magnetic field increases damping in the system.

The behavior of the MR damper can be characterized by its highly nonlinear motion [2–6]. The force versus velocity plot forms a hysteresis loop depicting the nonlinear character of the MR damper. There exist many different parametric models like Bingham model, Bingham Body model, Lee model, Spencer model, Bouc-Wen model, and Gamota-Filisko model that portray the behavior of the MR damper. These models are difficult to implement throughout the working range of the MR damper due to the hysteresis and jump type phenomena resulting from the specific properties, like viscoplasticity, viscoelastoplasticity, and viscoelasticity, of the MR fluid. It is also difficult to find a solution for the equations for some of the defined models due to the numerical and analytical complexity of the model equation. These drawbacks prevent the models from being implemented as

governing equations in real time controllers [7]. In the low speed range the Bingham model can predict the rigid plastic behavior better than the involution model when the MR damper is loaded with a sinusoidal input. When triangular loading is used on the MR damper, the involution model predicts the model better than the Bingham model [8]. The deformation in the hysteresis loop of the force-velocity and force-displacement graphs deviates from the Bouc-Wen model due to the force lag phenomenon in the MR damper. The modification of the Bouc-Wen model, the Bouc-Wen-Baber-Noori model, can to a certain extent describe the pinching hysteretic behavior [9]. The evolutionary variable equation for the modified Bouc-Wen model depends on 4 parameters A, β, γ, and n, where A, β, and γ control the shape and size of the hysteresis and the parameter n controls the smoothness of the transition from elastic to plastic region [10]. Genetic algorithm assisted inverse method and nonlinear-least square error optimization in MATLAB can be used to identify the parameters and develop a model [11, 12].

(a)

(b)

(c)

(d)

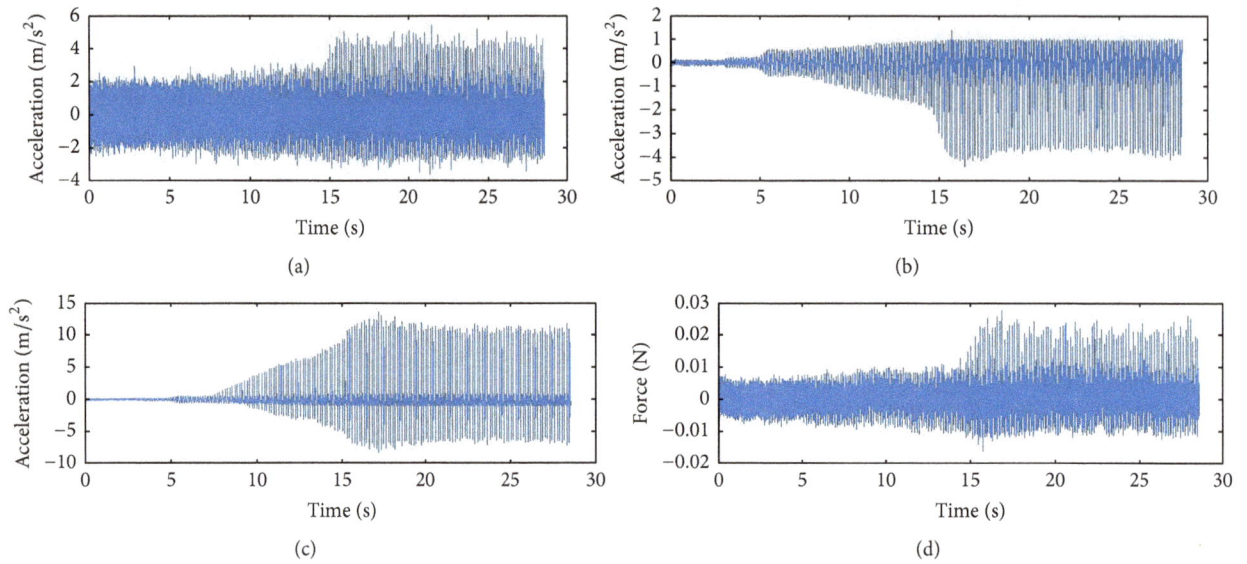

FIGURE 4: (a) Signatures of acceleration of road profile, (b) signatures of acceleration of lower plate, (c) signatures of acceleration of upper plate, and (d) signatures of force transmitted to upper plate.

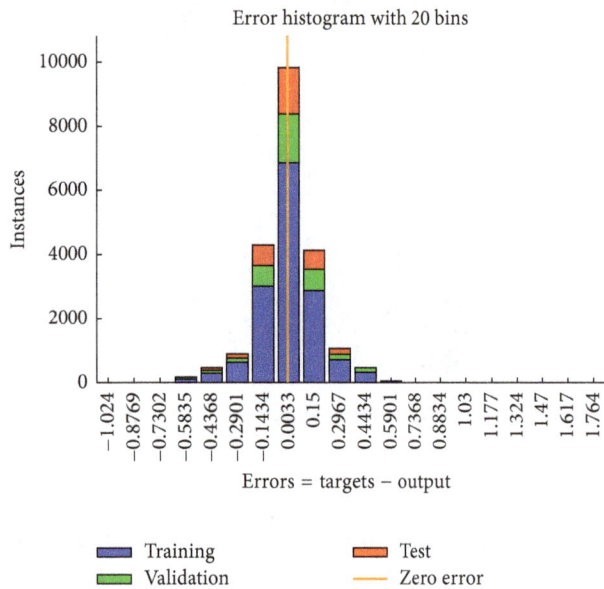

FIGURE 5: Error in the best model.

FIGURE 6: Best training performance.

2. Artificial Neural Networks

Artificial Neural Network is an important tool used to build models for dynamic systems. It includes a variety of modeling tools and models. This method has considerable advantages over other modeling tools. ANNs can work easily with nonlinearly separable data, and this makes them ideal for applications such as machine condition monitoring, where the training data are sparse, and the network will have to generalize well. Several applications have demonstrated that a neural network can successfully recognize and classify different faults in a number of different condition monitoring applications [13]. A good general introduction to neural networks is provided by Haykin [14] and also Rojas [15].

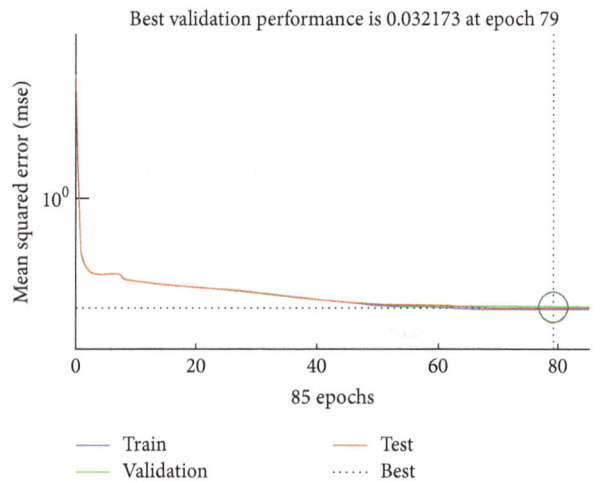

Multilayer perceptron (MLP) ANNs have been used for classification purposes in this experiment. The MLP used in this case consists of one hidden layer and an output layer, the hidden layer having a logistic activation function (see (1)), while the output layer uses a linear activation function (see (2)):

$$\varphi\left(v\right) = \frac{1}{\left(1 + \exp\left(-v\right)\right)} \tag{1}$$

$$\varphi\left(v\right) = v, \tag{2}$$

where v is the sum of the weighted outputs. Different sizes of the hidden layer were tested. The size of the output layer was set at two neurons for this particular application. Training of the ANN networks is carried out using the Levenberg-Marquardt algorithm discussed in Section 3. The network is

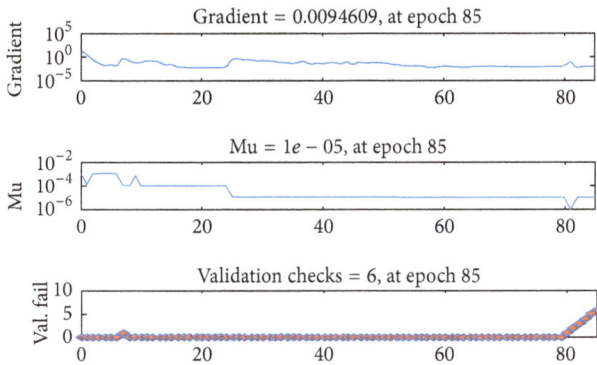

FIGURE 7: Training state of various parameters.

trained using a training and validation set, while testing is carried out using a further test feature set. Performance is measured in terms of the network's classification success on unseen data in the test set. Training is stopped when the classification performance of the validation set starts to diverge from that of the training set (i.e., the network becomes overtrained on the training set) [16].

The neural network architecture consists of fully connected feedforward network consisting of 100 hidden layers as shown in Figure 2. The optimization is achieved using the network growing approach where one hidden neuron is used initially and is increased until there is improvement in the performance. This optimization is carried out using the fitting tool in the neural network toolbox in Matlab.

3. Levenberg-Marquardt Algorithm

A mathematical description of the Levenberg-Marquardt (LM) neural network training algorithm has been presented by Hagan and Menhaj [17]. The LM algorithm was originally created [18, 19] to address the drawbacks of the Gauss-Newton (GN) method and gradient descent algorithm. The quadratic convergence properties of the GN algorithm make it very fast. However, the choice of initial weights is the key to success of this method and since these values are not readily available in real life problems, the GN method becomes slightly redundant in these cases. Since, in real-world problems, the prediction of an appropriate set of initial values is not always possible, the GN method is impractical for many applications. In comparison to the GN algorithm, the success of the gradient descent algorithm is less dependent on the initial choice of weight values. However, since the gradient descent algorithm approaches the minimum in a linear (first order) manner, its speed of convergence is normally low and, thus, does not always possess adequate convergence properties.

By merging the advantages of GN and gradient descent algorithms, a hybrid algorithm (LM algorithm) was created to cater to the drawbacks of GN and gradient descent algorithms. The LM algorithm possesses quadratic convergence (approximates the GN method) when it is in the vicinity of (but not too close to) a minimum [20]. LM uses gradient descent to improve on an initial guess for its parameters and transforms to the GN method as it approaches the minimum

value of the cost function. Once it approaches the minimum, it transforms back to the gradient descent algorithm to improve the accuracy. The LM algorithm is used for curve fitting [13–15] and many other optimization problems [21–24]. Due to its desirable convergence capabilities, in many optimization applications, the LM method is usually preferred over many other optimization techniques [25–29].

4. Experimental Setup

The experimental setup consists of three components: external actuation system, data acquisition system, and the controller. The external actuation for the system was fabricated to excite the damper with the help of the electrodynamic shaker. The system was designed to closely depict a quarter car model. The shaker transmits the road disturbances to the lower plate by means of a stringer as seen in Figure 3. The lower plate is equivalent to the tire of the vehicle in the quarter car model. The lower plate is coupled with the upper plate with a LORD manufactured MR damper RD-8040-1 [30] which is a part of the suspension system. The upper plate behaves as the chassis for vehicle. The MR damper is clamped to the upper and lower plates by means of an L clamp.

The electrodynamic shaker (Model 2110E) [31] with a load capacity of 489 N (sine-peak) is connected to the linear power amplifier (Model 2050E09) [32] which provides the shaker with the input profile that is given to the system using the Spider 81 data acquisition system [33] and the EDM software. The vibration signals, taken as inputs for genetic programming, are measured using sensors. The signal is measured on the shaker and the lower plate using accelerometers "352C34 LW 155857" of sensitivity 101.3 mV/g [34] and "352C68 SN 92017" of sensitivity 102.2 mV/g [35], respectively. To measure the force transmitted and the acceleration signals at the upper plate an impedance head "288D01 SN 3176" is used of sensitivity 98.73 mV/LBF for force and 101.3 mV/g for acceleration [36]. The MR damper is operated using an external power source. The current in the damper is varied using a potentiometer type device called Wonder Box RD-3002-03 [37]. The current is measured using a multimeter.

The signal parameters from these sensors are used as inputs for the algorithm and the current that is varied using the potentiometer is used as the output. This current can be directly provided to the control system to create a feedback loop.

5. Experimentation

The training set for the ANN is obtained from experimentation using the setup described in Section 3. In the experimental setup system there are 6 factors that are monitored to develop the training set. The experimentation is carried out by varying the frequency and current parameters. The data collected is divided into three sets; that is, 70% is training set, 15% is testing set, and the remaining 15% is validation. Using this to training set the final model is developed which best fits the data having minimum error. Table 1 provides the configuration of various experiments conducted by varying the parameters.

(a)

(b)

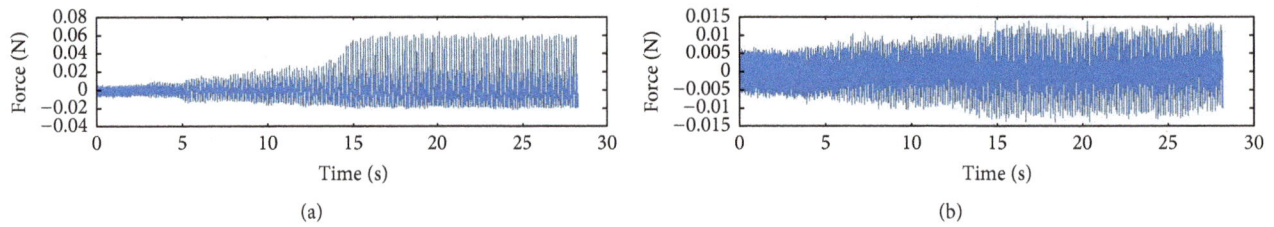

FIGURE 8: (a) Vibration signature without optimization; (b) signature with optimization.

TABLE 1: Different test configurations.

Serial number	Description	Frequency	Steps	Current
1	Dwell	5–7 Hz	0.1 Hz	0 A
2	Dwell	5–7 Hz	0.1 Hz	0.25 A
3	Dwell	5–7 Hz	0.1 Hz	0.50 A
4	Dwell	5–7 Hz	0.1 Hz	0.75 A
5	Dwell	5–7 Hz	0.1 Hz	1.0 A

The input parameters are recorded for 10 seconds of the response. These signatures are as shown in Figure 4.

In the first set of tests the current is kept constant and the frequency is increased from 5 Hz to 7 Hz in steps of 0.1 Hz. The data is collected at a sampling rate of 20.48 kHz. In the similar fashion the data set is obtained for all the current values. This data is then consolidated to form the training and testing matrix. There are 4 input parameters taken into consideration in this system:

(1) Frequency of excitation.

(2) Acceleration of the shaker.

(3) Relative acceleration of the upper level (chassis) with respect to the lower level (tire).

(4) Force transmitted to chassis.

The current in the damper coils is taken as the output parameter. Using these input-output parameters the algorithm is trained.

6. Results and Discussion

The best model selected from the models generated in the different stages is based on the performance of the model. The algorithm is trained using the Levenberg-Marquardt algorithm. The error histogram for the model obtained with 100 nodes in the hidden layer is shown in Figure 5. It can be seen from the histogram that the large centre peaks indicate very small errors or the output is very close to the target values. The absence of spikes towards the end implies that the there are no misclassifications in the model [38].

The best training performance for the different iterations of the model is shown in Figure 6.

The training state for each iteration for the various parameters is shown in Figure 7.

Figure 8 shows the vibration signature for the force exerted on the upper mass of the fabricated structure. From the figure it is visible that there is a massive reduction in the amplitude of the force exerted due to the optimization performed.

7. Conclusion

This work is the first of its kind that develops a model to predict the current value to adjust the damper force for the given input parameters, using actual experimental data, in order to maintain the driver acceleration. This proves the versatility and success of Artificial Neural Networks in highly nonlinear and hysteretic systems like MR damper. The optimization performed, based on the Artificial Neural Network, has reduced the amplitude of force transmitted to the driver. The magnitude of the force transmitted is approximately 1/10 that at the base of the vehicle. The current parameter determined based on this optimization has caused this reduction in the force transmitted to the passenger seated in the vehicle.

Competing Interests

The authors declare that they have no competing interests.

Acknowledgments

The authors acknowledge BITS-Pilani and Department of Science and Technology, FIST program, for providing funding for the experimental setup.

References

[1] W. Liu, W.-K. Shi, D.-W. Liu, and T.-Y. Yan, "Experimental modeling of magneto-rheological damper and PID neural network controller design," in *Proceedings of the 6th International Conference on Natural Computation (ICNC '10)*, pp. 1674–1678, Yantai, China, August 2010.

[2] F. Gandhi and I. Chopra, "A time-domain non-linear viscoelastic damper model," *Smart Materials and Structures*, vol. 5, no. 5, pp. 517–528, 1996.

[3] G. M. Kamath and N. M. Wereley, "Nonlinear viscoelastic-plastic mechanisms-based model of an electrorheological damper," *Journal of Guidance, Control, and Dynamics*, vol. 20, no. 6, pp. 1125–1132, 1997.

[4] R. A. Snyder, G. M. Kamth, and N. M. Werely, "Characterization and analysis of magnetorheological damper behaviour due to sinusoidal loading," in *Proceedings of the SPIE Symposium on Smart Materials and Structures*, vol. 3989 of *Proceedings of SPIE*, pp. 213–229, New Port Beach, Calif, USA, 2000.

[5] R. Stanway, N. D. Sims, and A. R. Johnson, "Modelling and control of a magnetorheological vibration isolator," in *Smart*

Structures and Materials: Damping and Isolation, vol. 3989 of *Proceedings of SPIE*, pp. 184–193, March 2000.

[6] N. M. Wereley, L. Pang, and G. M. Kamath, "Idealized hysteresis modeling of electrorheological and magnetorheological dampers," *Journal of Intelligent Material Systems and Structures*, vol. 9, no. 8, pp. 642–649, 1998.

[7] B. Sapiński and J. Filuś, "Analysis of parametric models of MR linear damper," *Journal of Theoretical and Applied Mechanics*, vol. 41, no. 2, pp. 215–240, 2003.

[8] H. Fujitani, H. Sodeyama, K. Hata et al., "Dynamic performance evaluation of 200kN magnetorheological damper," in *Proceedings of the SPIE's 7th Annual International Symposium*, vol. 3989, pp. 194–203, 2000.

[9] M. T. Braz-Cesar and R. C. Barros, "Experimental and numerical analysis of MR dampers," in *Proceedings of the 4th ECCOMAS Thematic Conference on Computational Methods in Structural Dynamics and Earthquake Engineering*, Kos Island, Greece, June 2013.

[10] G. R. Peng, W. H. Li, H. Du, H. X. Deng, and G. Alici, "Modelling and identifying the parameters of a magneto-rheological damper with a force-lag phenomenon," *Applied Mathematical Modelling*, vol. 38, no. 15-16, pp. 3763–3773, 2014.

[11] M. Giuclea, T. Sireteanu, D. Stancioiu, and C. W. Stammers, "Modelling of magnetorheological damper dynamic behaviour by genetic algorithms based inverse method," *Proceedings of the Romanian Academy, Series A*, vol. 5, no. 1, pp. 1–10, 2004.

[12] P. Prakash and A. K. Pandey, "Performance of MR damper based on experimental and analytical modelling," in *Proceedings of the 22nd International Congress of Sound and Vibration (ICSV '15)*, Florence, Italy, July 2015.

[13] T. J. Doyle, R. Pathak, J. S. Wolinsky, and P. A. Narayana, "Automated proton spectroscopic image processing," *Journal of Magnetic Resonance, Series B*, vol. 106, no. 1, pp. 58–63, 1995.

[14] S. Haykin, *Neural Networks: A Comprehensive Foundation*, Prentence Hall, Upper Saddle River, NJ, USA, 2nd edition, 1999.

[15] R. Rojas, *Neural Networks: A Systematic Introduction*, Springer Science & Business Media, Berlin, Germany, 2013.

[16] L. B. Jack and A. K. Nandi, "Fault detection using support vector machines and artificial neural networks, augmented by genetic algorithms," *Mechanical Systems and Signal Processing*, vol. 16, no. 2-3, pp. 373–390, 2002.

[17] M. T. Hagan and M. B. Menhaj, "Training feedforward networks with the Marquardt algorithm," *IEEE Transactions on Neural Networks*, vol. 5, no. 6, pp. 989–993, 1994.

[18] K. Levenberg, "A method for the solution of certain non-linear problems in least squares," *Quarterly of Applied Mathematics*, vol. 2, pp. 164–168, 1944.

[19] D. W. Marquardt, "An algorithm for least squares estimation of non-linear parameters," *Japan Journal of Industrial and Applied Mathematics*, vol. 11, no. 2, pp. 431–441, 1963.

[20] S. Kollias and D. Anastassiou, "An adaptive least squares algorithm for the efficient training of artificial neural networks," *IEEE Transactions on Circuits and Systems*, vol. 36, no. 8, pp. 1092–1101, 1989.

[21] J. Basu and L. Hazra, "Role of line search in least-squares optimization of lens design," *Optical Engineering*, vol. 33, no. 12, pp. 4060–4066, 1994.

[22] A. S. Deo and I. D. Walker, "Adaptive non-linear least squares for inverse kinematics," in *Proceedings of the IEEE International Conference on Robotics and Automation*, vol. 1, pp. 186–193, May 1993.

[23] V. I. Dmitriev and T. Y. Shameeva, "Solution of an inverse problem for estimation of ionospheric parameters," *Computational Mathematics and Modeling*, vol. 5, no. 2, pp. 157–161, 1994.

[24] M. Guarini, J. Urzua, A. Cipriano, and M. Matus, "Cardiac flow estimation using the radial arterial pressure waveform," in *Proceedings of the 25th Annual Summer Computer Simulation Conference*, pp. 1016–1019, 1993.

[25] J. L. del Alamo, "Comparison among eight different techniques to achieve an optimum estimation of electrical grounding parameters in two-layered earth," *IEEE Transactions on Power Delivery*, vol. 8, no. 4, pp. 1890–1899, 1993.

[26] P. G. Drennan, B. W. Smith, and D. Alexander, "Technique for the measurement of the in situ development rate," in *Proceedings of the Integrated Circuit Metrology, Inspection, and Process Control VIII*, vol. 2196 of *Proceedings of SPIE*, pp. 449–465, San Jose, Calif, USA, May 1994.

[27] P. Ojala, J. Saarinen, P. Elo, and K. Kaski, "Novel technology independent neural network approach on device modelling interface," *IEE Proceedings: Circuits, Devices and Systems*, vol. 142, no. 1, pp. 74–82, 1995.

[28] A. Swietlik, "Interactive approach to reconstruction of solids from range images," in *Applications of Artificial Intelligence: Machine Vision and Robotics*, vol. 1964 of *Proceedings of SPIE*, pp. 202–210, International Society for Optics and Photonics, 1993.

[29] B. G. Kermani, S. S. Schiffman, and H. T. Nagle, "Performance of the Levenberg-Marquardt neural network training method in electronic nose applications," *Sensors and Actuators B: Chemical*, vol. 110, no. 1, pp. 13–22, 2005.

[30] http://www.lordmrstore.com/lord-mr-products/rd-8040-1-mr-damper-short-stroke.

[31] http://www.modalshop.com/filelibrary/110lbf-Vibration-Shaker-Datasheet-(DS-0078).pdf.

[32] http://www.modalshop.co.in/filelibrary/2050E09-Linear-Power-Amplifier-Specifications-(PS-0064).pdf.

[33] http://www.hzrad.com/pdf/Spider-81-Brochure.pdf.

[34] http://www.pcb.com/contentstore/docs/pcb_corporate/vibration/products/manuals/352c34.pdf.

[35] http://www.pcb.com/contentstore/docs/PCB_Corporate/Vibration/products/Manuals/352C68.pdf.

[36] http://www.pcb.com/contentstore/docs/pcb_corporate/vibration/products/manuals/tla288d01.pdf.

[37] http://www.lordmrstore.com/lord-mr-products/wonder-box-device-controller-kit.

[38] J. M. Twomey and A. E. Smith, "Performance measures, consistency, and power for artificial neural network models," *Mathematical and Computer Modelling*, vol. 21, no. 1-2, pp. 243–258, 1995.

6

Experimental Investigation on Flutter Similitude of Thin-Flat Plates

I. P. G. Sopan Rahtika,[1,2] I. N. G. Wardana,[2] A. A. Sonief,[2] and E. Siswanto[2]

[1]*Department of Mechanical Engineering, Bali State Polytechnics, Badung, Bali 80361, Indonesia*
[2]*Department of Mechanical Engineering, Brawijaya University, Malang, East Java 65144, Indonesia*

Correspondence should be addressed to I. P. G. Sopan Rahtika; sopan_rahtika@yahoo.com

Academic Editor: Marc Thomas

This paper shows the experimental results of the flutter speed of thin-flat plates with free leading edge in axial flow as a function of plates' geometry, fluid densities, and viscosities, as well as natural frequencies of the plates. The experiment was developed based on similitude theory using dimensional analysis and Buckingham Pi Theorem. Dimensional analysis generates four dimensionless numbers. Experiment was conducted by placing the thin-flat plates in a laminar flow wind tunnel in order to obtain the relationship among those dimensionless numbers. The flutter speed was measured by varying the flow velocity until the instability occurred. The dimensional analysis gives a map of the flutter Reynolds number as a function of a new type of dimensionless number that is hereby called flutter fluid structure interaction number, thickness-to-length, and aspect ratios as the correcting factors. This map is a very useful tool for predicting the flutter speed of thin-flat plates in general. This investigation found that the flutter Reynolds number is very high at the region of high flutter fluid structure and thickness-to-length ratios numbers; however, it is very sensitive to the change of those two dimensionless numbers. The sensitivity is higher at lower aspect ratio.

1. Introduction

Flutter is a potentially damaging dynamic aeroelastic phenomenon where aerodynamic forces with the natural modes of vibration cause a periodic motion of a structure going unstable. In a certain fluid structure interaction, the aerodynamic forces serve as input energy to the structural vibration. When the system is not damped by the aerodynamic damping, the vibration amplitude will increase and eventually will lead to structural failure. Flutter can occur in a variety of structures, such as in aircraft wings and turbine blades or even on the bridge. Most previous studies on flutter were focused on the prediction of the flutter speed using numerical methods. Very few literatures are available that discuss flutter using experimental similitude method as a tool to predict flutter phenomenon, despite the fact that similitudes offer cost savings in the investigation of the fluid flow phenomena. This scarcity was one that motivated this study.

The flutter phenomenon observed in this research focused on the flutter of thin-flat plates with free leading edge in axial flow. There were numerous studies which have been done on the flutter of the thin-flat plates due to the canonical characteristics of the problem, such as researches by Chad Gibbs et al. [1], Tang et al. [2], Tang and Païdoussis [3], Tang and Dowell [4], Howell et al. [5], and Zhao et al. [6].

Chad Gibbs et al. [1] performed experimental and theoretical work on the flutter of a flat plate with a fixed leading edge using the three-dimensional vortex-lattice method. Gibbs' report comprehensively described the characteristics of plate flutter as a function of the mass ratio and the aspect ratio. Tang and Païdoussis [3] discussed the flutter of two flat plates which were positioned in parallel with the axial direction of the fluid flow.

The work of Tang and Dowell [4] was about nonlinear flutter and Limit Cycle Oscillation (LCO) of two-dimensional panels in low subsonic flow and later was extended to three-dimensional panels by Tang et al. [2]. The dynamics of the system was built to produce nonlinear models. Zhao et al. [6] discussed both theory and experiment of the flutter of a flat plate. The discussion emphasizes nonlinear analysis to produce Poincaré maps. The nonlinear analysis results were compared to experimental results. Howell et al. [5] used fluid flow interactions to discuss the nature of a cantilevered plate with ideal flow.

The aeroelastic instability of a flexible plate has been investigated using weakly nonlinear analyses by Eloy et al. [7]. Later, a deeper investigation was focused on the origin of the instability hysteresis [8].

Despite the detrimental effect of flutter, there are new research trends to utilize flutter for wind harvesting. The utilization of flutter phenomena for energy harvesting has been explored by Doaré and Michelin [9], Makihara and Shimose [10], and Dunmon et al. [11]. The other works on energy harvesting using a slender structure in the wake of a bluff body were also conducted by Allen and Smits [12]. As a matter of fact, this research is oriented as a theoretical base to explore this application further.

This research chose to explore the free leading edge instead of a fixed leading edge because the chosen config- uration experiences flutter at a lower wind speed. A free leading edge plate will experience flutter in its first mode shape as shown later on the experimental results, while the fixed leading edge will experience flutter at the second mode [1, 5, 7]. This advantage of free leading edge is the reason of the selection of this configuration in this research.

Furthermore, the utilization of thin-flat plates for wind harvester requires a map of flutter speed as a function of plates' geometry for design optimization. One significant part of this research is to generate a flutter speed map for wind harvester design optimization. The authors currently have an on-going development of a micropower generator utilizing flutter of a free leading edge configuration. This is new expansion of flutter energy conversion utilization into the field of microelectromechanical systems. This paper puts a benchmark for this research.

Flutter phenomena of plates continuously have been observed in the last several years. More recent studies on plate or panel flutter were also done by Fernandes and Mirzaeisefat [13], Cunha-Filho et al. [14], Peng and DeSmidt [15], and Yaman [16].

Similitude theory has been well developed and widely used in the field of fluid dynamic. The use of similitude theory continues to grow in the field of structural vibration, for example, Torkamani et al. [17], and acoustic, for example, De Rosa et al. [18]. A work has also been done on structural similitude for flutter of composite plates by Yazdi and Reza-eepazhand [19].

In this study, a testing method is developed using dimen- sional similitude analysis based on the Buckingham Pi The- orem. Dimensional analysis generated four dimensionless numbers. Experiment was conducted on thin-flat plate which is placed in a wind tunnel. The problem in this research was to formulate relationships between the flutter speed and the affecting parameters. Flutter speed is defined as the velocity of fluid flow at which flutter started happening. The plates' parameters that affect the flutter speed that were taken into account in this experiment were their length, width, thickness, and natural frequencies, while the fluid's parameters (in this case, fluid is air) that were taken into

account were its density and viscosity. By using the similitude principle, the experimental results were used to obtain the relationship between the dimensionless numbers Π_1, Π_2, Π_3, and Π_4.

Later on, it has been observed during the analysis that the first dimensionless number Π_1 is the Reynolds number measured during flutter and then called flutter Reynolds number. The second dimensionless number Π_2 is called fluid structure interaction number since it contains the interacting forces. The third Π_3 and fourth Π_4 numbers are the thickness- to-length and aspect ratios, consecutively. This study finally discovered the relationship of the flutter Reynolds number as a function of the fluid-structure interaction number and the geometric ratios.

2. Theory of Flutter and Similitude

Theoretically, this research was about implementing the similitude theory to observe the flutter phenomenon of thin-flat plates which then was used to generate a map for predicting the flutter speed. The conceptual theory of this research was developed from the theory of flutter and the similitude theory.

2.1. Flutter. Theoretical work on flutter has been recognized as early as 1878 by Rayleigh [20]. However, the practical work on flutter was later on reported by Theodorsen in 1934. Theodorsen explained the flutter mechanism theoretically and experimentally of the aircraft wing and also the combi- nation of wing-aileron-tab [21–23].

Flutter is an unstable fluid and structure interaction. The dynamics of a thin-flat plate's structure is stable by itself. However, when it is placed in a moving air, the aerodynamic forces shift the stability of the system. The system will become unstable when the air speed reaches a certain speed. This speed is called the flutter speed.

The dynamics of a thin-flat plate can be modeled as a matrix equation of motion where the plate's structural dynamics is subjected to aerodynamic force F_a.

$$M\ddot{U} + C\dot{U} + KU = F_a. \tag{1}$$

In this case, $U, \dot{U}, \ddot{U}$ are the element nodal displacement, velocity, and acceleration vectors, consecutively. M, C, and K are the mass, damping, and stiffness matrices.

The aerodynamic force F_a is nonlinear in nature. For the purpose of predicting the flutter speed, F_a is often linearized to be

$$F_a = C_a\dot{U} + K_aU. \tag{2}$$

The denotation "a" on C_a and K_a represents the aerodynamic contribution to the damping and stiffness matrices.

Substituting the linearized F_a into the full plate's aeroelas- tic equation of motion yields the plate's linearized equation of motion:

$$M\ddot{U} + (C - C_a)\dot{U} + (K - K_a)U = 0. \tag{3}$$

Equation (3) shows how the flutter can occur. The flutter will occur if (3) is unstable. The *eigenvalue* analysis of this linearized equation can give the value of the flutter speed of the plate.

2.2. Similitude Requirements for Modeling in Fluid Mechanics.

The similitude requirements for the fluid dynamic problems are already well developed. Similitude is usually used in analyzing fluid dynamic problems such as lift and drag forces. In this study, the similitude method is used for analyzing the fluid structure interaction.

Similitude deals with the similarity of the actual system with its lab-scaled model or prototype. The similitude requirements for fluid dynamic problems have been well defined by Wolowicz et al. [24]. Similarity in geometric configuration is a fundamental requirement. Prototype and actual objects have to be geometrically congruent.

Another requirement is kinematic similarity. Two flows are kinematically similar if the associated velocities at the same point are related to a constant factor in the direction and magnitude [24]. This means that two streams are equal in their kinematics when streamline pattern associated with a constant factor.

Further requirements that must be met are the dynamics similarity. Two flows are dynamically similar when the associated forces at the same point are related to a constant factor in the direction and magnitude.

For the purpose of this research, an extra similitude requirement should be considered. Flutter problem is a moving boundary problem. The shape of the boundary varies with time. Ideally, in a steady case, the change is periodic or constant in frequency spectrum. Flutter will occur in a certain shape that is associated with the structural natural mode. Due to this reason, it is required to add a fourth similarity requirement when similitude theory is implemented to flutter problem. This fourth requirement is that the mode shape of the actual and the prototype must be congruent.

2.3. Buckingham Pi Theorem.

The development of a similitude method relays very much on the Buckingham Pi Theorem. In this section, the Beckingham Pi Theorem is recalled as the base for the dimensional analysis.

For every physical phenomenon dependent parameter which can be expressed by a function of $n - 1$ independent parameters, we can express the relationship between the parameters to form

$$q_1 = f(q_2, q_3, \ldots, q_n) \tag{4}$$

which is dependent parameters q_1 and $q_2, q_3, \ldots, q_n$ is the $n - 1$ independent parameters. In mathematics the above functional relationship can be expressed by an equivalent function

$$g(q_1, q_2, q_3, \ldots, q_n) = 0, \tag{5}$$

where g is an unspecified function, different from f.

Buckingham Pi Theorem states that [25] if there are n parameters in the function

$$g(q_1, q_2, q_3, \ldots, q_n) = 0 \tag{6}$$

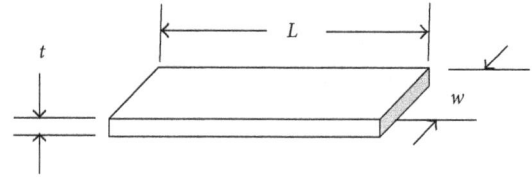

FIGURE 1: Plate's dimensions.

then n parameters can be grouped into different $n - m$ dimensionless ratio, or Π parameters, which can be expressed in the form of the function

$$G(\Pi_1, \Pi_2, \ldots, \Pi_{n-m}) = 0 \tag{7}$$

or

$$\Pi_1 = G_1(\Pi_2, \Pi_3, \ldots, \Pi_{n-m}). \tag{8}$$

Theorem does not predict the form of functional Gs or G_1. The form of the functional relationship between Π dimensionless independent parameters must be determined experimentally.

2.4. Π Group for Flutter Similitude Thin-Flat Plate.

In this experimental study, the dependent variable is the flutter speed V_f. Measurements were performed in SI units; thus, V_f unit is m/s. Variable definitions for the plate's dimensions can be referred to Figure 1. Independent parameters that are expected to affect the value of V_f are

L = plate length (m),

t = plate thickness,

w = plate width,

μ = fluid viscosity (N·s/m² = kg/(m·s)),

ρ = fluid mass density (kg/m³),

ω_n = natural frequency of the *thin-flat plate* (rad/s).

Relationship between the variables studied can be expressed by the symbolic function as follows:

$$V_f = f(L, \mu, \rho, \omega_n). \tag{9}$$

Based on this function then $n = 5$. Primary dimensional dimension used in this analysis is the mass, length, and time. The selected repeating parameters are L, μ, ρ. So $m = 3$. Analysis results obtained two Πs that are

$$\Pi_1 = \frac{\rho V_f L}{\mu}, \tag{10}$$

$$\Pi_2 = \frac{\rho \omega_n L^2}{\mu}, \tag{11}$$

$$\Pi_1 = f(\Pi_2). \tag{12}$$

The formulation of the dimensional analysis on (12) will work for geometrically similar plates. For the cases of thin-flat

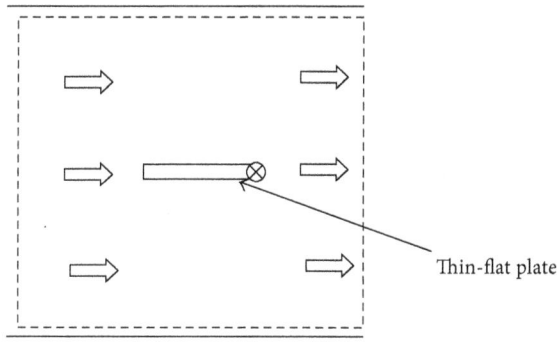

FIGURE 2: Experimental setup (top view).

plates, the thicknesses of the plates are relatively much smaller than the other geometric parameters and the variations in the thickness will not affect the geometric similarity of the flow significantly. Based on this reasoning—also approved by the experimental data—it is reasonable to do similitude analysis of the plates with various thicknesses but having the same aspect ratios. Allowing the variations in the plates' thickness requires the thickness (t) to be included in the dimensional analysis. An extra dimensionless number Π_3 is introduced to the analysis with the addition of thickness (t) to the dimensional analysis with the value:

$$\Pi_3 = \frac{t}{L}. \tag{13}$$

Hence, the experiment was used to obtain the relationship of Π_1 as a function of Π_2 and Π_3.

$$\Pi_1 = f\left(\Pi_2, \Pi_3\right) \tag{14}$$

or

$$\frac{\rho V_f L}{\mu} = f\left(\frac{\rho \omega_n L^2}{\mu}, \frac{t}{L}\right). \tag{15}$$

Consider the control volume of the fluid-structure interaction as shown in Figure 2. It is shown that the system is geometrically congruent from the top view projection. Then, it is possible to include the plates with different aspect ratio to analysis by considering the width to dimensional analysis. Inclusion of all plates with different aspect ratio to the analysis then introduces a fourth dimensionless number:

$$\Pi_4 = \frac{w}{L}, \tag{16}$$

$$\Pi_1 = f\left(\Pi_2, \Pi_3, \Pi_4\right), \tag{17}$$

$$\frac{\rho V_f L}{\mu} = f\left(\frac{\rho \omega_n L^2}{\mu}, \frac{t}{L}, \frac{w}{L}\right). \tag{18}$$

The form of first dimensionless number Π_1 can be viewed as a form of comparison between inertial forces to the viscous forces of the fluid. It is a kind of Reynolds number. Since it is calculated during the flutter speed, it can be called flutter Reynolds number. Dimensionless number Π_2 contains the inertia forces of both fluid and structure as well as the structural elastic forces and fluid viscous forces. Hence, Π_2 represents the fluid-structural interaction of the system.

Since the experiments allow the variation in the thickness and aspect ratios free from the requirement of structural similarity, the thickness factor Π_3 and aspect ratio Π_4 have to be included in the empirical equation (18).

2.5. Structural Analysis. The experiment shows the flutter of the plates in this free leading edge configuration occurring in their first natural frequencies. The calculation of the second dimensionless number Π_2 requires the value of the plate's natural frequency. The mode shapes and the natural frequencies of the plates can be approximated using Euler-Bernoulli beam model as in (19). Using the beam approximation for a plate, the displacement of the plate is restricted to the transverse direction $W(x,t)$ with its value as a function of the distance x from the end of the beam and time t:

$$\frac{\partial^2 W(x,t)}{\partial t^2} + c^2 \frac{\partial^4 W(x,t)}{\partial x^4} = 0 \tag{19}$$

with

$$c = \sqrt{\frac{EI}{\rho_s A}}. \tag{20}$$

The natural frequency of a clamped-free beam [26] can be derived from partial differential equation (19) to become the following formula:

$$\beta^4 = \frac{\rho_s A \omega_n^2}{EI}. \tag{21}$$

The calculated natural frequency ω_n from (21) is used to calculate the dimensionless number Π_2. In (20) and (21), E is plate's Young's modulus of elasticity; ρ_s is plate's mass density; A is plate's cross-sectional area; I is second moment of plate's cross-sectional area. The value of constant β depends on the boundary conditions and the length of the beam. For a cantilevered beam, the value of βL is 1.8751 for the first natural frequency.

3. Experimental Setup

The purpose of the experiment was to determine the function f as in (18) that describes the relationship of the three dimensionless numbers. Therefore, the test data should be taken with a plate that has a variety of different length and different natural frequencies. Form of the function f will be obtained through multivariable nonlinear regression analysis. All plates tested were copper plates with Young's modulus of elasticity of 110 GPa and mass density of 8960 kg/m^3.

Experimental setup can be seen in Figure 2. The flutter speed of every single thin-flat plate was measured one by one. The plate was placed on the wind tunnel that has maximum speed of 40 m/s. Each plate was positioned in vertical direction in the width direction. The trailing edge of the plate was clamped and the leading edge was set free.

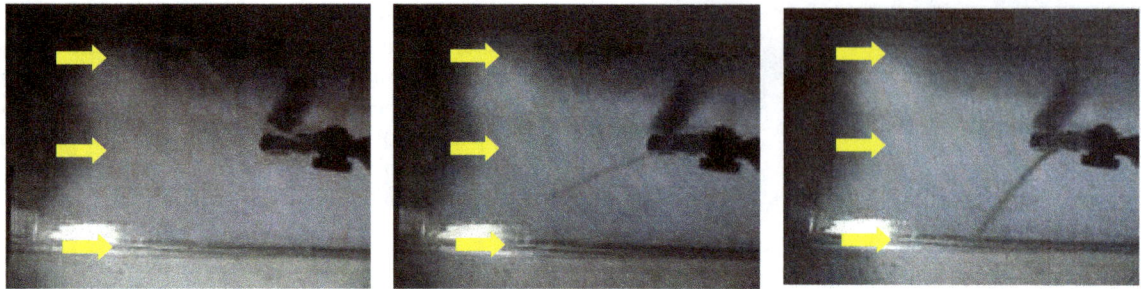

FIGURE 3: Snapshot of flutter mode shape taken by high speed camera (top view).

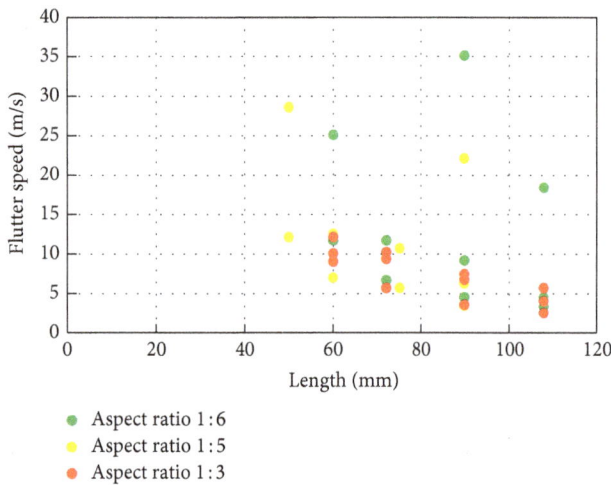

FIGURE 4: Flutter speed for various plates' dimensions.

Air was blown by a blower into the test section. Air velocity in the wind tunnel was set by regulating the air intake into the blower. The flow velocity was measured using a pitot tube.

For each plate, the testing is done by varying the air velocity from zero to flutter speed. The image of the plate motion was captured by high speed camera that was set on 210 frame-per-second speed as shown in Figure 3. The flutter speed determined from that image is the air speed when the plate was stopped at the infinite deflection. The results of measurements were presented in the form of the flutter speed as a function of plates' dimensions.

4. Result and Discussion

The experimental results are tabulated in two ways: the first way is the tabulation without dimensional analysis as in Table 1 and the second one is the tabulation with dimensional analysis as in Table 2. Table 1 is divided into three subtables according to the plates' aspect ratio. The list of flutter speed for various dimensions was obtained from the experiment and tabulated on Table 1. The natural frequency of each plate was also calculated using (21). The data from Table 1 is plotted on Figure 4 in the form of flutter speeds verses the lengths of the plates. In this graph, the data are also classified according

to the plates' aspect ratio: green dots for aspect ratio of $1:6$; yellow ones for that of $1:5$; and red for that of $1:3$.

Not all physical trends can be observed in Figure 4. One trend can be observed from Figure 4: that is, a longer plate has a lower flutter speed. The length of the plate relates to the area of the interaction surface between structure and the fluid. A longer plate has a larger area of interaction surface so that it is easier to transfer energy from fluid; thus, the flutter will occur at lower speed.

Similitude analysis on flutter speed offers a comprehensive view of all data trends. The values of dimensionless variable are calculated using the data from Table 1 and tabulated on Table 2 using (10), (11), (13), and (16). Table 2 is also divided into three subtables according to the value of Π_4. The calculated dimensionless numbers from Table 2 are plotted on 2-dimensional chart of Π_1 versus Π_2 as in Figure 5. In Figure 5, the data are also classified into three different aspect ratios: black dots are for $\Pi_4 = 1:6$; blue dots are for $\Pi_4 = 1:5$; and red dots are for $\Pi_4 = 1:3$.

Some physical trends can be observed from Figure 5. The dimensionless number Π_1 represents flutter speed and $\Pi2$ represents the plate's natural frequency. The relationship of the flutter speed with the plate's natural frequency can be described using energy scheme. The higher the frequency of a vibrating plate, the higher its vibration energy. A plate with higher natural frequency will need a higher energy from the fluid to reach flutter.

Also from Figure 5, a plate with relative higher thickness with respect to its length will tend to have a higher flutter Reynolds number, because thicker plates tend to be stiffer and have a higher value of structural stiffness matrices. By thinking in the static stability scheme, to invoke the instability requires a higher speed with the stiffer plate.

The trends of the data are not clearly identified if the data are plotted on 2-dimensional graph as in Figure 5 without the classification of their geometric ratios. This implies that the geometric ratios should be included in the analysis as the correcting factors as in (17). The data were apparently nonlinear and approximating the power trends. Then, a multivariable nonlinear regression analysis is conducted to obtain the form of function f in (17). The result of the regression analysis is shown on (22) and (23).

$$\Pi_1 = 1.51 \times 10^7 \Pi_2^{0.60} \Pi_3^{1.86} \Pi_4^{-1.63}. \qquad (22)$$

TABLE 1: Flutter speed for several plates with different dimensions and aspect ratios.

(a) Aspect ratio 1 : 6

Plate's dimension (mm)			ω_n	Flutter
Thickness	Width	Length	(rad/s)	(m/s)
0.06	10	60	59.27	11.71
0.08	10	60	79.03	25.09
0.06	12	72	41.16	6.69
0.08	12	72	54.88	11.71
0.06	15	90	26.34	4.52
0.08	15	90	35.12	9.20
0.13	15	90	57.08	35.12
0.06	18	108	18.29	3.34
0.08	18	108	24.39	4.52
0.13	18	108	39.64	18.40

(b) Aspect ratio 1 : 5

Plate's dimension (mm)			ω_n	Flutter
Thickness	Width	Length	(rad/s)	(m/s)
0.06	10	50	85.35	13.58
0.08	10	50	113.80	18.11
0.06	12	60	59.27	9.43
0.08	12	60	79.03	12.58
0.06	15	75	37.93	6.04
0.08	15	75	50.58	8.05
0.06	18	90	26.34	4.19
0.08	18	90	35.12	5.59
0.13	18	90	57.08	9.08

(c) Aspect ratio 1 : 3

Plate's dimension (mm)			ω_n	Flutter
Thickness	Width	Length	(rad/s)	(m/s)
0.06	20	60	59.27	9.43
0.08	20	60	79.03	12.58
0.13	20	60	128.42	20.44
0.06	24	72	41.16	6.55
0.08	24	72	54.88	8.73
0.13	24	72	89.18	14.19
0.06	30	90	26.34	4.19
0.08	30	90	35.12	5.59
0.13	30	90	57.08	9.08
0.06	36	108	18.29	2.91
0.08	36	108	24.39	3.88
0.13	36	108	39.64	6.31

Substituting the values of Π_1 as a function of Π_2, Π_3, and Π_4 into (23) yields

$$\frac{\rho V_f L}{\mu} = 1.51 \times 10^7 \left(\frac{\rho \omega_n L^2}{\mu} \right)^{0.60} \left(\frac{t}{L} \right)^{1.86} \left(\frac{w}{L} \right)^{-1.63}. \quad (23)$$

The illustration of trends body of (22) requires a 4-dimensional plot. The 4-dimensional plot is projected as iso-aspect-ratio surfaces on 3-dimensional space as shown in

TABLE 2: Π_1 as a function of Π_2, Π_3, and Π_4.

(a) $\Pi_4 = 1 : 6$

Π_1	Π_2	Π_3
45538.65	829.80	0.00100
97571.71	1106.40	0.00133
31219.84	995.76	0.00083
54646.38	1327.69	0.00111
26366.53	1244.70	0.00067
53666.39	1659.61	0.00089
204865.60	2696.86	0.00144
23379.88	1493.65	0.00056
31639.84	1991.53	0.00074
128799.33	3236.23	0.00120

(b) $\Pi_4 = 1 : 5$

Π_1	Π_2	Π_3
39291.69	691.50	0.00120
92611.40	922.00	0.00160
27222.08	829.80	0.00100
48696.34	1106.40	0.00133
27783.42	1037.25	0.00080
51978.02	1383.01	0.00107
20279.99	1244.70	0.00067
36522.25	1659.61	0.00089
129125.67	2696.86	0.00144

(c) $\Pi_4 = 1 : 3$

Π_1	Π_2	Π_3
35143.56	829.80	0.00100
39134.21	1106.40	0.00133
47150.03	1797.91	0.00217
26672.08	995.76	0.00083
43826.70	1327.69	0.00111
47712.47	2157.49	0.00181
21086.13	1244.70	0.00067
39448.54	1659.61	0.00089
43470.18	2696.86	0.00144
17892.18	1493.65	0.00056
28290.02	1991.53	0.00074
40008.12	3236.23	0.00120

Figure 6. The iso-aspect-ratio surfaces are the relationship of Π_1 as a function of Π_2 and Π_3 with constant value of Π_4.

Equation (23) is 4-dimensional map of the tendency of the flutter speed for various plates' dimensions. The projection of this map to 3-dimensional space on Figure 6 shows the characteristic of the flutter Reynolds numbers Π_1 as a function of flutter fluid structure interaction numbers Π_2, with the geometric ratios as the correcting factors. This map shows that flutter Reynolds number Π_1 is very high at the region of high flutter fluid structure interaction number Π_2 and thickness-to-length ratios number Π_3. However, it is very sensitive to the change of those two dimensionless numbers at the region. The sensitivity is higher at lower

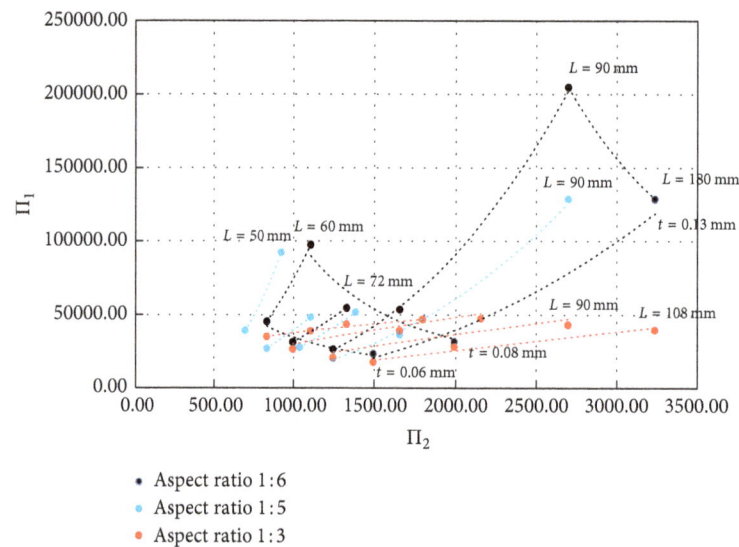

FIGURE 5: Flutter Reynolds number verses flutter fluid interaction number on 2-dimensional plot.

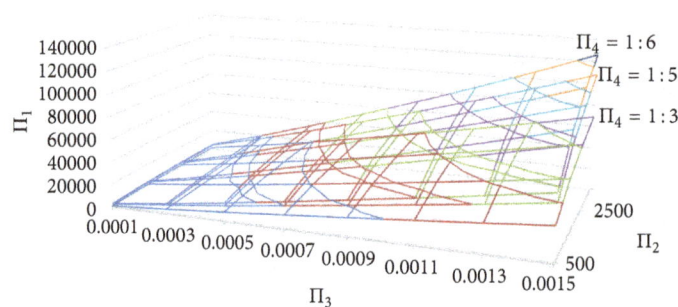

FIGURE 6: Iso-aspect-ratio trend surfaces showing flutter Reynolds number as a function of flutter fluid interaction number and thickness-to-length ratio on 3-dimensional plot.

aspect ratio Π_4 following the power trends. A lower aspect ratio means the plate is narrow and long (see (16)) so it has smaller interaction surface. The sensitivity of values of flutter Reynold number increases to the change of the values of fluid structure interaction number for the plate with lower aspect ratio. This happens as a result of decreasing contact surface for lower aspect ratio plates. Thus, the change of natural frequencies, which is also identical to the change of flutter fluid interaction number, has more dominant effect on flutter Reynolds number. On the contrary, at higher aspect ratio the contact surface is broader and it has dominant effect. As a result flutter Reynolds number is less sensitive to the change in natural frequency Π_2.

The map generated in this research is a very useful tool in optimizing the design of wind harvester given its range of operating wind speed. The map illustrates the flutter speed trends for various plates' dimensions. A specific plate's dimension will be optimum for a specific range of wind speed.

An experimental setup is often limited to a certain flutter Reynolds number due to the limited speed of the wind tunnel. Numerical investigations should have more freedom to exceed this limitation. The dimensional analysis and similitude method formulated in this paper can be used in

a future research as method of presenting a result numerical investigation to generate a broader map, of course with some minor drawbacks of the numerical method limitations in representing the real case.

The similitude analysis presented in this paper is focused on the free leading edge plates in axial flow with application orientation for wind harvesting. However, there are also possibilities of using this method for other applications such as for the flutter speed analysis of airfoil of an aircraft wings or turbine blades.

5. Conclusions

Similitude method has been developed to generate an empirical map to predict the flutter speed of a thin-flat plate. The experiment has showed that the trend of flutter speed has been traced in various geometric parameters. The relationship between the flutter speeds of thin flat-plates in a variety of sizes that are geometrically congruent and the parameters that affect them such as the plates' width, thickness, length, fluid density, fluid viscosity, and their natural frequencies has been described.

This research also has found an empirical relationship between the flutter Reynolds number and the fluid-structure interaction number with the geometric ratios as correcting factors.

Departing from the testing of thin-flat plate, this study provides a formulation in testing the similitude theory for a future study on flutter analysis of an airfoil that is used in the aircraft or turbomachinery industries. In addition, flutter similitude method can also be used to test the possibility of a detrimental effect of flutter on narrow and long bridges.

References

[1] S. Chad Gibbs, I. Wang, and E. Dowell, "Theory and experiment for flutter of a rectangular plate with a fixed leading edge in three-dimensional axial flow," *Journal of Fluids and Structures*, vol. 34, pp. 68–83, 2012.

[2] D. M. Tang, H. Yamamoto, and E. H. Dowell, "Flutter and limit cycle oscillations of two-dimensional panels in three-dimensional axial flow," *Journal of Fluids and Structures*, vol. 17, no. 2, pp. 225–242, 2003.

[3] L. Tang and M. P. Païdoussis, "The coupled dynamics of two cantilevered flexible plates in axial flow," *Journal of Sound and Vibration*, vol. 323, no. 3–5, pp. 790–801, 2009.

[4] D. Tang and E. H. Dowell, "Limit cycle oscillations of two-dimensional panels in low subsonic flow," *International Journal of Non-Linear Mechanics*, vol. 37, no. 7, pp. 1199–1209, 2002.

[5] R. M. Howell, A. D. Lucey, P. W. Carpenter, and M. W. Pitman, "Interaction between a cantilevered-free flexible plate and ideal flow," *Journal of Fluids and Structures*, vol. 25, no. 3, pp. 544–566, 2009.

[6] W. Zhao, M. P. Païdoussis, L. Tang, M. Liu, and J. Jiang, "Theoretical and experimental investigations of the dynamics of cantilevered flexible plates subjected to axial flow," *Journal of Sound and Vibration*, vol. 331, no. 3, pp. 575–587, 2012.

[7] C. Eloy, R. Lagrange, C. Souilliez, and L. Schouveiler, "Aeroelastic instability of cantilevered flexible plates in uniform flow," *Journal of Fluid Mechanics*, vol. 611, pp. 97–106, 2008.

[8] C. Eloy, N. Kofman, and L. Schouveiler, "The origin of hysteresis in the flag instability," *Journal of Fluid Mechanics*, vol. 691, pp. 583–593, 2012.

[9] O. Doaré and S. Michelin, "Piezoelectric coupling in energy-harvesting fluttering flexible plates: linear stability analysis and conversion efficiency," *Journal of Fluids and Structures*, vol. 27, no. 8, pp. 1357–1375, 2011.

[10] K. Makihara and S. Shimose, "Supersonic flutter utilization for effective energy-harvesting based on piezoelectric switching control," *Smart Materials Research*, vol. 2012, Article ID 181645, 10 pages, 2012.

[11] J. A. Dunnmon, S. C. Stanton, B. P. Mann, and E. H. Dowell, "Power extraction from aeroelastic limit cycle oscillations," *Journal of Fluids and Structures*, vol. 27, no. 8, pp. 1182–1198, 2011.

[12] J. J. Allen and A. J. Smits, "Energy harvesting eel," *Journal of Fluids and Structures*, vol. 15, no. 3-4, pp. 629–640, 2001.

[13] A. C. Fernandes and S. Mirzaeisefat, "Flow induced fluttering of a hinged vertical flat plate," *Ocean Engineering*, vol. 95, pp. 134–142, 2015.

[14] A. G. Cunha-Filho, A. M. G. de Lima, M. V. Donadon, and L. S. Leão, "Flutter suppression of plates using passive constrained viscoelastic layers," *Mechanical Systems and Signal Processing*, vol. 79, pp. 99–111, 2016.

[15] M. Peng and H. A. DeSmidt, "Stability analysis of a flutter panel with axial excitations," *Advances in Acoustics and Vibration*, vol. 2016, Article ID 7194764, 7 pages, 2016.

[16] K. Yaman, "Subsonic flutter of cantilever rectangular PC plate structure," *International Journal of Aerospace Engineering*, vol. 2016, Article ID 9212364, 10 pages, 2016.

[17] S. Torkamani, H. M. Navazi, A. A. Jafari, and M. Bagheri, "Structural similitude in free vibration of orthogonally stiffened cylindrical shells," *Thin-Walled Structures*, vol. 47, no. 11, pp. 1316–1330, 2009.

[18] S. De Rosa, F. Franco, X. Li, and T. Polito, "A similitude for structural acoustic enclosures," *Mechanical Systems and Signal Processing*, vol. 30, pp. 330–342, 2012.

[19] A. A. Yazdi and J. Rezaeepazhand, "Structural similitude for flutter of delaminated composite beam-plates," *Composite Structures*, vol. 93, no. 7, pp. 1918–1922, 2011.

[20] L. Rayleigh, "On the instability of jets," *Proceedings of the London Mathematical Society*, vol. s1-10, no. 1, pp. 4–13, 1878.

[21] T. Theodorsen, "General theory of aerodynamic instability and the mechanism of flutter," NACA Technical Report 469, 1934.

[22] T. Theodorsen and L. E. Garrick, "Mechanism of flutter: a theoretical and experimental investigation of the flutter problem," NACA Technical Report 685, 1940.

[23] T. Theodorsen and L. E. Garrick, "Nonstationary flow about a-wing-aeleron-tab combination including aerodynamic balance," NACA Technical Report 736, 1941.

[24] C. H. Wolowicz, J. S. Bowman Jr., and W. P. Gilbert, "Similitude requirement and scaling relationships as applied to model testing," NASA Technical Paper, 1979.

[25] R. W. Fox and A. T. McDonald, *Introduction to Fluid Mechanics*, John Wiley & Son, 1994.

[26] D. J. Inman, *Engineering Vibration*, Prentice-Hall, Upper Saddle River, NJ, USA, 1994.

7

Stability Analysis of a Flutter Panel with Axial Excitations

Meng Peng and Hans A. DeSmidt

Department of Mechanical, Aerospace and Biomedical Engineering, The University of Tennessee, 606 Dougherty Engineering Building, Knoxville, TN 37996-2210, USA

Correspondence should be addressed to Meng Peng; mpeng1@vols.utk.edu

Academic Editor: Marc Thomas

This paper investigates the parametric instability of a panel (beam) under high speed air flows and axial excitations. The idea is to affect out-of-plane vibrations and aerodynamic loads by in-plane excitations. The periodic axial excitation introduces time-varying items into the panel system. The numerical method based on Floquet theory and the perturbation method are utilized to solve the Mathieu-Hill equations. The system stability with respect to air/panel density ratio, dynamic pressure ratio, and excitation frequency are explored. The results indicate that panel flutter can be suppressed by the axial excitations with proper parameter combinations.

1. Introduction

Panel (beam) flutter usually occurs when high speed objects move in the atmosphere, such as flight wings [1] and ballute [2]. This phenomenon is a self-excited oscillation due to the coupling of aerodynamic load and out-of-plane vibration. Since flutter can cause system instability and material fatigue, many scholars have carried out theoretical and experimental analyses on this topic. Nelson and Cunningham [3] investigated flutter of flat panels exposed to a supersonic flow. Their model is based on small-deflection plate theory and linearized flow theory, and the stability boundary is determined after decoupling the system equations by Galerkin's method. Olson [4] applied finite element method to the two-dimensional panel flutter. A simply supported panel was calculated and an extremely accuracy approximation could be obtained using only a few elements. Parks [5] utilized Lyapunov technique to solve a two-dimensional panel flutter problem and used piston theory to calculate aerodynamic load. The results gave a valuable sufficient stability criterion. Dugundji [6] examined characteristics of panel flutter at high supersonic Mach numbers and clarified the effects of damping, edge conditions, traveling, and standing waves. The panel, Dugundji considered, is a flat rectangular one, simply supported on all four edges, and undergoes two-dimensional midplane compressive forces.

Dowell [7, 8] explored plate flutter in nonlinear area by employing Von Karman's large deflection plate theory. Zhou et al. [9] built a nonlinear model for the panel flutter via finite element method, including linear embedded piezoelectric layers. The optimal control approach for the linearized model was presented. Gee [10] discussed the continuation method, as an alternate numerical method that complements direct numerical integration, for the nonlinear panel flutter. Tizzi [11] researched the influence of nonlinear forces on flutter beam. In most cases, the internal force in panel or beam results from either external constant loads or geometric nonlinearities. Therefore, their models are time-invariant system.

Panel is usually excited by in-plane loads resulting from the vibrations generated and/or transmitted through the attached structures and dynamics components when experiencing aerodynamic loads. If the in-plane load is time dependent, the system becomes time-varying. The topic of dynamic stability of time-varying systems attracts many attentions. Iwatsubo et al. [12] surveyed parametric instability of columns under periodic axial loads for different boundary conditions. They used Hsu's results [13] to determine stability conditions and discussed the damping effect on combination resonances. Sinha [14] and Sahu and Datta [15, 16] studied the similar problem for Timoshenko beam and curved panel, respectively, and both models are classified as Mathieu-Hill

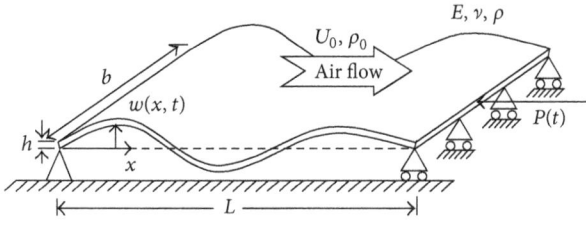

FIGURE 1: Simply supported panel (beam) subjected to an air flow and an axial excitation.

equations. Furthermore, the stability of the nonlinear elastic plate subjected to a periodic in-plane load was analyzed by Ganapathi et al. [17]. They solved nonlinear governing equations by using the Newmark integration scheme coupled with a modified Newton-Raphson iteration procedure. In addition, many papers have been published for dynamic stability analyses of shells under periodic loads [18–24]. Furthermore, Hagedorn and Koval [25] considered the effect of longitudinal vibrations and the space distributed internal force. The combination resonance was analyzed for Bernoulli-Euler and Timoshenko beams under the spatiotemporal force. Yang et al. [26] developed a vibration suppression scheme for an axially moving string under a spatiotemporally varying tension. Lyapunov method was employed to design robust boundary control laws, but the effect of parameters of the spatiotemporally varying tension on system stability has not been fully analyzed.

Nevertheless, the published investigations on the parametric stability of the flutter panel (beam) with the periodically time-varying system stiffness due to axial excitations are scarce. This paper is to explore the coactions of time-varying axial excitations and aerodynamic loads on panel (beam) and conduct parameter studies. The stability analysis is executed first by Floquet theory numerically and then by Hsu's method analytically for approximations.

2. System Description and Model

The configuration considered herein is an isotropic thin panel (beam) with constant thickness and cross section. As shown in Figure 1, the panel is simply supported at both ends and a periodic axial excitation acts on the right end. The panel's upper surface is exposed to a supersonic flow, while the air beneath the lower surface is assumed not to affect the panel dynamics. Another assumption made here is the axial strain from the lateral displacement is very small so that it can be ignored. The system model is based on the coupled effects from the out-of-plane (lateral) displacement of the panel, aerodynamic loads, and time-varying in-plane (axial) excitation forces.

The total kinetic energy of the penal due to lateral displacements is

$$T = \frac{1}{2}\rho A \int_0^L \dot{w}(x,t)^2 \, dx, \quad (1)$$

where $w(x,t)$ is the lateral displacement of the panel measured in the ground fixed coordinate frame, ρ the panel

density, A the panel cross-sectional area, and "·" the differentiation with respect to time t.

The total potential energy of the panel due to lateral displacements is

$$V = \frac{1}{2}E^* I \int_0^L w''(x,t)^2 \, dx + \frac{1}{2}P(t)\int_0^L w'(x,t)^2 \, dx, \quad (2)$$

where $E^* = E/(1 - v^2)$ for panel and $E^* = E$ for beam with Young's modulus E and Poisson's ratio v; the moment of inertia is given by $I = bh^3/12$ with panel width b and panel thickness h; $P(t)$ is the periodic axial excitation with the frequency ω and "$\prime$" indicates differentiation with respect to axial position x.

For material viscous damping, a Rayleigh dissipation function is defined as

$$R = \frac{1}{2}\xi E^* I \int_0^L \dot{w}''(x,t)^2 \, dx. \quad (3)$$

Here, ξ is the material viscous loss factor of the panel.

The aerodynamic load is expressed by using the classic quasisteady first-order piston theory [4, 6, 7, 9, 11]:

$$p(x,t) = \frac{2q_0}{\beta}\left[\frac{\partial w(x,t)}{\partial x} + \frac{(\mathrm{Ma}^2 - 2)}{(\mathrm{Ma}^2 - 1)}\frac{1}{U_0}\frac{\partial w(x,t)}{\partial t}\right], \quad (4)$$

where $q_0 = \rho_0 U_0^2/2$ is the dynamic pressure, ρ_0 is the undisturbed air flow density, U_0 is the flow speed at infinity, Ma is Mach number, and $\beta = (\mathrm{Ma}^2 - 1)^{1/2}$.

The flow goes against the lateral vibrations of the panel, so the nonconservative virtual work from aerodynamic load is negative and its expression is

$$\delta W_{\mathrm{nc}} = -\int_0^L p(x,t)\, b\delta w(x,t)\, dx. \quad (5)$$

For a simply supported panel, the modal expansion of $w(x,t)$ can be assumed in the form

$$w(x,t) = \sum_{n=1}^{\infty} \eta_n(t)\sin\left(\frac{n\pi x}{L}\right), \quad n = 1,2,3,\ldots, \quad (6)$$

where n is the positive integer and $\eta_n(t)$ the generalized coordinate. After substituting (6) into all energy expressions, the system equations-of-motion are obtained via Lagrange's Equations

$$\frac{d}{dt}\left(\frac{\partial T}{\partial \dot{q}}\right) - \frac{\partial T}{\partial q} + \frac{\partial V}{\partial q} + \frac{\partial D}{\partial \dot{q}} = Q_{\mathrm{nc}} \quad (7)$$

with the generalized force $Q_{\mathrm{nc}} = \partial \delta W_{\mathrm{nc}}/\partial \delta q$ and generalized coordinates vector

$$q(t) = [\eta_1(t) \ \ \eta_2(t) \ \ \eta_3(t) \ \cdots]^T. \quad (8)$$

Finally, the system equations-of-motion are given as

$$\mathbf{M}\ddot{q}(t) + \left(\mathbf{C}^d + \mathbf{C}^a\right)\dot{q}(t) + \left[\mathbf{K}^e + \mathbf{K}^a + \mathbf{K}^P(t)\right]q(t) = 0, \quad (9)$$

where the elements of coefficient matrices are

$$\mathbf{M}_{mn} = \rho A \int_0^L \sin\left(\frac{m\pi x}{L}\right) \sin\left(\frac{n\pi x}{L}\right) dx,$$

$$\mathbf{C}_{mn}^d = \xi E^* I \frac{m^2 n^2 \pi^4}{L^4} \int_0^L \sin\left(\frac{m\pi x}{L}\right) \sin\left(\frac{n\pi x}{L}\right) dx,$$

$$\mathbf{C}_{mn}^a$$

$$= \frac{2bq_0}{\beta U_0} \frac{(Ma^2 - 2)}{(Ma^2 - 1)} \int_0^L \sin\left(\frac{m\pi x}{L}\right) \sin\left(\frac{n\pi x}{L}\right) dx, \quad (10)$$

$$\mathbf{K}_{mn}^e = E^* I \frac{m^2 n^2 \pi^4}{L^4} \int_0^L \sin\left(\frac{m\pi x}{L}\right) \sin\left(\frac{n\pi x}{L}\right) dx,$$

$$\mathbf{K}_{mn}^a = \frac{2bq_0}{\beta} \frac{n\pi}{L} \int_0^L \sin\left(\frac{m\pi x}{L}\right) \cos\left(\frac{n\pi x}{L}\right) dx,$$

$$\mathbf{K}_{mn}^P(t) = \frac{mn\pi^2}{L^2} P(t) \int_0^L \cos\left(\frac{m\pi x}{L}\right) \cos\left(\frac{n\pi x}{L}\right) dx.$$

The stiffness matrix $\mathbf{K}^P(t)$ resulting from the periodic axial excitation introduces a periodically time-varying item into the system, so (9) is identified as a set of coupled Mathieu-Hill equations. Subsequently, the system equations-of-motion are transformed into the nondimensional (N.D.) form

$$\overline{\mathbf{M}}\, \overset{**}{\overline{q}}(\tau) + \left(\overline{\mathbf{C}}^d + \overline{\mathbf{C}}^a\right) \overset{*}{\overline{q}}(\tau)$$
$$+ \left[\overline{\mathbf{K}}^e + \overline{\mathbf{K}}^a + \overline{\mathbf{K}}^P(\tau)\right] \overline{q}(\tau) = 0 \quad (11)$$

with the dimensionless parameters and coordinates:

$$\overline{x} = \frac{x}{L},$$

$$\overline{q} = \frac{q(t)}{h},$$

$$\Omega = \frac{\pi^2 h}{L^2} \sqrt{\frac{E^*}{12\rho}},$$

$$\tau = t\Omega,$$

$$\mu = \frac{\rho_0}{\rho},$$

$$\overline{\xi} = \xi\Omega,$$

$$P_{cr} = \frac{E^* I \pi^2}{L^2},$$

$$f(\tau) = \frac{P(t)}{P_{cr}},$$

$$\sigma = \frac{q_0}{E^*},$$

$$\alpha = \frac{h}{L},$$

$$\overline{\omega} = \frac{\omega}{\Omega},$$

$$\overset{*}{(\,)} = \frac{d(\,)}{d\tau}.$$

$$(12)$$

The elements of the N.D. coefficient matrices in (11) are

$$\overline{\mathbf{M}}_{mn} = \begin{cases} 1 & \text{if } m = n \\ 0 & \text{if } m \neq n, \end{cases}$$

$$\overline{\mathbf{C}}_{mn}^d = \begin{cases} \overline{\xi} n^4 & \text{if } m = n \\ 0 & \text{if } m \neq n, \end{cases}$$

$$\overline{\mathbf{C}}_{mn}^a = \begin{cases} \dfrac{2\sqrt{6\mu\sigma}\,(Ma^2 - 2)}{\pi^2 \alpha^2 (Ma^2 - 1)^{3/2}} & \text{if } m = n \\ 0 & \text{if } m \neq n, \end{cases}$$

$$\overline{\mathbf{K}}_{mn}^e = \begin{cases} n^4 & \text{if } m = n \\ 0 & \text{if } m \neq n, \end{cases}$$

$$\overline{\mathbf{K}}_{mn}^a = \begin{cases} 0 & \text{if } m = n \\ \dfrac{48\sigma mn\left[1 - (-1)^{m+n}\right]}{\sqrt{Ma^2 - 1}\,(m^2 - n^2)\,\pi^4 \alpha^3} & \text{if } m \neq n, \end{cases}$$

$$\overline{\mathbf{K}}_{mn}^P(\tau) = \begin{cases} n^2 f(\tau) & \text{if } m = n \\ 0 & \text{if } m \neq n. \end{cases}$$

Here, $\overline{\mathbf{M}}, \overline{\mathbf{C}}^d, \overline{\mathbf{C}}^a$, and $\overline{\mathbf{K}}^e$ are constant symmetric matrices; $\overline{\mathbf{K}}^a$ is a constant skew-symmetric matrix; $\overline{\mathbf{K}}^P$ is a symmetric time-varying matrix with a period of $2\pi/\overline{\omega}$.

3. Mathematical Methods for Stability Analysis

Due to the periodic axial excitations, (11) becomes a periodically linear time-varying system. Floquet theory is able to assess the stability of this type of systems through evaluating the eigenvalues of the Floquet transition matrix (FTM) numerically [24, 27–32]. The FTM method can obtain all unstable behaviors of a system but at the cost of intensively numerical computations, so the perturbation method originally developed by Hsu [13, 33] is modified in this paper to approximate the system stability boundary in an efficient way. The results from the perturbation (analytical) method will be compared with those from the FTM (numerical) method.

To implement Hsu's perturbation method, the time-invariant part of the system stiffness matrix, $\overline{\mathbf{K}}^e + \overline{\mathbf{K}}^a$, needs to be diagonalized by its left and right eigenvectors [34]. The resulting equations-of-motion are

$$\overset{**}{\overline{q}}(\tau) + \widetilde{\mathbf{C}}\overset{*}{\overline{q}}(\tau) + \left[\widetilde{\mathbf{K}} + \widetilde{\mathbf{K}}^P(\tau)\right]\overline{q}(\tau) = 0 \quad (14)$$

with

$$\widetilde{\mathbf{C}} = \mathbf{X}_L^T \left(\overline{\mathbf{C}}^d + \overline{\mathbf{C}}^a \right) \mathbf{X}_R$$

$$\widetilde{\mathbf{K}} = \mathbf{X}_L^T \left(\overline{\mathbf{K}}^e + \overline{\mathbf{K}}^a \right) \mathbf{X}_R \qquad (15a)$$

$$\widetilde{\mathbf{K}}^P (\tau) = \mathbf{X}_L^T \overline{\mathbf{K}}^P (\tau) \mathbf{X}_R,$$

$$\left(\overline{\mathbf{K}}^e + \overline{\mathbf{K}}^a \right) \mathbf{X}_R = \lambda \mathbf{X}_R$$

$$\left(\overline{\mathbf{K}}^e + \overline{\mathbf{K}}^a \right)^T \mathbf{X}_L = \lambda \mathbf{X}_L \qquad (15b)$$

$$\mathbf{X}_L^T \mathbf{X}_R = \mathbf{I},$$

where λ is the eigenvalues of $\overline{\mathbf{K}}^e + \overline{\mathbf{K}}^a$ and $\mathbf{X}_L$ and $\mathbf{X}_R$ are the corresponding left and right eigenvectors, respectively, and orthonormal to each other. $\mathbf{I}$ is the identity matrix.

The damping matrix and the time-varying stiffness matrix resulting from the axial excitation are assumed to be small quantities relative to the time-invariant system stiffness for better predictability through Hsu's method. The standard form in Hsu's method is obtained by separating the regular and perturbed items in (14) and then expanding the periodic time-varying stiffness matrix into Fourier series,

$$\overset{**}{\overline{q}} (\tau) + \widetilde{\mathbf{K}} \overline{q} (\tau) = -\widetilde{\mathbf{C}} \overset{*}{\overline{q}} (\tau) - \widetilde{\mathbf{K}}^P (\tau) \overline{q} (\tau), \qquad (16)$$

where

$$\widetilde{\mathbf{K}}^P (\tau) = \widetilde{\mathbf{K}}^c \cos (\overline{\omega}\tau) + \widetilde{\mathbf{K}}^s \sin (\overline{\omega}\tau). \qquad (17)$$

With (16), the stability criteria given in [13, 33] can be applied by setting epsilon to one.

4. Stability Boundaries for the Panel under Flow

For the simple demonstrations of the model and the solving process developed above, only the first two modes are considered so that the closed-form stability solutions can be obtained. The axial excitation force considered here is a single frequency cosine function:

$$\overset{**}{\overline{q}} (\tau) + \begin{bmatrix} \omega_1^2 & 0 \\ 0 & \omega_2^2 \end{bmatrix} \overline{q} (\tau)$$

$$= - \begin{bmatrix} c_{11} & c_{12} \\ c_{21} & c_{22} \end{bmatrix} \overset{*}{\overline{q}} (\tau) - \begin{bmatrix} k_{11} & k_{12} \\ k_{21} & k_{22} \end{bmatrix} \cos (\overline{\omega}\tau) \overline{q} (\tau), \qquad (18)$$

$$f(\tau) = F \cos (\overline{\omega}\tau),$$

where

$$\omega_1 = \sqrt{\frac{17}{2} - \frac{15}{2} \sqrt{1 - \frac{\sigma^2}{\sigma_c^2}}},$$

$$\omega_2 = \sqrt{\frac{17}{2} + \frac{15}{2} \sqrt{1 - \frac{\sigma^2}{\sigma_c^2}}}, \qquad (19a)$$

$$c_{11} = \overline{\xi} \left(\frac{17}{2} - \frac{15}{2} \frac{\sigma_c}{\sqrt{\sigma_c^2 - \sigma^2}} \right)$$

$$+ \frac{2\sqrt{6} \left(\mathrm{Ma}^2 - 2 \right) \sqrt{\mu\sigma}}{\pi^2 \alpha^2 \left(\mathrm{Ma}^2 - 1 \right)^{3/2}},$$

$$c_{22} = \overline{\xi} \left(\frac{17}{2} + \frac{15}{2} \frac{\sigma_c}{\sqrt{\sigma_c^2 - \sigma^2}} \right) \qquad (19b)$$

$$+ \frac{2\sqrt{6} \left(\mathrm{Ma}^2 - 2 \right) \sqrt{\mu\sigma}}{\pi^2 \alpha^2 \left(\mathrm{Ma}^2 - 1 \right)^{3/2}},$$

$$c_{12} = \frac{15 \overline{\xi} \sigma}{2 \sqrt{\sigma_c^2 - \sigma^2}},$$

$$c_{21} = -c_{12},$$

$$k_{11} = F \left(\frac{5}{2} - \frac{3\sigma_c}{2\sqrt{\sigma_c^2 - \sigma^2}} \right),$$

$$k_{22} = F \left(\frac{5}{2} + \frac{3\sigma_c}{2\sqrt{\sigma_c^2 - \sigma^2}} \right), \qquad (19c)$$

$$k_{12} = F \frac{3\sigma}{2\sqrt{\sigma_c^2 - \sigma^2}},$$

$$k_{21} = -k_{12},$$

$$\sigma_c = \frac{15}{128} \pi^4 \alpha^3 \sqrt{\mathrm{Ma}^2 - 1}. \qquad (19d)$$

The material viscous loss factor is set to zero for the eigen-analyses in this paper. The stability boundaries can be obtained via solving the following equations:

1st principle resonance: $\overline{\omega} = 2\omega_1 + \Delta\omega$

$$\Delta\omega = \pm \sqrt{\frac{k_{11}^2}{4\omega_1^2} - c_{11}^2}$$

2nd principle resonance: $\overline{\omega} = 2\omega_2 + \Delta\omega$

$$\Delta\omega = \pm \sqrt{\frac{k_{22}^2}{4\omega_2^2} - c_{22}^2} \qquad (20)$$

Combination resonance: $\overline{\omega} = \omega_2 - \omega_1 + \Delta\omega$

$$\Delta\omega = \pm \sqrt{\frac{k_{12}^2}{4\omega_1\omega_2} - c_{11}c_{22}}.$$

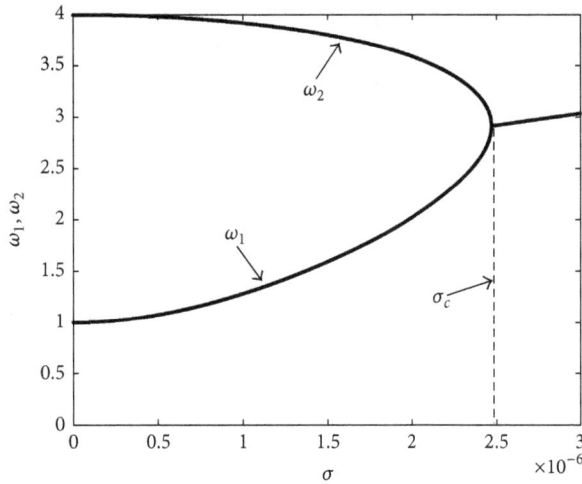

FIGURE 2: N.D. natural frequency variations with respect to dynamic pressure ratio: Ma = 2, α = 0.005, and $\sigma_c = 2.47 \times 10^{-6}$.

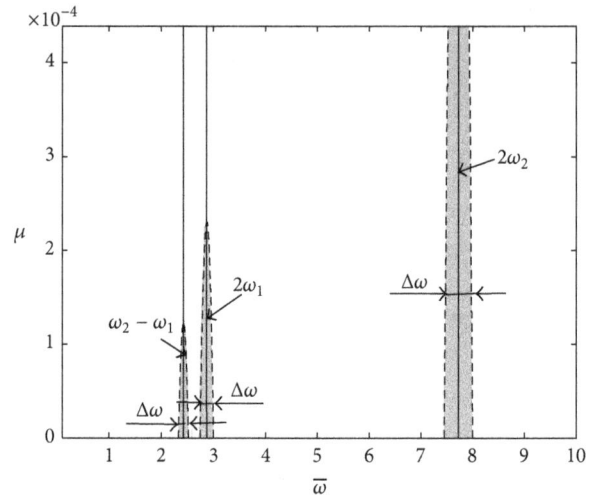

FIGURE 4: Stability plot for the flutter panel with respect to axial excitation frequency and air/panel density ratio: Ma = 2, α = 0.005, $\sigma_c = 2.47 \times 10^{-6}$, $\sigma = 1.27 \times 10^{-6}$, and $F = 0.5$.

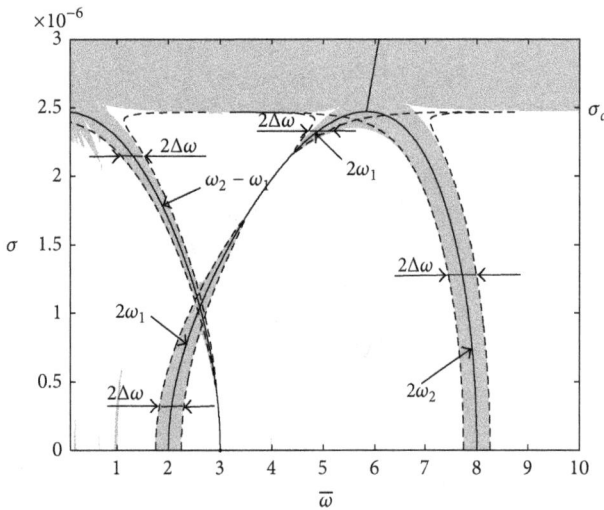

FIGURE 3: Stability plot for the flutter panel with respect to axial excitation frequency and dynamic pressure ratio: Ma = 2, α = 0.005, $\sigma_c = 2.47 \times 10^{-6}$, $\mu = 4.39 \times 10^{-5}$, and $F = 0.5$.

5. Numerical Results

The N.D. natural frequencies due to materials and aerodynamic loads are plotted in Figure 2 with respect to the dynamic pressure ratio, σ. Once the dynamic pressure ratio exceeds the critical value that equals σ_c, two natural frequencies merge together, which means they are conjugate pairs and the system instability occurs.

The system stability with respect to the axial excitation frequency and the dynamic pressure ratio is shown in Figure 3. In this paper, the gray regions in stability plots indicate the instabilities computed by the numerical FTM method and the black dash lines are the stability boundaries calculated from the analytical perturbation method. The black solid lines in Figure 3 represent the N.D. natural frequencies. With the same observations as in Figure 2, the system flutters for the entire axial excitation frequency range calculated here

when the dynamic pressure ratio passes its critical value. The principle resonances and combination resonance given by the perturbation method successfully match those by the FTM method with a lot of computation savings.

The system stability with respect to the axial excitation frequency and the air/panel density ratio is shown in Figure 4. Since $\sigma < \sigma_c$, the combination resonance and principal resonances are clearly separated. Again, the results from both the FTM method and the perturbation method match each other very well.

It can be observed in Figures 3 and 4 that the principal resonance of the second mode causes instabilities for all dynamic pressure ratio and air/panel density ratio values explored here. However, it could be stabilized by different axial excitation frequencies. The instabilities around the first principal resonance and combination resonance can be suppressed for some dynamic pressure ratio and air/panel density ratio values. Their axial excitation frequency stability boundaries are calculated by solving (20) for $\Delta\omega = 0$ and the results are plotted in Figure 5. The system stability depends on both resonances in each area divided by those boundaries. The panel system is only stable when both resonances are stable, shown in the white area in Figure 5.

6. Summary and Conclusions

This paper investigates the parametric stability of the panel (beam) under both aerodynamic loads and axial excitations. The dimensionless equation-of-motion is derived, including material viscous damping, axial excitation, and aerodynamic load. The eigen-analyses based on the first two modes are taken as examples to explore the stability properties of the flutter panel system with an axial single frequency cosine excitation. Both numerical FTM method and analytical perturbation method solve the problem and their results match each other very well.

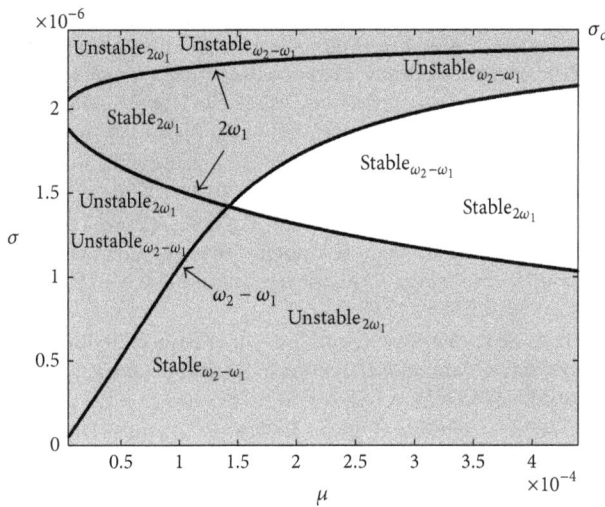

FIGURE 5: Stability plot for the flutter panel with respect to air/panel density ratio and dynamic pressure ratio: Ma = 2, $\alpha = 0.005$, $\sigma_c = 2.47 \times 10^{-6}$, and $F = 0.5$.

The panel may flutter under high-speed air flows when its out-of-plane dynamics couples with the aerodynamic loads. The parameter study was conducted for the system instability zones with respect to axial excitation frequency, air/panel density ratio, and air/panel dynamic pressure ratio. Different from the static axial force, this paper introduces a periodic axial excitation that brings the system into the time-varying domain. The axial excitation force could increase the panel stiffness locally to overcome aerodynamic loads when interacting with the out-of-plane vibrations. The study results in this paper indicate that the system is stable under the combinations of the proper excitation frequency and certain air/panel density ratio and dynamic pressure ratio.

The perturbation method developed in this paper saves lots of computations, which can help understand the flutter phenomenon of the panel with axial excitations more efficiently.

Competing Interests

The authors declare that they have no competing interests.

References

[1] F. Liu, J. Cai, Y. Zhu, H. M. Tsai, and A. S. F. Wong, "Calculation of wing flutter by a coupled fluid-structure method," *Journal of Aircraft*, vol. 38, no. 2, pp. 334–342, 2001.

[2] J. L. Hall, "A review of ballute technology for planetary aerocapture," in *Proceedings of the 4th IAA Conference on Low Cost Planetary Missions*, pp. 1–10, Laurel, Md, USA, May 2000.

[3] H. C. Nelson and H. J. Cunningham, "Theoretical investigation of flutter of two-dimensional flat panels with one surface exposed to supersonic potential flow," NACA Technical Note 3465/Report 1280, 1955.

[4] M. D. Olson, "Finite elements applied to panel flutter," *AIAA Journal*, vol. 5, no. 12, pp. 2267–2270, 1967.

[5] P. C. Parks, "A stability criterion for panel flutter via the second method of Liapunov," *AIAA Journal*, vol. 4, no. 1, pp. 175–177, 1966.

[6] J. Dugundji, "Theoretical considerations of panel flutter at high supersonic Mach numbers.," *AIAA Journal*, vol. 4, no. 7, pp. 1257–1266, 1966.

[7] E. H. Dowell, "Nonlinear oscillations of a fluttering plate," *AIAA Journal*, vol. 4, no. 7, pp. 1267–1275, 1966.

[8] E. H. Dowell, "Nonlinear oscillations of a fluttering plate II," *AIAA Journal*, vol. 5, no. 10, pp. 1856–1862, 1967.

[9] R. C. Zhou, C. Mei, and J.-K. Huang, "Suppression of nonlinear panel flutter at supersonic speeds and elevated temperatures," *AIAA Journal*, vol. 34, no. 2, pp. 347–354, 1996.

[10] D. J. Gee, "Numerical continuation applied to panel flutter," *Nonlinear Dynamics*, vol. 22, no. 3, pp. 271–280, 2000.

[11] S. Tizzi, "Influence of non-linear forces on beam behaviour in flutter conditions," *Journal of Sound and Vibration*, vol. 267, no. 2, pp. 279–299, 2003.

[12] T. Iwatsubo, Y. Sugiyama, and S. Ogino, "Simple and combination resonances of columns under periodic axial loads," *Journal of Sound and Vibration*, vol. 33, no. 2, pp. 211–221, 1974.

[13] C. S. Hsu, "On the parametric excitation of a dynamic system having multiple degrees of freedom," *ASME Journal of Applied Mechanics*, vol. 30, pp. 367–372, 1963.

[14] S. K. Sinha, "Dynamic stability of a Timoshenko beam subjected to an oscillating axial force," *Journal of Sound and Vibration*, vol. 131, no. 3, pp. 509–514, 1989.

[15] S. K. Sahu and P. K. Datta, "Parametric instability of doubly curved panels subjected to non-uniform harmonic loading," *Journal of Sound and Vibration*, vol. 240, no. 1, pp. 117–129, 2001.

[16] S. K. Sahu and P. K. Datta, "Dynamic stability of curved panels with cutouts," *Journal of Sound and Vibration*, vol. 251, no. 4, pp. 683–696, 2002.

[17] M. Ganapathi, B. P. Patel, P. Boisse, and M. Touratier, "Nonlinear dynamic stability characteristics of elastic plates subjected to periodic in-plane load," *International Journal of Non-Linear Mechanics*, vol. 35, no. 3, pp. 467–480, 2000.

[18] K. Nagai and N. Yamaki, "Dynamic stability of circular cylindrical shells under periodic compressive forces," *Journal of Sound and Vibration*, vol. 58, no. 3, pp. 425–441, 1978.

[19] C. Massalas, A. Dalamangas, and G. Tzivanidis, "Dynamic instability of truncated conical shells, with variable modulus of elasticity, under periodic compressive forces," *Journal of Sound and Vibration*, vol. 79, no. 4, pp. 519–528, 1981.

[20] V. B. Kovtunov, "Dynamic stability and nonlinear parametric vibration of cylindrical shells," *Computers & Structures*, vol. 46, no. 1, pp. 149–156, 1993.

[21] K. Y. Lam and T. Y. Ng, "Dynamic stability of cylindrical shells subjected to conservative periodic axial loads using different shell theories," *Journal of Sound and Vibration*, vol. 207, no. 4, pp. 497–520, 1997.

[22] T. Y. Ng, K. Y. Lam, K. M. Liew, and J. N. Reddy, "Dynamic stability analysis of functionally graded cylindrical shells under periodic axial loading," *International Journal of Solids and Structures*, vol. 38, no. 8, pp. 1295–1309, 2001.

[23] F. Pellicano and M. Amabili, "Stability and vibration of empty and fluid-filled circular cylindrical shells under static and periodic axial loads," *International Journal of Solids and Structures*, vol. 40, no. 13-14, pp. 3229–3251, 2003.

[24] M. Ruzzene, "Dynamic buckling of periodically stiffened shells: application to supercavitating vehicles," *International Journal of Solids and Structures*, vol. 41, no. 3-4, pp. 1039–1059, 2004.

[25] P. Hagedorn and L. R. Koval, "On the parametric stability of a Timoshenko beam subjected to a periodic axial load," *Ingenieur-Archiv*, vol. 40, no. 3, pp. 211–220, 1971.

[26] K.-J. Yang, K.-S. Hong, and F. Matsuno, "Robust adaptive boundary control of an axially moving string under a spatiotemporally varying tension," *Journal of Sound and Vibration*, vol. 273, no. 4-5, pp. 1007–1029, 2004.

[27] Z. Viderman, F. P. J. Rimrott, and W. L. Cleghorn, "Parametrically excited linear nonconservative gyroscopic systems," *Mechanics of Structures and Machines*, vol. 22, no. 1, pp. 1–20, 1994.

[28] V. V. Bolotin, *The Dynamic Stability of Elastic Systems*, Holden-Day, 1964.

[29] C. S. Hsu, "On approximating a general linear periodic system," *Journal of Mathematical Analysis and Applications*, vol. 45, pp. 234–251, 1974.

[30] P. P. Friedmann, "Numerical methods for determining the stability and response of periodic systems with applications to helicopter rotor dynamics and aeroelasticity," *Computers and Mathematics with Applications*, vol. 12, no. 1, pp. 131–148, 1986.

[31] O. A. Bauchau and Y. G. Nikishkov, "An implicit Floquet analysis for rotorcraft stability evaluation," *Journal of the American Helicopter Society*, vol. 46, no. 3, pp. 200–209, 2001.

[32] H. A. DeSmidt, K. W. Wang, and E. C. Smith, "Coupled torsion-lateral stability of a shaft-disk system driven throuoh a universal joint," *Journal of Applied Mechanics, Transactions ASME*, vol. 69, no. 3, pp. 261–273, 2002.

[33] C. S. Hsu, "Further results on parametric excitation of a dynamic system," *ASME Journal of Applied Mechanics*, vol. 32, pp. 373–377, 1965.

[34] L. Meirovitch, *Computational Methods in Structural Dynamics*, Sijthoff & Noordhoff, 1980.

8

Kaybob Revisited: What We Have Learned about Compressor Stability from Self-Excited Whirling

Edgar J. Gunter[1,2] and Brian K. Weaver[3]

[1]*RODYN Vibration Analysis, Inc., Charlottesville, VA, USA*
[2]*Rotor Dynamics Laboratory, Mechanical and Aerospace Engineering Department, University of Virginia, Charlottesville, VA, USA*
[3]*Rotating Machinery and Controls Laboratory, Mechanical and Aerospace Engineering Department, University of Virginia, Charlottesville, VA, USA*

Correspondence should be addressed to Brian K. Weaver; bkw3q@virginia.edu

Academic Editor: Lars Hakansson

The Kaybob compressor failure of 1971 was an excellent historic example of rotordynamic instability and the design factors that affect this phenomenon. In the case of Kaybob, the use of poorly designed bearings produced unstable whirling in both the low and high pressure compressors. This required over five months of vibration troubleshooting and redesign along with over 100 million modern U.S. dollars in total costs and lost revenue. In this paper, the history of the Kaybob compressor failure is discussed in detail including a discussion of the ineffective bearing designs that were considered. Modern bearing and rotordynamic analysis tools are then employed to study both designs that were considered along with new designs for the bearings that could have ultimately restored stability to the machine. These designs include four-pad, load-between-pad bearings and squeeze film dampers with a central groove. Simple relationships based on the physics of the system are also used to show how the bearings could be tuned to produce optimum bearing stiffness and damping of the rotor vibration, producing insights which can inform the designers as they perform more comprehensive analyses of these systems.

1. Introduction

The rotordynamic stability of high-speed compressors and turbines has been a critical element of their design for decades. With the ever-increasing demand for greater output through an increased number of stages, higher speeds, and higher pressures, these machines are continuously being pushed to the design limits of the previous generation of machines. As they are being pushed to these new extremes of operation the designs must also adapt to accommodate the consequences of these extremes including increased rotor flexibility and an increased likelihood of subsynchronous whirl. This whirl can be excited by a number of sources in turbomachinery including seals and even the bearings supporting the rotor. The stability of that whirl can then be reduced by destabilizing cross-coupled stiffness forces from the bearings, interstage seals, balance pistons, and Alford-type aerodynamic cross-coupled forces around blades.

The increasing use of tilting pad journal bearings in the 1960s and 70s brought about new opportunities for compressor design as these bearings eliminated the self-exciting oil whirl commonly found in fixed geometry bearings. They also essentially eliminated the destabilizing cross-coupled stiffness forces produced by these bearings. However, despite the fact that a number of important papers were being published on these topics by Lund [2] and others, it was not until 1974 that Lund published his landmark paper on damped eigenvalue solutions for stability analysis [3]. It was because of this lack of accurate design and analysis tools that a significant number of high-speed machines were not designed appropriately for the conditions in which they were intended to operate.

The Kaybob compressor failure of 1971 was a historic case of rotordynamic instability from self-excited, subsynchronous whirl. The many months of troubleshooting that followed this failure served as a valuable lesson to the team

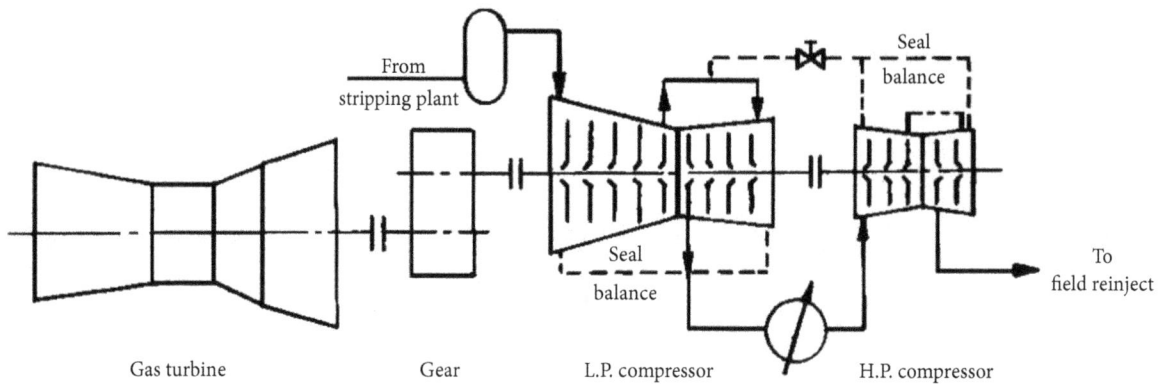

FIGURE 1: Kaybob compressor train schematic from Smith [1].

involved but even to this day this classic example serves as an important motivator for understanding the underlying principles that govern the design of these machines, many of which that were developed after the incident and troubleshooting took place. With the advent of advanced bearing and rotordynamic analysis tools, designers now have the ability to fine tune their designs to avoid these failures; however it is critically important that the underlying principles and physical insights gained from the analysis of these machines are properly considered as even modern bearing technologies can contribute to stability problems when not implemented correctly.

In this study, modern bearing and rotordynamic analysis tools are applied to the Kaybob compressor instability case to highlight the effects of the design choices made with this machine and how they contributed to its ultimate failure. By combining useful design principles with these analysis tools it also showed not only why the costly solution to this failure worked at the time, but also how a number of alternative and much simpler design changes to the support structure could have restored stability to the machine including the use of properly designed tilting pad journal bearings and squeeze film dampers. It is the goal of this study to provide practical information and design approaches that designers can then use to avoid costly failures like the one experienced in Kaybob.

2. Stage Case History: Kaybob Plant, Alberta

The case of the Kaybob compressor instability began in November of 1971, a detailed summary of which is provided by Smith of Cooper-Bessemer [1]. Three natural gas reinjection trains were being commissioned at the Kaybob South Beaverhill Lake plant in Fox Creek, Alberta, operated by Chevron-Standard.

These duplicate trains (Figure 1) consist of a gas turbine drive, a speed-increasing gearbox, a 9-stage low pressure compressor, and a 5-stage high pressure compressor. Both compressors have back-to-back impeller arrangements with centrally located balance pistons to control thrust and are supported by tilting pad thrust and radial bearings.

The tilting pad journal bearings were designed as five-pad, load-on-pad bearings, a design that has been commonly employed in compressors over the years. Floating oil film ring seals were employed in both compressor casings. The trains were designed for a rated speed of 10,200 rpm with a maximum continuous operating speed of 11,400 rpm.

As site testing began, a significant vibration problem became apparent in the low pressure compressor. A first train was operated at speeds up to 10,000 rpm with inlet pressures of 700–750 psi and discharge pressures of 1,300–1,500 psi. Under these conditions a significant vibration was present at a frequency of approximately 4,250 cpm, producing 3 to 4 mil peak-to-peak amplitudes in one of the bearings which had a 6 mil clearance.

A second train was then tested to rule out possible unexplained sources of resonance from the first train. This machine was ran up to its maximum speed of 11,440 rpm, followed by increases to the inlet and discharge pressures to 1,120 psi and 3,300 psi.

At this point the machine became highly unstable, resulting in a violent 5,100 cpm whirl of 9+ mils peak-to-peak amplitude. Figure 2 shows the 20-second time lapse of the instability onset in the thru-drive end bearing along with the orbit shape. Smith [1] notes that "the rotor assembly and internal labyrinths were severely damaged as a result of this violent, high-energy motion" and that the "vibration could be heard and felt in the control room some 50 feet away from the unit."

The five months that followed included a thorough investigation into the causes of this instability as well as tests of various potential solutions to the problem including investigations into both seals and journal bearings. The seal investigation involved the consideration of four different seal configurations to determine whether seal lockup or effective damping had any significant effects on the instability. It was concluded, though, that the seals were in fact floating and their configuration overall had little effect on the problem.

A number of design modifications were considered for the tilting pad journal bearings as well. These modifications included increasing the specific load of the bearing, introducing asymmetry to the bearing design, different pad load

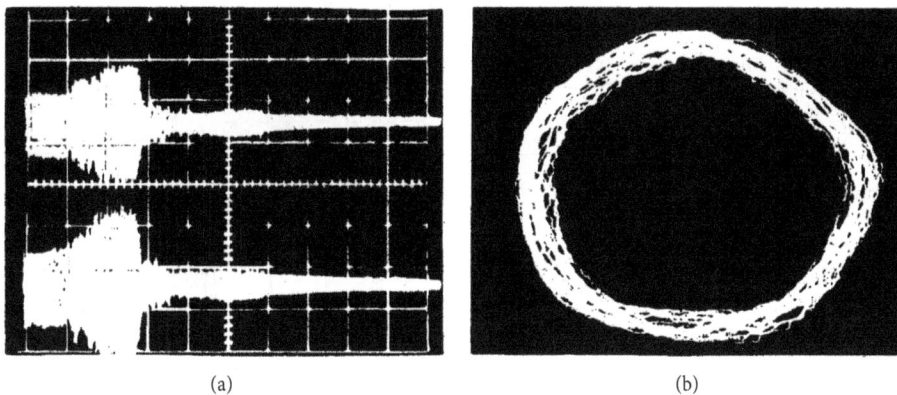

FIGURE 2: (a) Time lapse and (b) orbit of the instability onset. From Smith [1].

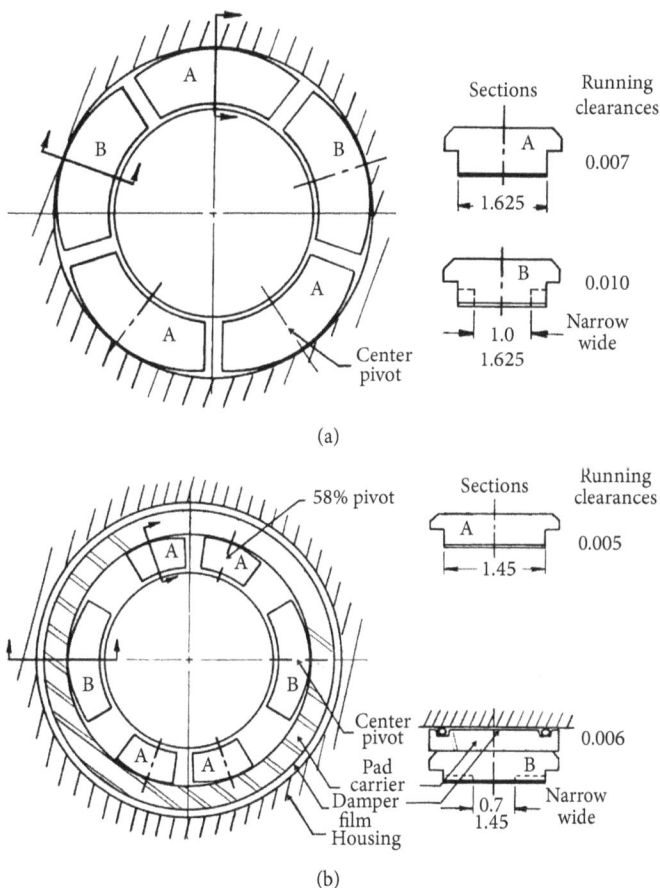

FIGURE 3: (a) Asymmetric tilt pad bearing from Smith [1]. (b) Squeeze film damper design from Smith [1].

configurations, offset pads, variable preloads, the introduction of squeeze film dampers, and combinations thereof. Two of the configurations tested are shown in Figure 3. Overall it was found that all of these bearing design-based solutions produced marginal improvements in rotordynamic stability. Additional analysis and discussion of these and other bearing designs is presented in the following sections.

Note that, in Figure 3(a), the side pads, B, are reduced in width in an attempt to incorporate some bearing asymmetry.

This modification had a negligible influence on stability. In Figure 3(b), the A pads are reduced in arc length and a squeeze film damper was added without a central groove. This design results in a damper with excessive damping and stiffness.

The ultimate solution to this vibration problem was a redesign of the rotor to a shorter, stiffer configuration with increased diameters beneath the impellers. This stiffer shaft would thereby accommodate the stiff bearing designs that

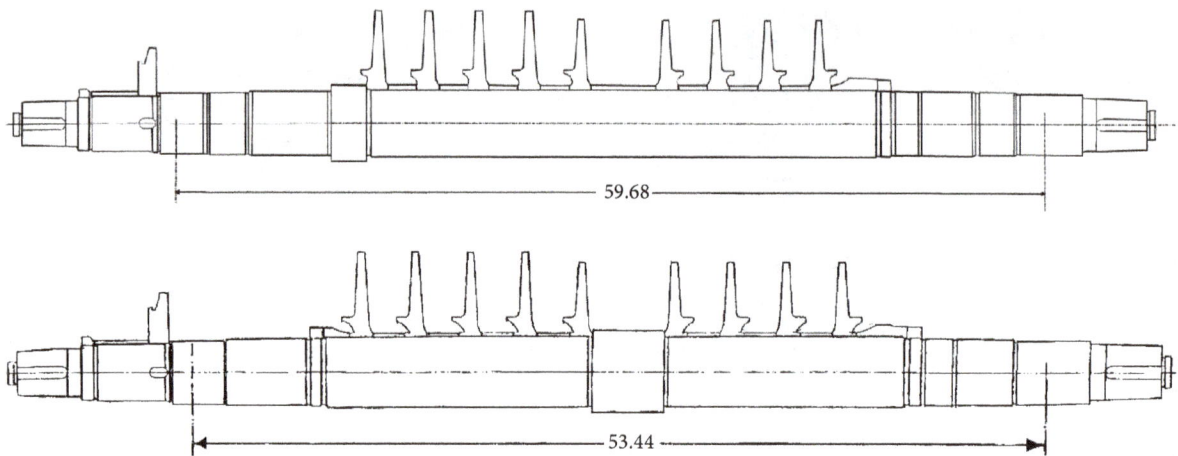

FIGURE 4: Rotor redesign from Smith [1].

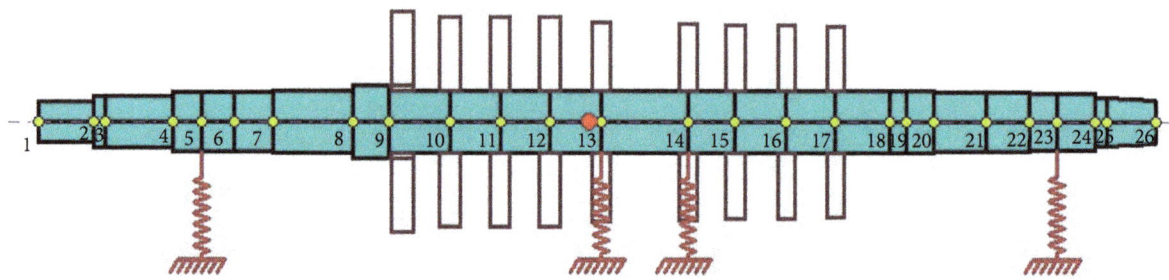

FIGURE 5: Kaybob 9-stage compressor model with aerodynamic cross-coupling at stations 13 and 14.

were used and that will be demonstrated in Section 9. Figure 4 shows the initial and final shaft designs. Overall, these significant changes resulted in a redesign of most of the primary machine components, a process which would cost the plant over 100 million modern U.S. dollars.

In the original design of the Kaybob rotor, the bearing span was 59.688 inches. A modal analysis of the rotor indicated that the shaft first critical speed modal stiffness was approximately 115,000 lb/in. This fundamental shaft modal stiffness is extremely important for the proper tuning of the bearings to the shaft. From previous research studies of the optimum bearing stiffness, it has been determined that the optimum bearing stiffness is approximately one half the rotor shaft stiffness. When the bearing stiffness greatly exceeds this fundamental design value, the first mode damping is drastically reduced causing the rotor to be very susceptible to self-excited whirl effects.

The fundamental stiffness of a uniform team is given by the following equation:

$$K_{\text{shaft}} = 48\frac{EI}{L^3} \text{ lb/in.} \quad (1)$$

In the final redesign as shown in the lower rotor of Figure 4, the bearing span has been reduced to 53.438 inches. In addition to the reduced bearing span, the center section diameter has also been increased. This then leads to an improved shaft modal stiffness of 200,000 lb/in.

3. Rotordynamic Characteristics

In revisiting the design and troubleshooting of the Kaybob compressor, the first step was to establish the overall rotordynamic characteristics of the system. A finite element model of the compressor was first developed in the Dyrobes software suite used for this analysis and is presented to scale in Figure 5. The rotor spans 78 inches in total length with a bearing span of 60 inches. The model consists of 25 elements and 26 nodes with 9 rigid disks representing the mass and inertial properties of the 9 compressor stages, the details of which are provided in Table 1. The total rotor weight is 400 lbs. The bearings are represented as linear stiffness and damping coefficients at nodes 5 and 23.

After building the shaft model, the analysis began with an assessment of the rotor critical speeds in the operating range of the compressor (Figure 6). Two modes were found in the operating range: a first bending mode at approximately 4,600 rpm (Figure 7) which proved to be the unstable mode for the machine and a well-damped conical mode at approximately 8,500 rpm. Overlaying the critical speed map is the vertical stiffness K_{yy} of the original 5-pad load-on-pad tilting pad bearings supporting the rotor. Its position on the curve for the first bending mode provides an early indication of a relatively stiff support structure for the compressor.

In order to evaluate the stability characteristics of this compressor, a nominal amount of cross-coupled stiffness of 25,000 lb/in was assumed acting at stations 13 and 14. This

TABLE 1: Kaybob compressor model disk properties.

Disk #	Station	Mass (lbm)	Transverse inertia (lbm·in^2)	Polar inertia (lbm·in^2)	Length (in)	Inner diameter (in)	Outer diameter (in)
1	9	15	760	1400	1.5	4.36	15.0
2	10	15	760	1400	1.5	4.36	14.3
3	11	15	760	1400	1.4	4.36	14.3
4	12	15	760	1400	1.4	4.36	14.3
5	13	5	760	1400	1.3	4.36	13.6
6	14	5	400	750	1.3	4.36	13.3
7	15	5	400	750	1.3	4.36	13.1
8	16	5	400	750	1.2	4.36	13.1
9	17	5	400	750	1.2	4.36	13.0

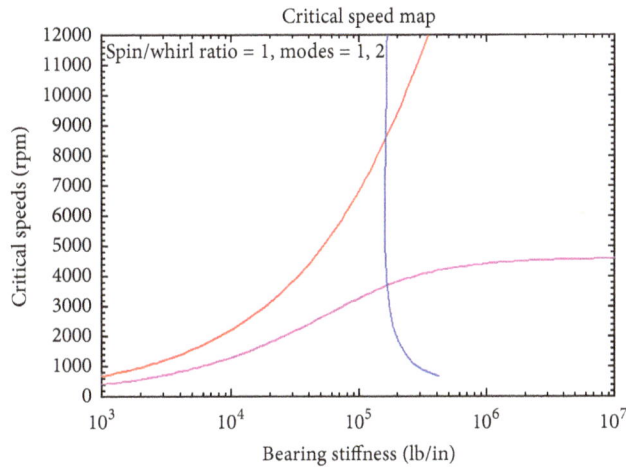

FIGURE 6: Compressor critical speed map.

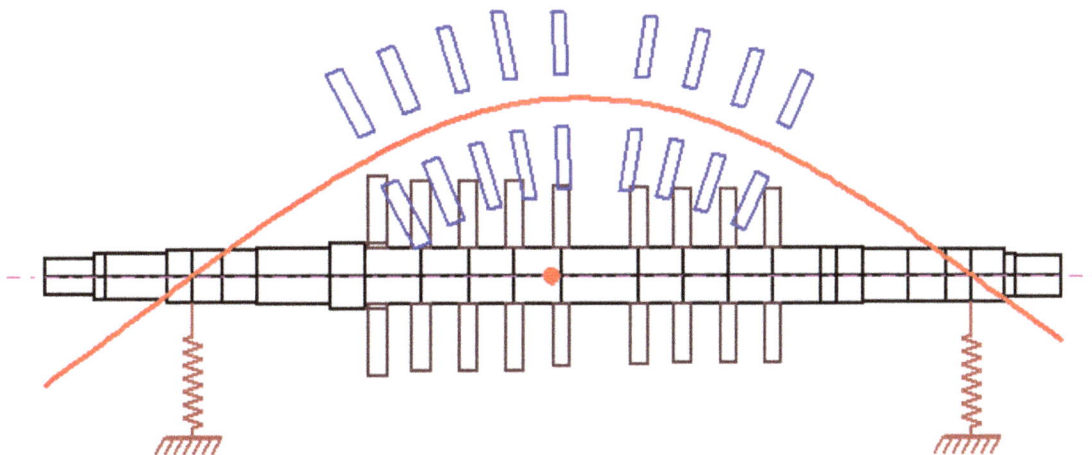

FIGURE 7: First bending mode of the compressor.

value was estimated based on experience and the description of the stability characteristics of the machine. These cross-coupling effects are known as the Alford effect [4] and can occur in all gas turbines and high pressure compressors. These effects are often not observed on a test stand when the unit is operated under low pressure ratios. When operated in the field under high pressure conditions, the Alford effects become apparent resulting in self-excited whirling which can then lead to damage of seals and labyrinths. Due to a lack of information on the design of the seals these were not modeled

Critical speed mode shape, mode number = 1
Spin/whirl ratio = 1, stiffness: K_{xx}
Critical speed = 4612 rpm = 76.86 Hz

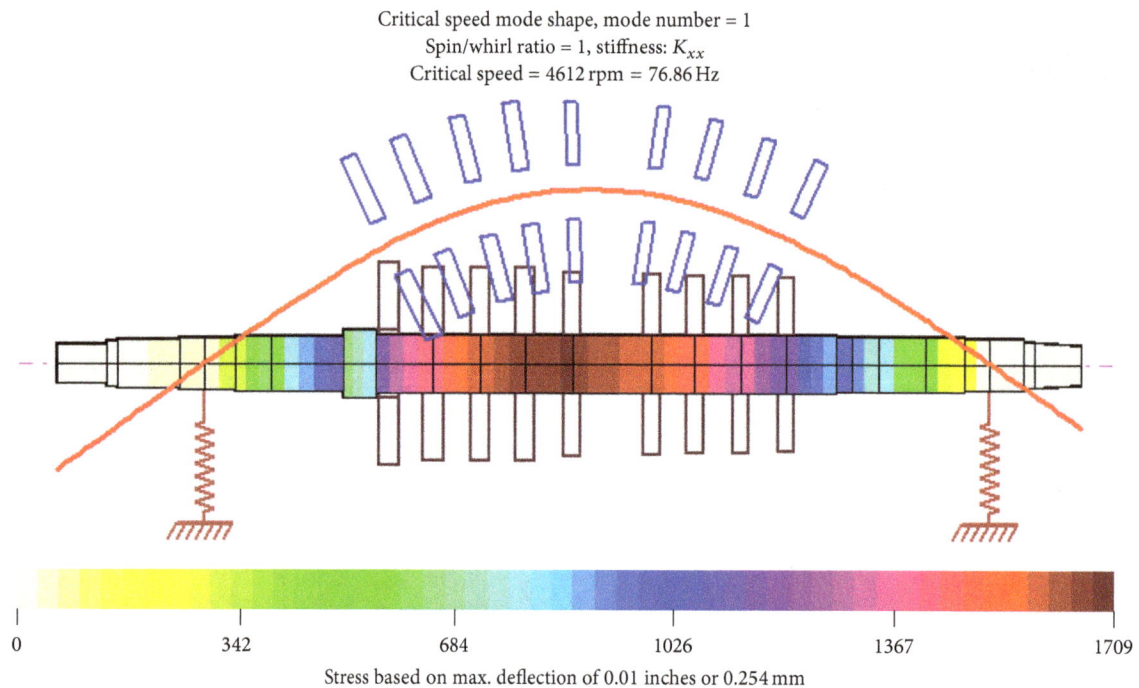

Stress based on max. deflection of 0.01 inches or 0.254 mm

| 0 | 342 | 684 | 1026 | 1367 | 1709 |

FIGURE 8: Shaft modal stresses of the first bending mode.

separately, with their contribution to the destabilization of the compressor being included in the cross-coupled stiffness applied.

The shaft modal stresses and stiffness were also assessed for the first bending mode assuming rigid bearings with a stiffness of 10^8 lb/in (Figure 8). Peak stresses of 1,709 psi were predicted along with a shaft modal stiffness of 115,850 lb/in. This magnitude of stiffness will prove to be very important to the stability of the machine as it is compared to the stiffness of the bearings in the following sections. It was presented by Barrett et al. [5] that the optimum ratio of twice the bearing stiffness to the shaft modal stiffness for maximizing modal stability is one.

To further support this important relationship in rotor dynamics, a plot was created showing the amplification factors of the first bending mode under various sets of bearing stiffness and damping (Figure 9). It is shown that the amplification factors are reduced as the bearing stiffness approaches this ideal ratio. The plot also highlights the importance of designing the bearings in such a manner that underdamped and overdamped conditions can also be avoided through proper design. The following sections on various bearing designs for the compressor will further show how various design variables affect these bearing properties and the stability of the compressor.

Figure 9 shows the importance of tuning the bearing stiffness to the rotor shaft stiffness. For example, with the original rotor design, it is seen that, with a vertical bearing stiffness of 100,000 lb/in, the optimum damping is about 250 lb-sec/in with an amplification factor of two. As the bearing stiffness increases, the required optimum damping increases

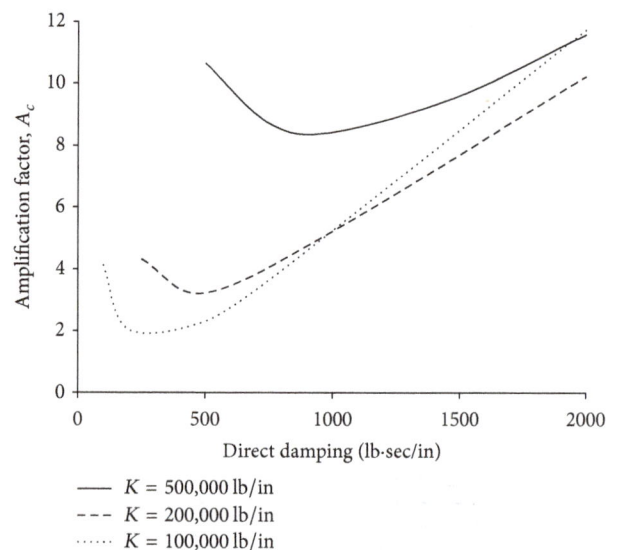

FIGURE 9: Amplification factors of the first bending mode.

to 500 lb-sec/in and the minimum amplification factor is over three.

For the case of very high stiffness bearing of 500,000 lb/in, it is seen that the optimum bearing damping is over 800 lb-sec/in and the amplification factor is now over eight. Therefore it is very apparent that it is critical to select the bearing stiffness in relationship to the predicted first modal shaft stiffness. This concept will often lead to the requirement of an extended width bearing in order to have a reduced vertical

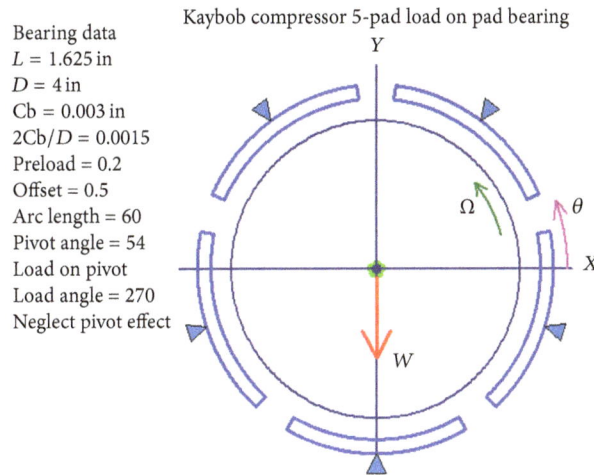

Bearing data
$L = 1.625$ in
$D = 4$ in
$Cb = 0.003$ in
$2Cb/D = 0.0015$
Preload = 0.2
Offset = 0.5
Arc length = 60
Pivot angle = 54
Load on pivot
Load angle = 270
Neglect pivot effect

FIGURE 10: Original compressor bearing design.

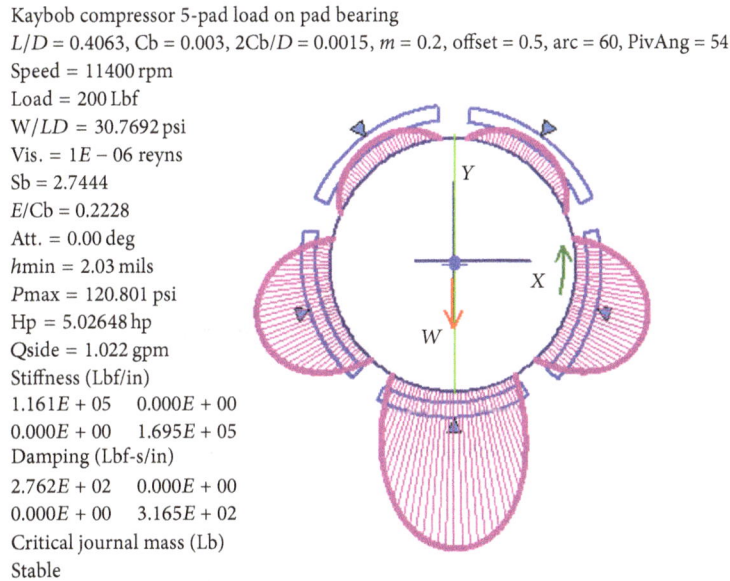

Kaybob compressor 5-pad load on pad bearing
$L/D = 0.4063$, $Cb = 0.003$, $2Cb/D = 0.0015$, $m = 0.2$, offset = 0.5, arc = 60, PivAng = 54
Speed = 11400 rpm
Load = 200 Lbf
$W/LD = 30.7692$ psi
Vis. = $1E - 06$ reyns
Sb = 2.7444
$E/Cb = 0.2228$
Att. = 0.00 deg
$h$min = 2.03 mils
Pmax = 120.801 psi
Hp = 5.02648 hp
Qside = 1.022 gpm
Stiffness (Lbf/in)
$1.161E + 05 \quad 0.000E + 00$
$0.000E + 00 \quad 1.695E + 05$
Damping (Lbf-s/in)
$2.762E + 02 \quad 0.000E + 00$
$0.000E + 00 \quad 3.165E + 02$
Critical journal mass (Lb)
Stable

FIGURE 11: Original compressor bearing analysis results.

4. Original Bearing Design

The analysis of the Kaybob compressor continued with the development of various bearing models to better understand how the design of the bearings would impact the stability of the machine. The original design of the compressor bearings was a 5-pad load-on-pad configuration with an L/D of approximately 0.4, depicted in Figure 10. This model utilizes a two-dimensional Reynolds equation solver that for the current analysis neglects thermal effects and deformation for simplicity. All bearing analyses were performed at the maximum continuous operating speed of 11,400 rpm.

The results of this initial bearing analysis are shown in Figure 11. It was found that for this configuration the tight clearance, preload, and load-on-pad configuration together produced a relatively stiff bearing, resulting in a vertical bearing stiffness of 169,500 lb/in. Recall from the previous section that the ideal ratio of twice the bearing stiffness to the shaft modal stiffness is unity. When comparing twice the vertical stiffness of the original compressor bearing to the calculated shaft modal stiffness, this results in a ratio of over 2.9. These results alone are indicative of an overly stiff support structure for the compressor that could contribute to an unstable whirl motion.

The next step in analyzing the original rotor-bearing system was to perform a stability analysis of the compressor with the original bearings. The results of the damped eigenvalue analysis are presented in Figure 12. It is shown that the poor choices made in the bearing design process resulted in an unstable forward bending mode in the operating range. The combination of high stiffness and damping in the bearings

bearing stiffness in order to properly tune the bearings to the fundamental shaft stiffness.

Mode number = 3 unstable forward precession
Shaft rotational speed = 11400 rpm
Whirl speed (damped natural freq.) = 4486 rpm, log decrement = −0.1641

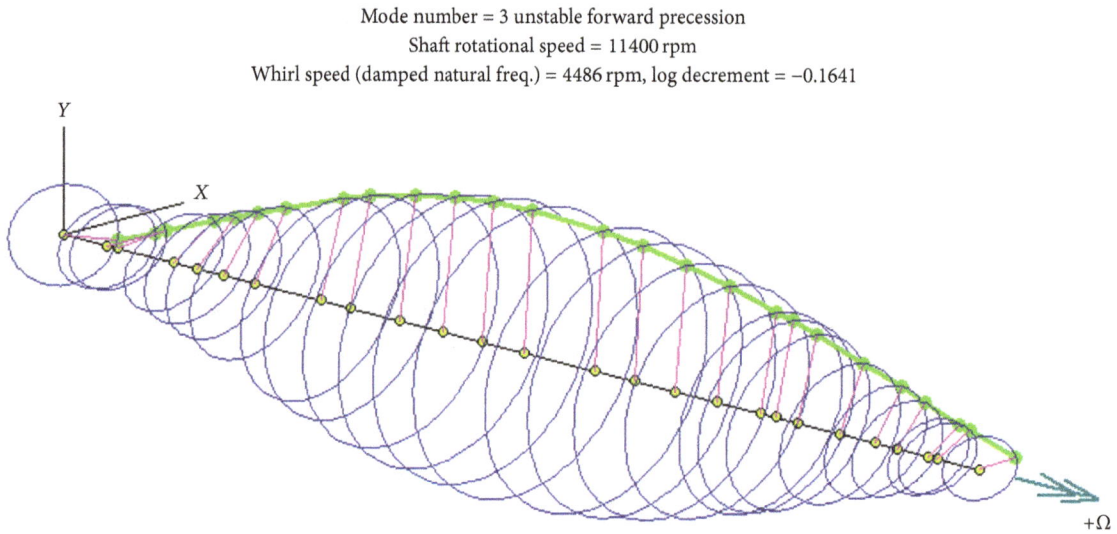

FIGURE 12: Stability results for the original bearing design.

Mode number = 1, critical speed = 4065 rpm = 67.75 Hz
Potential energy distribution (s/w = 1)
Overall: shaft (S) = 51.86%, bearing (Brg) = 48.14%

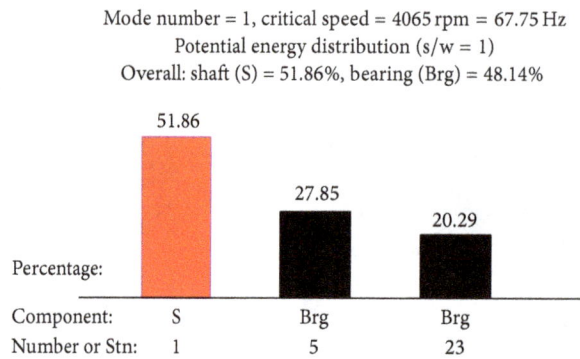

Percentage:	51.86	27.85	20.29
Component:	S	Brg	Brg
Number or Stn:	1	5	23

FIGURE 13: First mode shaft and bearings energy distribution with the original bearing design.

resulted in less shaft motion at the bearings and therefore a greater distribution of the system energy into the shaft, resulting in the unstable mode. A critical speed analysis using the bearing vertical stiffness reveals the relative energy distribution in the shaft and bearings (Figure 13).

5. A Lesson in Bearing Asymmetry

A second bearing design considered when troubleshooting the Kaybob compressor was an asymmetric bearing. Smith [1] cited two papers by Gunter [6, 7] as evidence that introducing bearing asymmetry could lead to an improvement in rotor stability. However, this evidence was misinterpreted as asymmetry has little effect in cases of high bearing stiffness and large quantities of aerodynamic cross-coupling [8]. This is demonstrated in the below analysis of an asymmetric bearing design similar to that considered by the team at Cooper-Bessemer.

The model (Figure 14) shows how the bearing was clocked to a load-between-pad position. The bearing clearance was also increased in the upper-left and upper-right pads from a 3.5 mil radial clearance to 5 mils. These two pads were also

narrowed from an axial length of 1.625″ to a length of 1″, though this modification is not captured by the model. This proves to have little effect, however, as shown in the results (Figure 15). The pads with the increased clearance show no participation in controlling the shaft motion, resulting in predicted vertical stiffness values that are still quite large when compared to the shaft stiffness. A significant drop in damping is also produced as a result of this design.

A stability analysis of the bearing-rotor system (Figure 16) in fact shows a decrease in the log decrement of the first bending mode when compared to the original bearing design. This is due to the fact that not only is the bearing stiffness still too large, but now the system is also likely underdamped with this bearing design.

The team at Cooper-Bessemer tried multiple bearing designs that utilized bearing asymmetry, but with such high levels of bearing stiffness and aerodynamic cross-coupling they were unable to achieve any significant increases in modal stability.

Figure 16 represents the damped eigenvalue analysis of the nine-stage Kaybob compressor with the aerodynamic cross-coupling included at stations 13 and 14. It has also been

Kaybob compressor 5-pad load between pad asymmetric bearings

Bearing data
$L = 1.625\,\text{in}$
$D = 4\,\text{in}$
$Cb = 0.0035\,\text{in}$
$2Cb/D = 0.0018$
Fixed $Cp = 0.0050\,\text{in}$
$Cb = 0.0035{-}0.0050$
PivAng = 378, 90, 162, 234, 306
Preload = −0, 0.3, −0, 0.3, 0.3
Offset = 0.5
Arc length = 60
Load between pivots
Load angle = 270
Neglect pivot effect

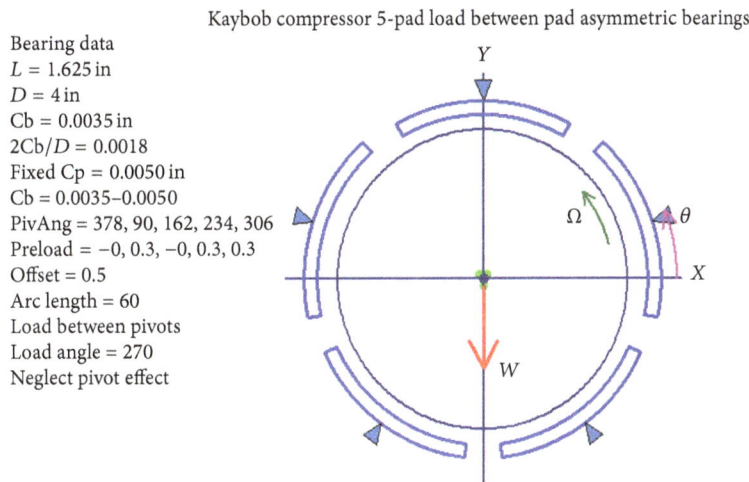

FIGURE 14: Asymmetric bearing model.

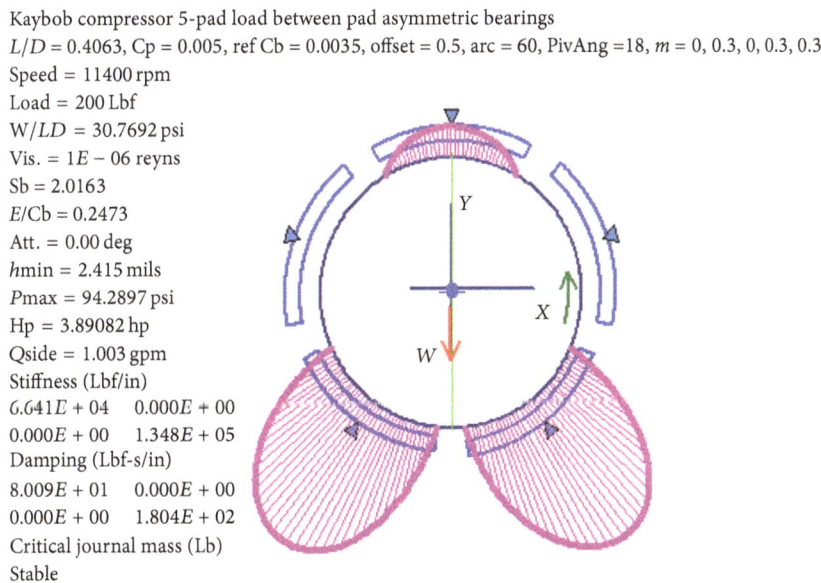

Kaybob compressor 5-pad load between pad asymmetric bearings
$L/D = 0.4063$, $Cp = 0.005$, ref $Cb = 0.0035$, offset = 0.5, arc = 60, PivAng =18, $m = 0, 0.3, 0, 0.3, 0.3$
Speed = 11400 rpm
Load = 200 Lbf
$W/LD = 30.7692\,\text{psi}$
Vis. = $1E - 06$ reyns
$Sb = 2.0163$
$E/Cb = 0.2473$
Att. = 0.00 deg
$h\text{min} = 2.415\,\text{mils}$
$P\text{max} = 94.2897\,\text{psi}$
$Hp = 3.89082\,\text{hp}$
$Q\text{side} = 1.003\,\text{gpm}$
Stiffness (Lbf/in)
$6.641E + 04 \quad 0.000E + 00$
$0.000E + 00 \quad 1.348E + 05$
Damping (Lbf-s/in)
$8.009E + 01 \quad 0.000E + 00$
$0.000E + 00 \quad 1.804E + 02$
Critical journal mass (Lb)
Stable

FIGURE 15: Results for the asymmetric bearing.

shown that, in addition to turbines and compressors developing aerodynamic cross-coupling, a center span balance piston can also create aerodynamic cross-coupling effects. Therefore small values of aerodynamic cross-coupling of 25,000 lb/in were placed at the center stations 13 and 14 at the balance piston location.

It is seen that, at the operating speed of 11,400 rpm, there now occurs a self-excited whirl component at 3733 cpm. The log decrement shows a value of −0.51 which represents a highly unstable system. Noncontact probes are often placed at the bearings to observe the rotor motion. The motion observed at the bearing locations is often quite small as compared to the large orbital motion occurring at the rotor center due to the added whirl component. This large motion occurring at the center causes extensive rubbing and seals to be extensively damaged.

6. 4-Pad Bearing Design

A design not considered by the team at Cooper-Bessemer was a 4-pad tilting pad bearing. These bearings, particularly when oriented in a load-between-pad position, are known to be beneficial from a rotordynamic standpoint due to their ability to more evenly distribute stiffness forces in the vertical and horizontal directions. To see how this design might impact the Kaybob compressor, an initial 4-pad load-between-pad bearing design was analyzed with other design parameters that were kept constant from the original 5-pad bearing including the axial length, bearing clearance, and preload as shown in Figure 17.

The results of the bearing analysis (Figure 18) show that the 4-pad design does produce equal values of vertical and horizontal stiffness. When compared to the original 5-pad

Mode number = 2 unstable forward precession
Shaft rotational speed = 11400 rpm
Whirl speed (damped natural freq.) = 3733 rpm, log decrement = −0.5112

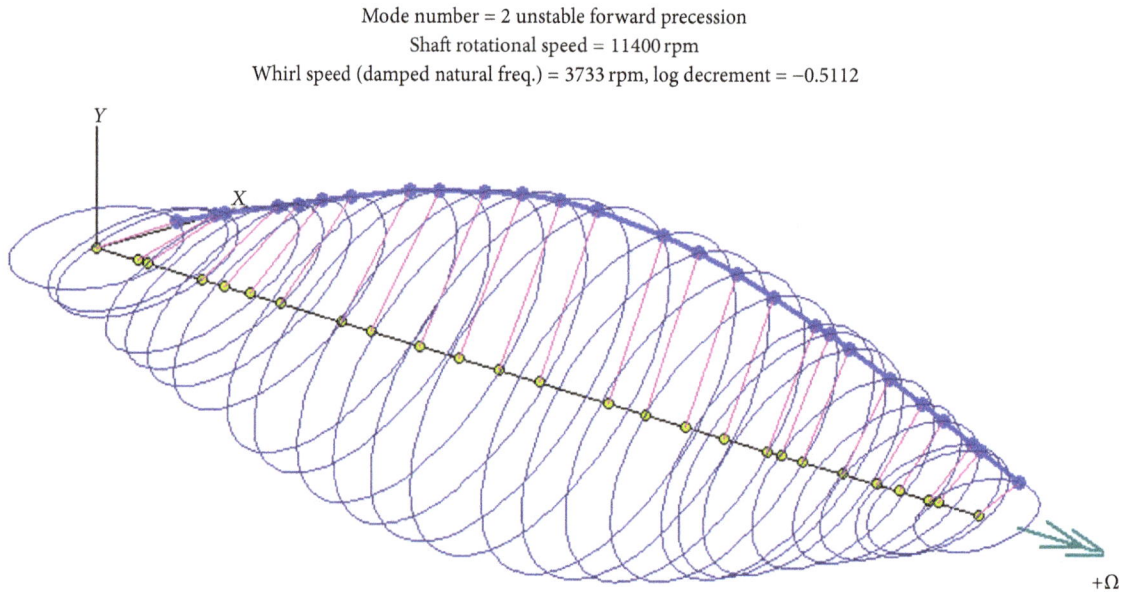

FIGURE 16: Stability results for the asymmetric bearing.

Kaybob compressor 4-pad load between pad bearings

Bearing data
L = 1.625 in
D = 4 in
Cb = 0.003 in
2Cb/D = 0.0015
Preload = 0.2
Offset = 0.5
Arc length = 80
Pivot angle = 45
Load between pivots
Load angle = 270
Neglect pivot effect

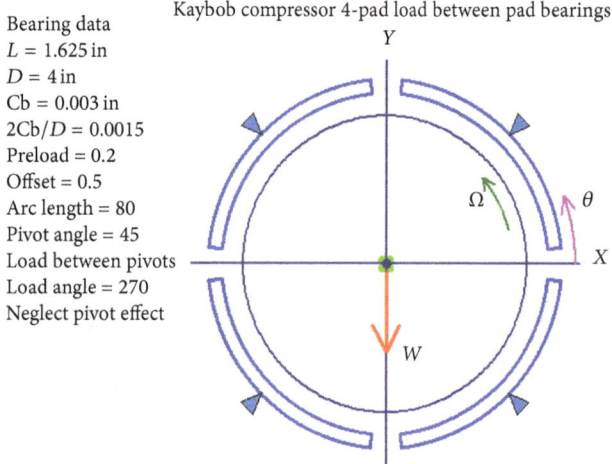

FIGURE 17: Initial four-pad bearing design.

bearing, the vertical stiffness is slightly lower along with an increase in direct damping. However, in general this design is still very stiff for this machine.

A stability analysis (Figure 19) reveals a reduction in the log decrement of the first bending mode when compared to the original design, likely due to increases in damping and horizontal stiffness preventing the bearings from absorbing the shaft vibrational energy. These results indicate that more drastic changes to the bearing design are necessary in order to restore stability to the system.

The design for the 4-pad bearing is superior to the asymmetric 5-pad bearing in that the log decrement has improved. However the dynamic analysis of the compressor with the 4-pad design as shown in Figure 19 is still unstable. The system has been improved over the asymmetric 5-pad

design with a reduction of the log decrement from −0.51 for the 5-pad case as compared to −0.233 with the 4-pad design.

In order to improve the rotor dynamics with the 4-pad bearings, it will be necessary to further reduce the vertical bearing stiffness in order to bring the values more in line with the fundamental shaft stiffness value. This requires a longer pad with a reduced bearing preload.

The paradox of this type of design is that the noncontact probes monitoring the bearing motion indicate higher amplitudes of motion, although the shaft center motion has been reduced. This phenomenon led one chief engineer of a compressor company to remove the redesigned 4-pad bearings and replace them with the original 5-pad design because of the increased motion observed at the bearings. Total rotor failure occurred shortly after the 5-pad bearings were installed!

7. Improved 4-Pad Bearing

In an effort to determine what bearing design choices were necessary in order to restore stability to Kaybob compressor, a number of modifications were made to the previously analyzed 4-pad bearing. Because it has been revealed that all of the designs considered thus far have produced bearings that are overly stiff, a number of design parameters must be considered to reduce the stiffness of the bearing while maintaining damping so that the bearings can better absorb the energy of the first bending mode.

Design parameters including the bearing clearance, pad preload, load orientation, and lubricant properties have all been shown by many authors [9–11] to affect the bearing stiffness. Therefore, for this design (Figure 20), it was decided to increase the bearing clearance from 3 mils radial to 6 mils and reduce the pad preload from 0.2 to 0.1 while maintaining the load-between-pad orientation to reduce the stiffness of

Kaybob compressor 4-pad load between pad bearings
$L/D = 0.4063$, Cb = 0.003, 2Cb/D = 0.0015, $m = 0.2$, offset = 0.5, arc = 80, PivAng = 45
Speed = 11400 rpm
Load = 200 Lbf
$W/LD = 30.7692$ psi
Vis. = $1E - 06$ reyns
Sb = 2.7444
E/Cb = 0.1871
Att. = 0.00 deg
$h$min = 2.221 mils
Pmax = 104.715 psi
Hp = 5.31435 hp
Qside = 1.396 gpm
Stiffness (Lbf/in)
$1.636E + 05$ $0.000E + 00$
$0.000E + 00$ $1.636E + 05$
Damping (Lbf-s/in)
$3.458E + 02$ $0.000E + 00$
$0.000E + 00$ $3.458E + 02$
Critical journal mass (Lb)
Stable

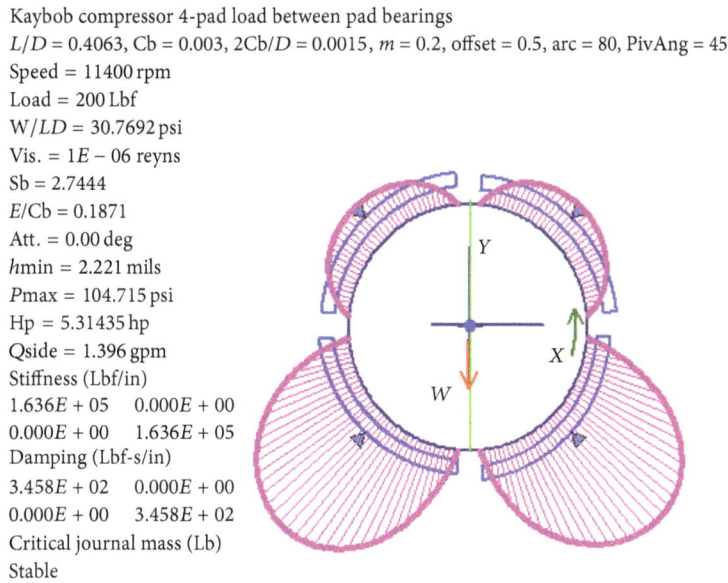

FIGURE 18: Results of the four-pad bearing analysis.

Mode number = 4 unstable forward precession
Shaft rotational speed = 11400 rpm
Whirl speed (damped natural freq.) = 4652 rpm, log decrement = −0.2329

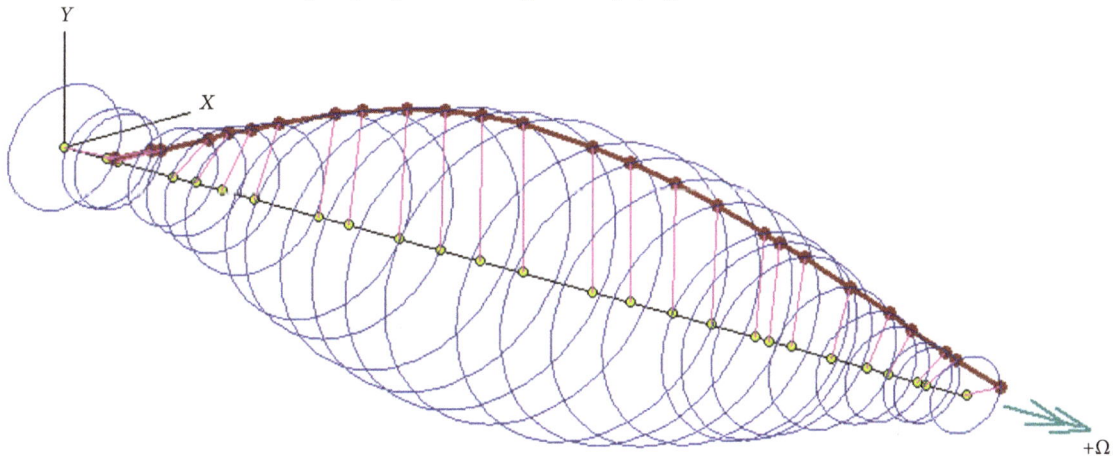

FIGURE 19: Stability results for a four-pad bearing design.

the bearing. In order to maintain adequate damping the bearing axial length was also increased from 1.625 inches to 4 inches, resulting in an increase in the L/D ratio from 0.4 to 1.

The results from the bearing analysis show a significant change in bearing performance when compared to the previous designs (Figure 21). The bearing stiffness has now been reduced to 48,960 lb/in. When comparing twice the bearing stiffness to the shaft modal stiffness this now results in a ratio of 0.85, much closer to the desired value of unity. The increase in bearing axial length also results in reasonable levels of damping. A stability analysis of this system shows a drastic increase in the log decrement of the first bending mode (Figure 22), resulting in a fully stable mode for the compressor. An energy distribution plot

(Figure 23) shows how the improved bearing design does a much better job of allowing the bearings to absorb the vibrational energy, resulting in an increase of 32% of the energy being transmitted to the bearings when compared to the original 5-pad design.

8. Squeeze Film Dampers

Another solution considered by the team at Cooper-Bessemer was a squeeze film damper. Smith [1] describes the overall design as a long damper surrounding a number of different bearing designs that were considered including a unique design that involved six tilting pads with asymmetries in the pad clearance, axial length, and arc length. However,

Kaybob compressor improved 4-pad load between pad bearings

Bearing data
$L = 4\,\text{in}$
$D = 4\,\text{in}$
$Cb = 0.006\,\text{in}$
$2Cb/D = 0.0030$
Preload = 0.1
Offset = 0.5
Arc length = 80
Pivot angle = 45
Load between pivots
Load angle = 270
Neglect pivot effect

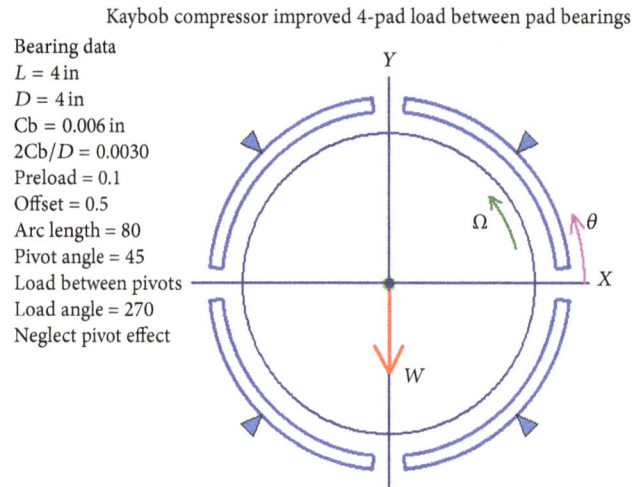

FIGURE 20: Improved four-pad bearing design.

Kaybob compressor improved 4-pad load between pad bearings
$L/D = 1$, Cp = 0.006, $2Cb/D = 0.003$, $m = 0.1$, offset = 0.5, arc = 80, PivAng = 45
Speed = 11400 rpm
Load = 200 Lbf
$W/LD = 12.5\,\text{psi}$
Vis. = $1E - 06$ reyns
Sb = 1.6889
$E/\text{Cb} = 0.1229$
Att. = 0.00 deg
$h\text{min} = 4.944\,\text{mils}$
Pmax = 37.308 psi
Hp = 6.49116 hp
Qside = 1.898 gpm
Stiffness (Lbf/in)
$4.896E + 04 \quad 0.000E + 00$
$0.000E + 00 \quad 4.896E + 04$
Damping (Lbf-s/in)
$2.825E + 02 \quad 0.000E + 00$
$0.000E + 00 \quad 2.825E + 02$
Critical journal mass (Lb)
Stable

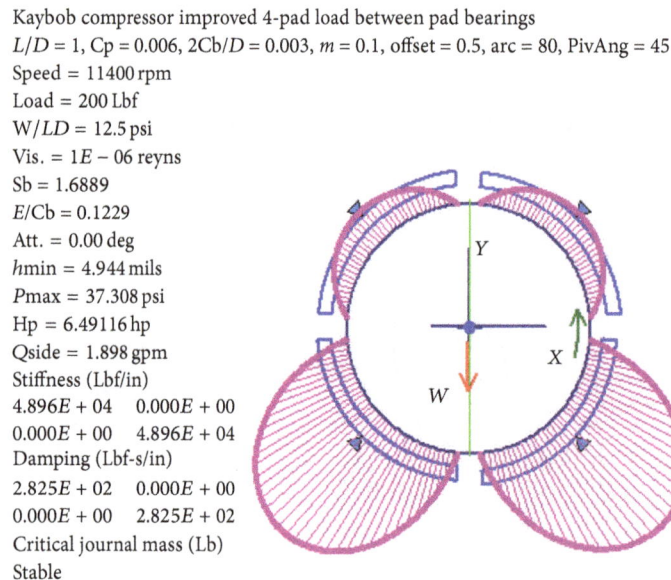

FIGURE 21: Analysis of the improved four-pad bearing.

none of these designs produced an ultimately stable solution for the machine and the use of a long squeeze film damper was likely a key factor in this.

To further make this point, the analysis was performed on two damper designs surrounding the original 5-pad bearing (Figure 24): a long damper design and a damper with a central groove of 0.15 inches in axial length. The nongrooved damper is treated as a long bearing, but because an added groove to the damper results in a different pressure distribution within the damper the grooved damper is modeled as a short bearing.

With all other inputs equal it can be seen that the groove has a significant effect on the predicted stiffness and damping of the damper. The longer damper design—similar to the one used by Cooper-Bessemer—produces very high levels of stiffness and damping while the grooved design produces much more reasonable values.

Stability analyses of these two support scenarios also reveal the significant effect of the central groove (Figures 25 and 26). It is shown that the nongrooved damper produces only a very slight increase in the log decrement of the unstable mode when compared to the original bearing design.

The grooved damper, however, provides a substantial increase in the log decrement, resulting in a marginally stable system. Had the team at Cooper-Bessemer considered adding a central groove to their squeeze film damper design they may have actually avoided the costly rotor redesign discussed in Section 9.

Mode number = 5 stable forward precession
Shaft rotational speed = 11400 rpm
Whirl speed (damped natural freq.) = 4061 rpm, log decrement = 1.7522

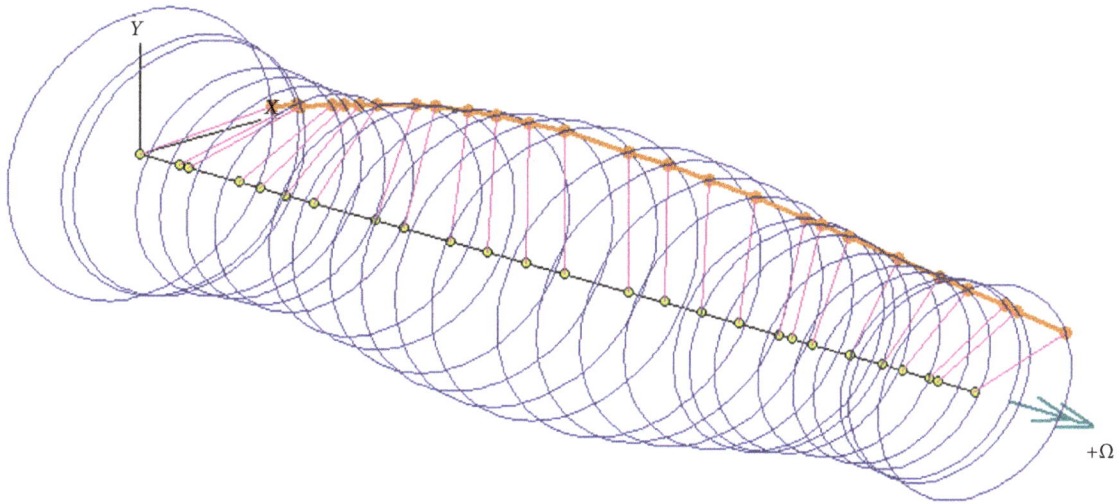

FIGURE 22: Stability analysis of the improved four-pad bearing.

Mode number = 1, critical speed = 2706 rpm = 45.10 Hz
Potential energy distribution (s/w = 1)
Overall: shaft (S) = 19.75%, bearing (Brg) = 80.25%

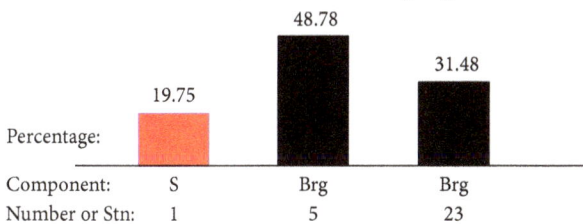

Percentage:	19.75	48.78	31.48
Component:	S	Brg	Brg
Number or Stn:	1	5	23

FIGURE 23: Energy distribution in the shaft and bearings for the improved four-pad bearing design.

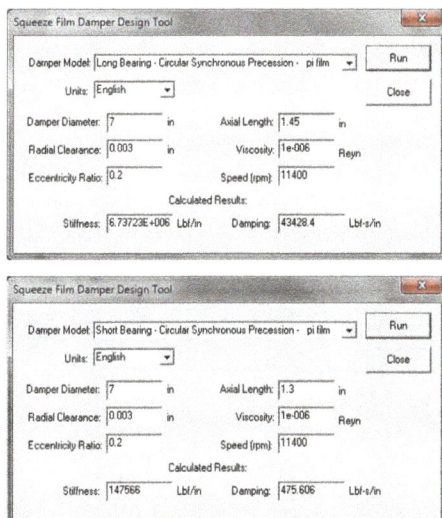

FIGURE 24: Nongrooved and grooved squeeze film damper analysis.

9. Rotor Redesign

To produce a stable operating environment the team at Cooper-Bessemer ultimately had to redesign the rotor. This involved multiple design iterations that resulted in a final shaft design with a bearing span reduced from about 60 inches to approximately 53 inches, modified shrink fits, an integral center seal section, and an increased shaft diameter under the impellers (Figure 4). This design, along with either a squeeze film damper or a 5-pad tilting pad bearing with no preload, was enough to restore the full operational capabilities of the compressor.

To show how these shaft design modifications made such a difference in the performance of the machine a shaft model was developed to analyze with the original bearing design and the original design with a long squeeze film damper (Figure 27). A critical speed analysis assuming rigid bearings with a stiffness of 10^8 lb/in predicted shaft stresses at the first bending mode of 2,371 psi (Figure 28), an increase of 39% when compared to the original rotor.

The shaft modal stiffness was calculated to be 201,780 lb/in, a 74% increase when compared to the original rotor. This significant increase in shaft modal stiffness made the new rotor design much more suited to the high stiffness bearings utilized.

A stability analysis of the original bearing design with the new shaft design (Figure 29) demonstrates how the new rotor accommodates the stiffer bearing design, resulting in a stable first bending mode. Note that the ratio of twice the bearing stiffness to the shaft modal stiffness in this case is 1.7, a significant improvement from the original design.

Since the long squeeze film damper was also considered along with the new rotor design, this case was analyzed for stability as well (Figure 30). Similar to before, however, the

Mode number = 3 unstable forward precession
Shaft rotational speed = 11400 rpm
Whirl speed (damped natural freq.) = 4477 rpm, log decrement = −0.1537

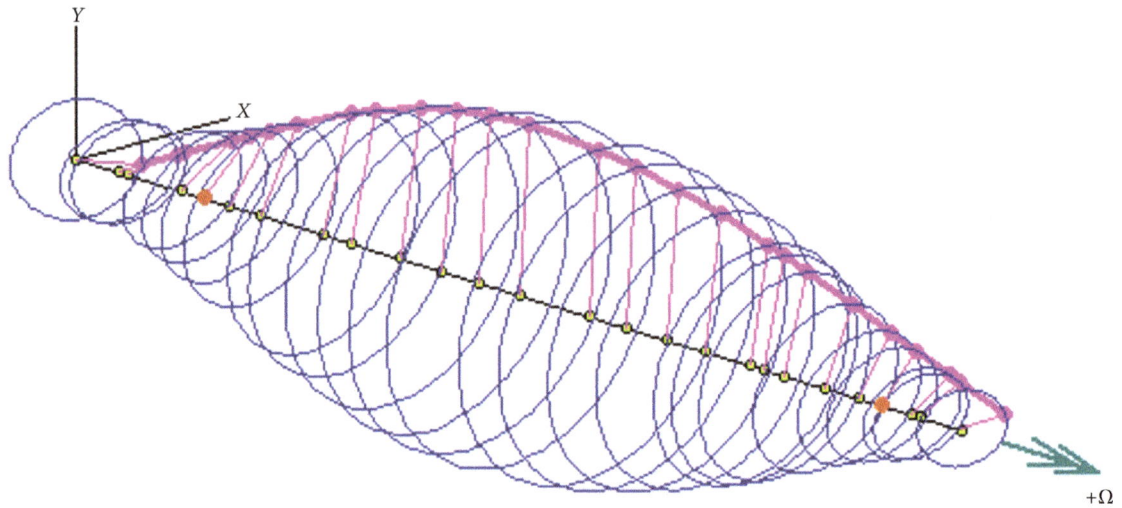

FIGURE 25: Stability analysis of the nongrooved squeeze film damper supported system.

Mode number = 3 stable forward precession
Shaft rotational speed = 11400 rpm
Whirl speed (damped natural freq.) = 3651 rpm, log decrement = 0.0669

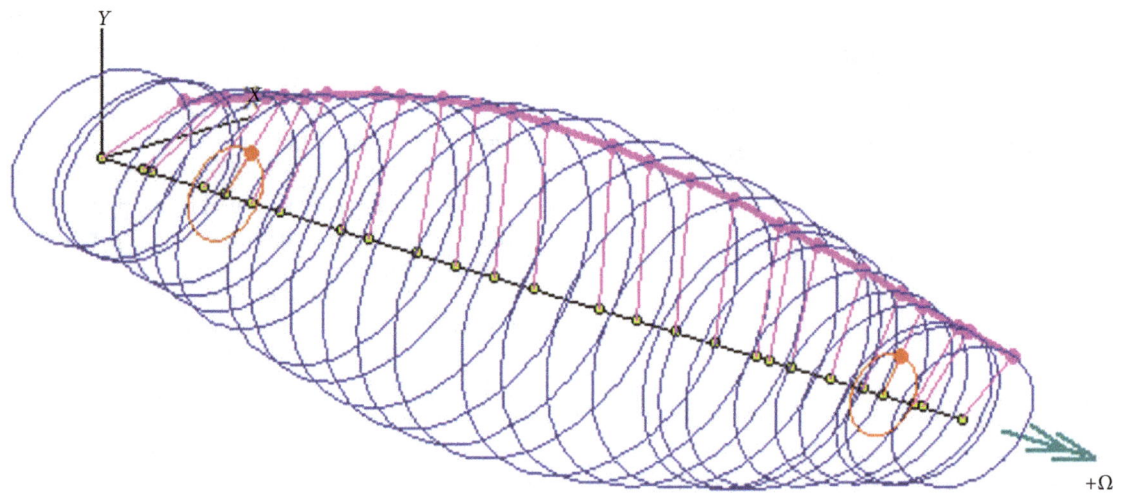

FIGURE 26: Stability analysis of the grooved squeeze film damper supported system.

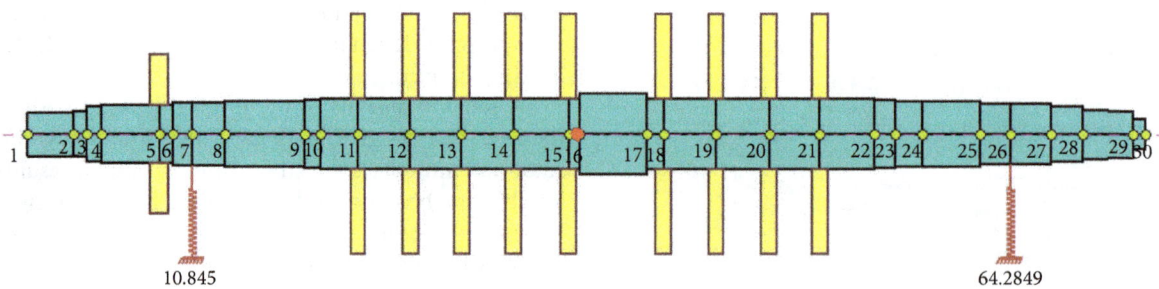

FIGURE 27: Reduced bearing span shaft model.

Critical speed mode shape, mode number = 1
Spin/whirl ratio = 1, stiffness: K_{xx}
Critical speed = 5444 rpm = 90.73 Hz

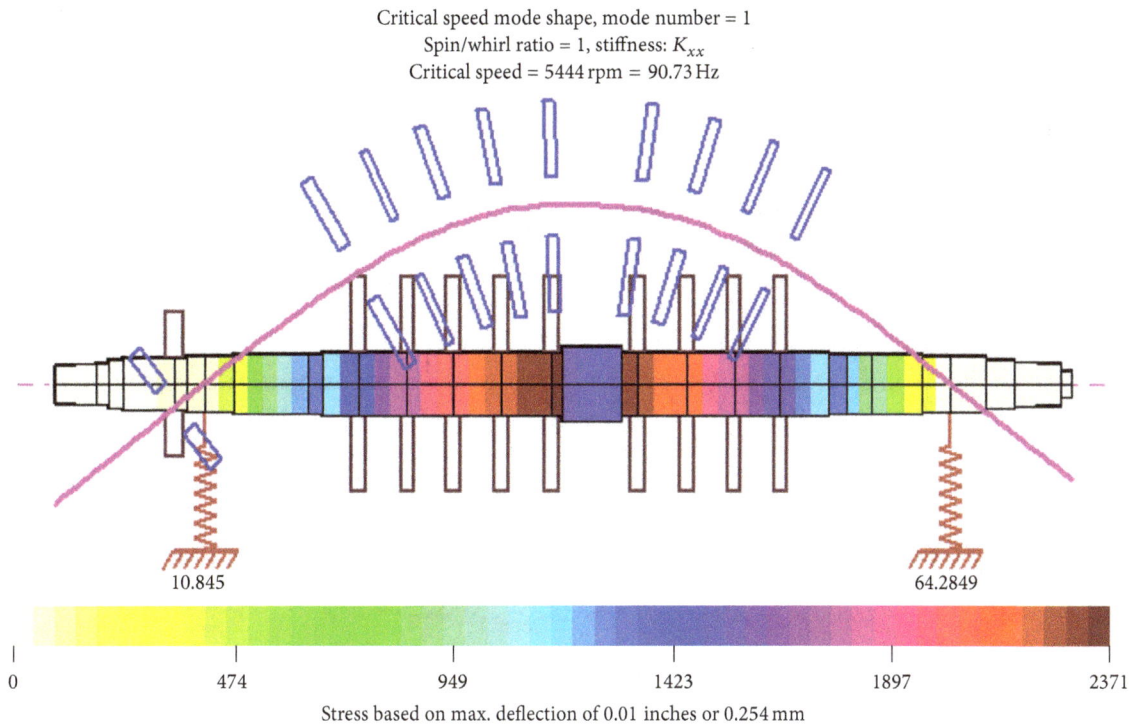

10.845   64.2849

| 0 | 474 | 949 | 1423 | 1897 | 2371 |

Stress based on max. deflection of 0.01 inches or 0.254 mm

FIGURE 28: Shaft modal stresses of the first bending mode.

Mode number = 3 stable forward precession
Shaft rotational speed = 11400 rpm
Whirl speed (damped natural freq.) = 4229 rpm, log decrement = 0.1238

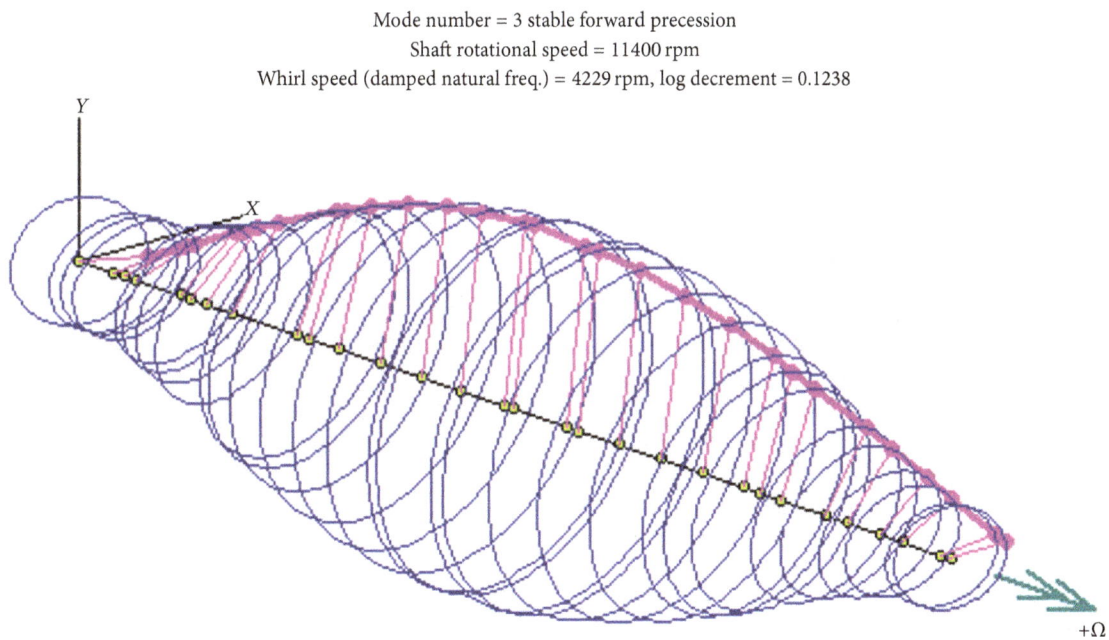

FIGURE 29: Stability analysis of the new rotor design with the original five-pad LOP bearings.

addition of the damper without a central groove proves to have little effect on the stability of the system.

10. Conclusions

The Kaybob compressor failure of 1971 was a historic case of rotordynamic instability and has taught us many things about the design of these machines. Tilting pad journal bearings have become a new normal in compressor stabilization due to their innate capacity for avoiding self-excited oil whirl; however great care must be taken in the design of these bearings for both ensuring a stable machine and meeting API specifications. Many have been tempted to try and avoid critical speeds and reduce motion at the bearings by

Mode number = 3 stable forward precession
Shaft rotational speed = 11400 rpm
Whirl speed (damped natural freq.) = 4217 rpm, log decrement = 0.1355

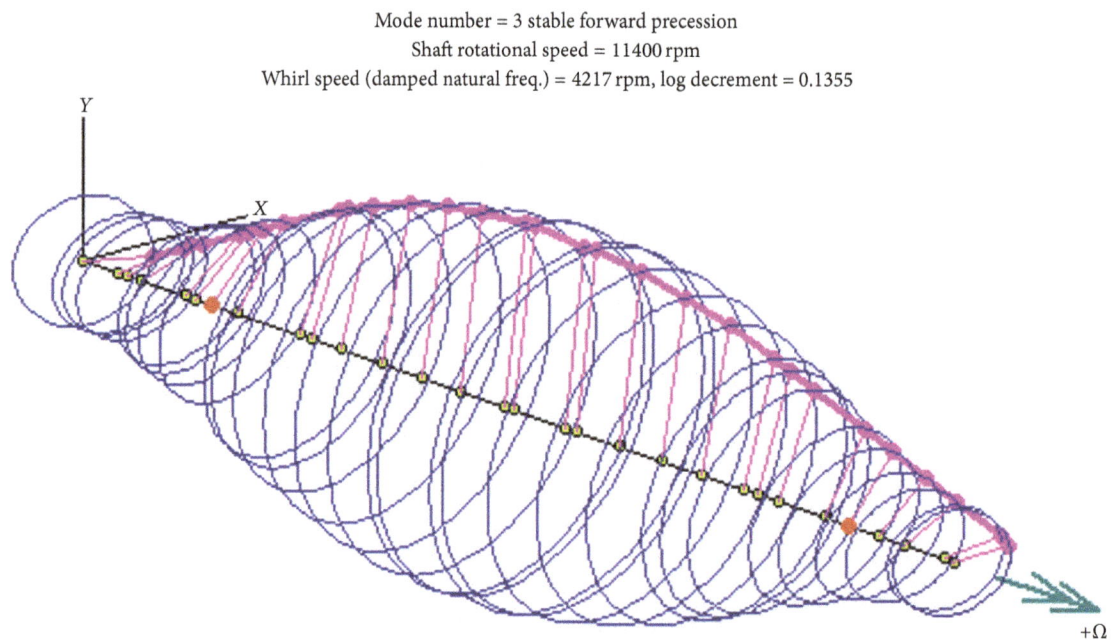

FIGURE 30: Stability analysis of the new rotor design with the original bearing and a long squeeze film damper.

increasing bearing stiffness; however this approach can and has often created a stability disaster as demonstrated by the Kaybob case. By designing for stability, however, these modes can also be eliminated from the operating range to the extent that they are not even detected because they are so well-damped.

In particular the Kaybob case has revealed the importance of designing bearings such that the bearing stiffness is tuned in to be in proper proportion to the shaft modal stiffness. By using this simple relationship and designing for proper damping as well, the bearings can properly absorb shaft vibrational energy and produce highly stable modes. While bearing asymmetry has been shown in past cases to produce some increases in stability—particularly in the presence of internal shaft friction or moderate levels of aerodynamic cross-coupling—the bearing stiffness must still be considered as a primary design variable as asymmetry will not benefit a system with a large bearing to shaft modal stiffness ratio. The presented design cases for 5- and 4-pad bearings demonstrate the importance of these concepts.

Squeeze film dampers have also been a significant source of compressor stabilization in decades past and will continue to do so when properly designed. It was shown in the context of the Kaybob failure that important features such as central grooves can make a large difference in the stiffness and damping characteristics of the damper, which can ultimately have as large an effect on compressor stability as the design of the bearing itself. While the redesign of the compressor rotor also produced a stable machine, this solution was far more costly than the simple use of a properly designed bearing or squeeze film damper. However, the team at Cooper-Bessemer cannot be faulted for these errors as they did not have the luxury of high-speed analysis tools that are available to designers today. Overall, the failure of the Kaybob

compressor and their work to investigate the instability has helped create the knowledge and methods necessary to avoid these kinds of failures in the future.

Competing Interests

The authors declare that they have no competing interests.

Acknowledgments

This work was supported by the Rotating Machinery and Controls Laboratory and industrial consortium at the University of Virginia and Rodyn Vibration Analysis, Inc.

References

[1] K. J. Smith, "An operation history of fractional frequency whirl," in *Proceedings of the 4th Turbomachinery Symposium*, pp. 115–125, Texas A&M University, 1975.

[2] J. W. Lund, "Spring and damping coefficients for the tilting-pad journal bearing," *ASLE Transactions*, vol. 7, no. 4, pp. 342–352, 1964.

[3] J. W. Lund, "Stability and damped critical speeds of a flexible rotor in fluid-film bearings," *Journal of Engineering for Industry*, vol. 96, no. 2, pp. 509–517, 1974.

[4] J. S. Alford, "Protecting turbomachinery from self–excited rotor whirl," *Journal of Engineering for Power*, vol. 87, no. 4, pp. 333–334, 1965.

[5] L. E. Barrett, E. J. Gunter, and P. E. Allaire, "Optimum bearing and support damping for unbalance response and stability of rotating machinery," *Journal of Engineering for Power*, vol. 100, no. 1, pp. 89–94, 1978.

[6] E. J. Gunter, "Dynamic stability of rotor-bearing systems," NASA SP-113, 1966.

[7] E. J. Gunter and P. R. Trumpler, "The influence of internal friction on the stability of high speed rotors with anisotropic supports," *Journal of Engineering for Industry*, vol. 91, no. 4, pp. 1105–1113, 1969.

[8] E. J. Gunter, "The influence of bearing asymmetry on rotor stability," in *Proceedings of the Vibration Institute Training Conference*, pp. 1–28, Williamsburg, Va, USA, June 2012.

[9] J. C. Nicholas and R. G. Kirk, "Selection and design of tilting pad and fixed lobe journal bearings for optimum turborotor dynamics," in *Proceedings of the 8th Turbomachinery Symposium*, pp. 43–57, Texas A&M University, College Station, Tex, USA, 1979.

[10] J. C. Nicholas and R. G. Kirk, "Four pad tilting pad bearing design and application for multistage axial compressors," *Journal of Lubrication Technology*, vol. 104, no. 4, pp. 523–529, 1982.

[11] B. K. Weaver, Y. Zhang, A. F. Clarens, and A. Untaroiu, "Nonlinear analysis of rub impact in a three-disk rotor and correction via bearing and lubricant adjustment," *Journal of Engineering for Gas Turbines and Power*, vol. 137, no. 9, Article ID 092504, 2015.

9

Mixed Skyhook and FxLMS Control of a Half-Car Model with Magnetorheological Dampers

Piotr Krauze and Jerzy Kasprzyk

Institute of Automatic Control, Silesian University of Technology, Akademicka 16, Gliwice, Poland

Correspondence should be addressed to Piotr Krauze; piotr.krauze@polsl.pl

Academic Editor: Mohammad Tawfik

The problem of vibration attenuation in a semiactive vehicle suspension is considered. The proposed solution is based on usage of the information about the road roughness coming from the sensor installed on the front axle of the vehicle. It does not need any preview sensor to measure the road roughness as other preview control strategies do. Here, the well-known Skyhook algorithm is used for control of the front magnetorheological (MR) damper. This algorithm is tuned to a quarter-car model of the front part of the vehicle. The rear MR damper is controlled by the FxLMS (Filtered-x LMS) taking advantage of the information about the motion of the front vehicle axle. The goal of this algorithm is to minimize pitch of the vehicle body. The strategy is applied for a four-degree-of-freedom (4-DOF) vehicle model equipped with magnetorheological dampers which were described using the Bouc-Wen model. The suspension model was subjected to the road-induced excitation in the form of a series of bumps within the frequency range 1.0–10 Hz. Different solutions are compared based on the transmissibility function and simulation results show the usefulness of the proposed solution.

1. Introduction

Damping of mechanical vibrations in vehicles using passive solution has a very long history and is widely used in practice. The conventional passive suspension system consists of a combination of springs and dampers. The characteristics of such suspension elements cannot be altered during operation. In recent years, much research has been carried out in the design of active and semiactive suspensions of vehicles for control of vibrations. However, active vibration damping entails considerable cost and difficulty in implementation of systems of this type; thus in the second half of the twentieth century a compromise between passive and active systems has been proposed. The semiactive suspension is an alternate to the active suspension [1]. Between different semiactive devices, magnetorheological (MR) dampers have received considerable interest in the last two decades [2]. These dampers comprise MR fluid belonging to the class of smart materials. The essential feature of MR fluid is its ability to reversibly change the state from a viscous to a semisolid with controllable yield strength when it is subjected to an external magnetic field. Varying electric current flowing through coils mounted in the piston of the damper allows for changing dynamical properties of the MR fluid. Inherent stability and low power consumption is favoured over active suspension force generators, so MR dampers are widely used in vehicle suspension systems [3–5].

Skyhook algorithm is one of the most widely used feedback control schemes dedicated to semiactive dampers firstly proposed in [1] and analysed based on a quarter-car model. The Skyhook control is favoured for its robustness and low complexity which is especially important in case of real time applications, for example, [3]. Numerous scenarios of Skyhook control were already published including control of quarter-car model [6] or plants modelled as quarter-car, for example, passenger's seat [7] or bicycle [8]. The classical Skyhook algorithm can be also used for separate control of each vehicle suspension part of a half-car experimental model [9] or a real-size vehicle [4]. Other variants of Skyhook control are related to coupled control of vehicle body vibration modes, that is, heave, pitch, and roll modes [10], or Skyhook control accompanied with procedure of its parameters adaptation [11].

The performance of the semiactive suspension system can be improved by knowing the future information about the road input, which is referred to as *preview control*. The semiactive suspension system with preview control using the linear quadratic theory was applied to quarter-car [12] and half-car [13] vehicle models. Stochastic optimal preview control of a vehicle suspension was presented in [14] and control using the H_{∞} theory in [15]. The other approach, based on the FxLMS algorithm modified for semiactive devices, was tested in [16] for a quarter-car model and in [17] for a half-car model. In all these references it is assumed that some kind of information about the road roughness is available in advance. However, in practice, the problem of the appropriate measurement of the road profile that can be available and useful for the control algorithm in real-time is very difficult. It can be obtained using specialized vision systems or laser scanners aimed in front of the vehicle. These devices are expensive and measurements are subjected to different disturbances or distortions; for example, quality of the information depends on lighting conditions [18]. Thus, they require sophisticated algorithms of measurement data processing [19]. In [20] it was proposed to use the lead vehicle response to generate preview functions for active suspension of convoy vehicles, but, of course, it is not the solution for normal civil vehicles. In this paper it is proposed to improve the quality of vibration control in a semiactive suspension by *adaptive semipreview* control. In this approach it is assumed that the front part of the suspension will be controlled without any information about the road, for example, using the Skyhook algorithm [1], whereas adaptive control of the rear part of the suspension can take advantage of the road roughness obtained due to the information about the movement of the front part. We proposed to use the FxLMS algorithm to control pitch of the vehicle body.

The paper is organized as follows. Section 2 refers to an experimental setup and defines a half-car model used in the research. In Section 2.2 behaviour of the MR damper is analysed and the required models of dynamics are obtained. Section 3 introduces the proposed semiactive vibration control algorithm. In Section 4 simulation results are presented and discussed as well as concluded in Section 5.

2. Modelling of Experimental Setup

Presented studies are related to the model of the experimental all-terrain vehicle equipped with magnetorheological dampers and vibration control system [21]. The vehicle is 2 meters long and 1 meter wide and it weighs 340 kilograms. The original shock absorbers in the vehicle were replaced with the suspension MR dampers produced by Lord Corporation. The measurement part of the system consists of numerous sensors, such as accelerometers, gyroscopes, and vehicle speed sensors, which track the vehicle motion during its ride. The motion of the vehicle body and underbody parts is measured using eight 3-axis accelerometers located in the vehicle

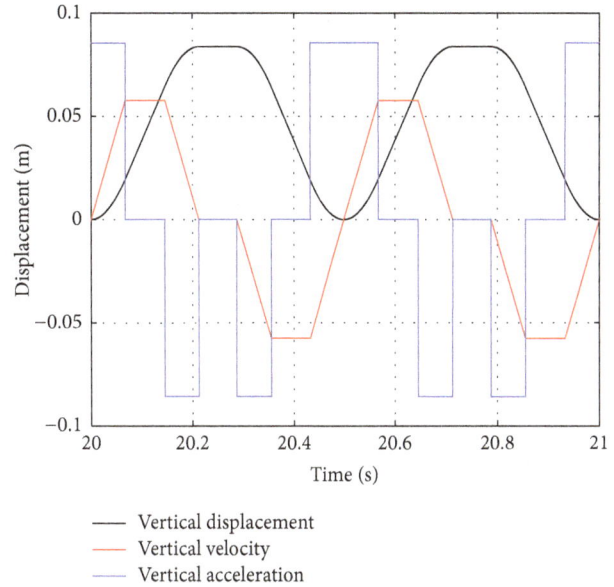

FIGURE 1: Simulated series of road bumps.

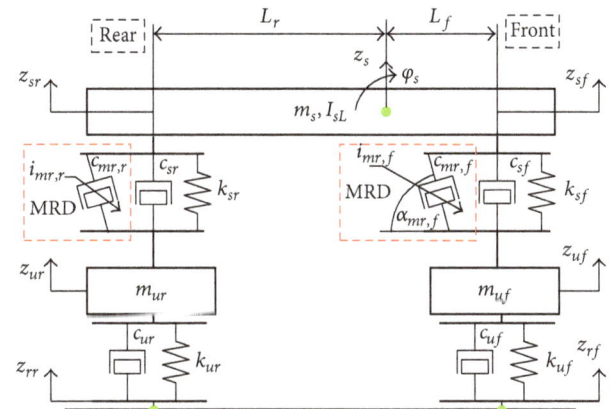

FIGURE 2: Half-car model of vehicle dynamics.

body as well as installed near the vehicle wheels. The vehicle speed is validated using the Hall effect sensors installed near the wheels. Moreover, vehicle suspension deflection sensors and inertial measurement units are applied.

2.1. Half-Car Model of Vehicle Dynamics. Generally, a full-car vehicle model which exhibits seven degrees of freedom [22] is used for the vibration control analysis. Such model maps heave, pitch, and roll motion of the vehicle body part as well as vertical motion of each wheel. The vehicle under consideration can be assumed with a high accuracy as longitudinally symmetrical. Vibrations of the right and left part of the vehicle are symmetrically excited using the simulated road-induced excitation, composed of a series of road bumps (see Figure 1). Therefore, a roll angle motion can be safely neglected in the analysis and it is sufficient to consider only a half-car model with 4 DOFs (see Figure 2).

In the particular case vibrations of the vehicle body part could be ideally decoupled which means that road excitation of the front vehicle part induces vibration of only front vehicle part and similarly for the rear vehicle part. Thus, double quarter-car models were also considered for simulation studies. However, generally, such assumption is not valid for the real vehicles. Furthermore, in the case of the experimental vehicle to which the presented studies are dedicated, vibration coupling between the front and the rear vehicle parts can be clearly measured. Thus, the following analysis is limited to the heave and pitch of the vehicle body as well as to the vertical motion of the wheels.

Dynamics of the road vehicle can be modelled by the following differential equations:

$$m_s \ddot{z}_s = -k_{sf}\left(z_{sf} - z_{uf}\right) - c_{sf}\left(\dot{z}_{sf} - \dot{z}_{uf}\right) + F_{mr,f}$$
$$\cdot \sin\left(\alpha_{mr,f}\right) - k_{sr}\left(z_{sr} - z_{ur}\right) - c_{sr}\left(\dot{z}_{sr} - \dot{z}_{ur}\right)$$
$$+ F_{mr,r}\sin\left(\alpha_{mr,r}\right),$$

$$I_{sL}\ddot{\varphi}_s = L_f\left[k_{sf}\left(z_{sf} - z_{uf}\right) + c_{sf}\left(\dot{z}_{sf} - \dot{z}_{uf}\right)\right.$$
$$\left. - F_{mr,f}\sin\left(\alpha_{mr,f}\right)\right] - L_r\left[k_{sr}\left(z_{sr} - z_{ur}\right)\right.$$
$$\left. + c_{sr}\left(\dot{z}_{sr} - \dot{z}_{ur}\right) - F_{mr,r}\sin\left(\alpha_{mr,r}\right)\right],$$

$$m_{uf}\ddot{z}_{uf} = -k_{uf}\left(z_{uf} - z_f\right) - c_{uf}\left(\dot{z}_{uf} - \dot{z}_f\right) - F_{mr,f}$$
$$\cdot \sin\left(\alpha_{mr,f}\right) + k_{sf}\left(z_{sf} - z_{uf}\right) + c_{sf}\left(\dot{z}_{sf} - \dot{z}_{uf}\right),$$

$$m_{ur}\ddot{z}_{ur} = -k_{ur}\left(z_{ur} - z_r\right) - c_{ur}\left(\dot{z}_{ur} - \dot{z}_r\right) - F_{mr,r}$$
$$\cdot \sin\left(\alpha_{mr,r}\right) + k_{sr}\left(z_{sr} - z_{ur}\right) + c_{sr}\left(\dot{z}_{sr} - \dot{z}_{ur}\right),$$

$$(1)$$

where subscripts f and r denote the front and rear part of the suspension, respectively. The sprung and unsprung parts of the vehicle are denoted using subscripts s and u, respectively. In the further analysis linear and angular velocities will be denoted as v and ω, respectively; namely, $v_s = \dot{z}_s$ and $\omega_s = \dot{\varphi}_s$. The linear and angular accelerations will be denoted as a and ε, respectively; namely $a_s = \ddot{z}_s$ and $\varepsilon_s = \ddot{\varphi}_s$. Location of the vehicle body's centre of gravity is described by distances L_f and L_r. Vertical displacements of the vehicle body z_{sf} and z_{sr} can be defined as

$$z_{sf} = z_s - L_f\varphi_s,$$
$$z_{sr} = z_s + L_r\varphi_s.$$

$$(2)$$

The road-induced displacements exciting the front and rear part of the vehicle are denoted as z_{rf} and z_{rr}. Vibrations propagate into the unsprung and sprung parts resulting in vertical displacement of wheels denoted as z_{uf} and z_{ur} and vertical displacement of the front and rear vehicle body as z_{sf} and z_{sr}, respectively. Parameters of the half-car model are listed as below.

Parameters of the Simulated 4 DOFs Half-Car Model

$$L_f = 1.116\,\text{m},$$
$$L_r = 1.232\,\text{m},$$
$$m_s = 348\,\text{kg},$$

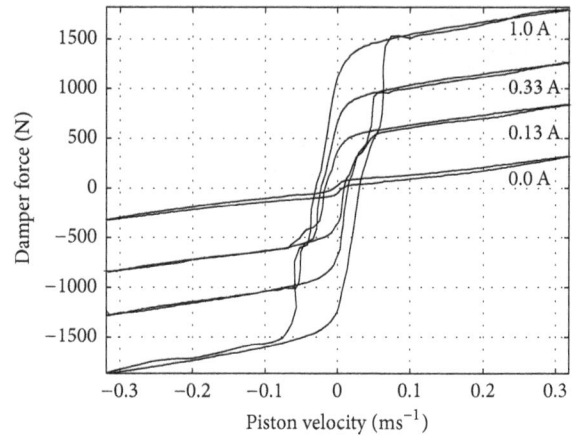

FIGURE 3: MR damper velocity-force characteristics.

$$I_{sL} = 175.2\,\text{kgm}^2,$$
$$k_{sf} = 28.0\,\text{kNm}^{-1},$$
$$k_{sr} = 30.4\,\text{kNm}^{-1},$$
$$c_{sf} = 971\,\text{Nsm}^{-1},$$
$$c_{sr} = 2464\,\text{Nsm}^{-1},$$
$$m_{uf} = 23.7\,\text{kg},$$
$$m_{ur} = 64.3\,\text{kg},$$
$$k_{uf/ur} = 169\,\text{kNm}^{-1},$$
$$c_{uf/ur} = 883\,\text{Nsm}^{-1},$$
$$\alpha_{mr,f} = 63°,$$
$$\alpha_{mr,r} = 63°.$$

The response of the vehicle to the bump excitation is mainly influenced by stiffness and viscous damping of the vehicle tires and the suspension system, denoted here by k and c. Stiffness of the tire corresponds to the pressure of compressed air therein. Passive invariant stiffness of the vehicle suspension describes behaviour of the shock absorbers. Viscous damping of the suspension is mainly related to passive damping of the shock absorbers, but it also corresponds to the overall and distributed suspension damping.

2.2. MR Damper Model. Axial force generated by the MR damper can change from hundreds to thousands of newtons in a couple of milliseconds [23]. However, due to the complex composition of the MR fluid inside the damper, this device exhibits strongly nonlinear behaviour. The velocity-force characteristics reveal saturation regions and hysteretic loop [3]. An exemplary characteristic for different current levels obtained for the MR damper produced by Lord Corporation is presented in Figure 3.

Numerous models are known in the literature that deal with these phenomena [24]. The Bingham model [25] includes viscous damping and Coulomb friction components. Extension of the Bingham model which additionally applies Kelvin-Voight body and Hooke body models is known as Gamota-Filisko model [2]. An accurate MR damper model needs to take into account numerous dependencies of the resultant force on control current, excitation magnitude, and frequency which was extensively reviewed in [26]. The proposed model was additionally validated and applied in control [27].

Among the presented models the Bouc-Wen model [2] is to be one of the most accurate. The Bouc-Wen model consists of hysteresis component defined using a nonlinear differential equation accompanied with linear viscous damping and stiffness components. Thus, simulation of the Bouc-Wen model requires more complex solvers comparing to static MR damper models. Furthermore, nonlinearity of such model makes its identification complex and not unique. It was also reported that the Bouc-Wen model exhibits greater inaccuracy for modelling force saturation in the postyield [24]. However, the Bouc-Wen model is defined in a compact form which map major features of MR damper behaviour, that is, hysteresis loops, force saturation, and additional viscous damping and stiffness indicated by MR damper behaviour. Furthermore, the hysteresis component can more accurately, in comparison to, for example, Tanh model, map hysteretic behaviour for nonharmonic and wideband piston excitations, apart from sinusoidal excitations. Parameters of the Bouc-Wen obtained for different control currents were additionally approximated using higher-degree polynomial in the presented paper in order to more accurately fit variations of parameters.

According to the Bouc-Wen model, force yield by the MR damper is defined as follows:

$$F_{mr,bw} = -\left[c_{bw}v_{mr} + k_{bw}\left(z_{mr} - x_{bw}\right) + \alpha_{bw}p\right], \quad (3)$$

where v_{mr} denotes relative velocity of MR damper piston and c_{bw}, k_{bw}, α_{bw}, and x_{bw} denote parameters of the model. The hysteretic displacement p is described by the first-order nonlinear differential equation:

$$\dot{p} = -\gamma_{bw} \cdot |v_{mr}| \cdot p \cdot |p|^{n-1} - \beta_{bw}v_{mr}|p|^{n} + A_{bw}v_{mr}, \quad (4)$$

where symbols γ_{bw}, β_{bw}, and A_{bw} denote the additional parameters of the Bouc-Wen model.

Commonly, parameters of the Bouc-Wen model are estimated separately for different levels of the control current denoted as i_{mr}. Since the relationship between the current and the model parameters is strongly nonlinear, it is generally approximated using the higher-order polynomials. It was shown in [28] that two parameters α_{bw} and c_{bw} can be related to the control current using the third-order polynomials as follows:

$$\alpha_{bw} = \sum_{j=0}^{3} \alpha_{j,bw} \cdot i_{mr}^{j},$$

$$\qquad\qquad\qquad\qquad\qquad (5)$$

$$c_{bw} = \sum_{j=0}^{3} c_{j,bw} \cdot i_{mr}^{j}.$$

Other parameters can be assumed as constant. Parameters of the Bouc-Wen model (see the list below) were estimated based on the identification results presented in [16].

Parameters of the MR Damper Models

Bouc-Wen Model of the MR Damper

$$i_{mr} \in (0.0; 1.33)\,\text{A},$$

$$t_{mr} = 12\,\text{ms},$$

$$n = 2,$$

$$k_{bw} = 0.001,$$

$$x_{bw} = 1.5,$$

$$[\alpha_{0,bw}, \alpha_{1,bw}, \alpha_{2,bw}, \alpha_{3,bw}] \quad = \quad [93506, 888021, 27374, -294583],$$

$$[c_{0,bw}, c_{1,bw}, c_{2,bw}, c_{3,bw}] = [792, 4195, -6390, 2565],$$

$$\gamma_{bw} = 987288,$$

$$\beta_{bw} = 983237,$$

$$A_{bw} = 7.979.$$

Tanh Based Model of the MR Damper

$$\alpha_{0,th} = -23.05,$$

$$\alpha_{1,th} = 1215,$$

$$\beta_{th} = 36.47,$$

$$\gamma_{th} = 1.6,$$

$$c_{th} = 1203,$$

$$k_{th} = 1297.$$

Furthermore, it was shown in [29] that the response time of the force yield by the MR damper depends on both the control current level and the kinematic excitation of the damper piston, and it can vary from 20 to 40 ms. Thus, the first-order filter $S_{F_{mr}}$ with a time constant t_{mr} equal to 12 ms was included at the output of the Bouc-Wen model to simulate this delay. Within the simulation environment, the MR damper force F_{mr} is processed using $S_{F_{mr}}$ filter resulting in the modified force denoted as F_{mr}^{*}. In this paper the Bouc-Wen model is treated as a reference model included in the vehicle suspension model (see Figure 4).

3. Control Algorithm

Here, the goal of control is to increase the ride comfort, that is, simultaneous mitigation of the heave and pitch vibrations. Results obtained for mixed Skyhook and FxLMS control will be compared with two other algorithms of the semiactive vibration control, that is, the Skyhook algorithm related to the quarter-car model and the two-dimensional Skyhook control optimized with respect to either heave or pitch. All these algorithms belong to the group of the feedback control; they use several response signals available from the vehicle. Mainly, for vibration control, heave and pitch velocities as well as front and rear body velocities are assumed as error signals. Furthermore, the relative displacement and velocity of the MR damper pistons are required for application of the MR damper inverse model. The MR damper displacement can be commonly measured in the experimental vehicle whereas the velocity needs to be estimated.

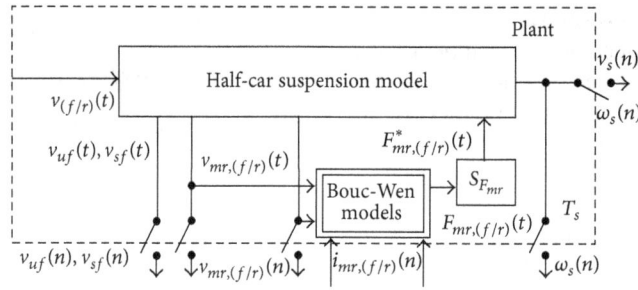

FIGURE 4: Block diagram of the vehicle simulator.

3.1. Inverse Tanh MR Damper Model. Common semiactive vibration control algorithms require the inverse MR damper model in order to partly linearise nonlinear relationships of the signal path from the input control current to the output MR damper force. Because this model should be used in a control system and solution of the Bouc-Wen inverse model cannot be obtained in the embedded controller in real-time, so for the purpose of vibration control it is proposed to use the hyperbolic tangent function [30] to model nonlinear behaviour of the MR damper as follows:

$$F_{mr,th}\left(i_{mr}, z_{mr}, v_{mr}\right)$$

$$= -\left(\alpha_{0,th} + \alpha_{1,th}\sqrt{i_{mr}}\right)\tanh\left[\beta_{th}v_{mr} + \gamma_{th}\operatorname{sign}\left(z_{mr}\right)\right] \quad (6)$$

$$- c_{th}v_{mr} - k_{th}z_{mr},$$

where symbols $\alpha_{0,th}$, β_{th}, γ_{th}, c_{th}, and k_{th} denote parameters of the model not related to the control current. The parameter $\alpha_{1,th}$ maps the influence of the control current on the MR damper force using a square root operator. Procedure of the *tanh* model identification is presented in [30] and model parameters are listed in Parameters of the MR Damper Models in Section 2.2.

Thus, the inverse model can be easily evaluated from (6) as follows:

$$i_{mr,th}$$

$$= \frac{1}{\alpha_{1,th}^2}\left(\frac{-F_{alg} - c_{th}v_{mr} - k_{th}z_{mr}}{\tanh\left[\beta_{th}v_{mr} + \gamma_{th}\operatorname{sign}\left(z_{mr}\right)\right]} - \alpha_{0,th}\right)^2, \quad (7)$$

where $i_{mr,th}$ stands for the desired control current, which should be generated by the control system in order to make the MR damper generate the desired force F_{alg}.

Additionally, the impact of modelling error on the final result of the damping system was tested. Model identification was repeated for the data contaminated by noise added to force values, and the model thus obtained was used in the control algorithm. It turned out that the results of damping did not differ substantially from the initial inverse model, which may indicate good tolerance to modelling errors.

3.2. Skyhook Control. Control force desired by the Skyhook algorithm related to the quarter-car model is expressed as follows [1]:

$$F_{alg,(f/r)}\left(n\right) = -g_{v_{s(f/r)}} \cdot v_{s(f/r)}\left(n\right), \quad (8)$$

where $g_{v_{s(f/r)}}$ denotes the gain of Skyhook control for the front or rear part of the vehicle, respectively.

The Skyhook approach can be generalised and extended to the two-dimensional clipped LQ (linear quadratic) control related to the half-car model [5, 31], where the following cost function is minimized:

$$J = \int_0^\infty \left[\mathbf{x}^T\left(\tau\right)\mathbf{Q}\mathbf{x}\left(\tau\right) + \mathbf{F_{alg}}^T\left(\tau\right)\mathbf{R}\mathbf{F_{alg}}\left(\tau\right)\right]d\tau. \quad (9)$$

The diagonal cost matrices $\mathbf{Q}$ and $\mathbf{R}$ related to the vectors $\mathbf{x}$ and $\mathbf{F_{alg}}$ of state and control variables, respectively, are evaluated according to Bryson's rule [32]:

$$\mathbf{Q} = \mathbf{S_x}^{-2},$$

$$\mathbf{R} = \mathbf{S_{F_{alg}}}^{-2}, \quad (10)$$

where diagonal elements of matrices $\mathbf{S_x}$ and $\mathbf{S_{F_{alg}}}$ are the limitations set on the state- $\mathbf{x}$ and control-related $\mathbf{F_{alg}}$ vectors. Herein, the analysis is focused only on two limitations included in matrix $\mathbf{S_x}$, strictly s_{v_s} and s_{ω_s} related to the v_s and ω_s state variables. The vectors of state and control variables are defined as follows:

$$\mathbf{x} = \left[\left(z_{uf} - z_{rf}\right) \ v_{uf} \ \left(z_{ur} - z_{rr}\right) \ v_{ur} \ \left(z_{sf} - z_{uf}\right) \ \left(z_{sr} - z_{ur}\right) \ v_s \ \omega_s\right]^T,$$

$$\mathbf{F_{alg}} = \left[F_{alg,f}, F_{alg,r}\right]^T. \quad (11)$$

The vector $\mathbf{x}$ consists of the vehicle suspension deflections and velocities related to the vehicle body heave and pitch as well as the deflections of tires and absolute velocities of the wheels. In the case of the clipped LQ control forces generated by the MR dampers are assumed as control signals included in the vector $\mathbf{F_{alg}}$.

Reformulation of a set of differential equations (1) and (2) gives the following state matrix equation of the linear half-car model:

$$\dot{\mathbf{x}} = \mathbf{A} \cdot \mathbf{x} + \mathbf{B_F} \cdot \mathbf{F_{alg}} + \mathbf{B_r} \cdot \mathbf{u_r}, \tag{12}$$

$$\mathbf{a_1} = \left[0 \quad -\frac{k_{uf}}{m_{uf}} \quad 0 \ 0 \ 0 \ 0 \ 0 \ 0 \right]^T,$$

$$\mathbf{a_2} = \left[1 \quad \left(-\frac{c_{uf}}{m_{uf}} - \frac{c_{sf}}{m_{uf}}\right) \quad 0 \ 0 \ -1 \ 0 \ \frac{c_{sf}}{m_s} \quad -\frac{L_f c_{sf}}{I_{sL}} \right]^T,$$

$$\mathbf{a_3} = \left[0 \ 0 \ 0 \quad -\frac{k_{ur}}{m_{ur}} \quad 0 \ 0 \ 0 \ 0 \right]^T,$$

$$\mathbf{a_4} = \left[0 \ 0 \ 1 \quad \left(-\frac{c_{ur}}{m_{ur}} - \frac{c_{sr}}{m_{ur}}\right) \quad 0 \ -1 \ \frac{c_{sr}}{m_s} \quad \frac{L_r c_{sr}}{I_{sL}} \right]^T,$$

$$\mathbf{a_5} = \left[0 \quad \frac{k_{sf}}{m_{uf}} \quad 0 \ 0 \ 0 \ 0 \quad -\frac{k_{sf}}{m_s} \quad \frac{L_f k_{sf}}{I_{sL}} \right]^T, \tag{14}$$

$$\mathbf{a_6} = \left[0 \ 0 \ 0 \quad \frac{k_{sr}}{m_{ur}} \quad 0 \ 0 \quad -\frac{k_{sr}}{m_s} \quad -\frac{L_r k_{sr}}{I_{sL}} \right]^T,$$

$$\mathbf{a_7} = \left[0 \quad \frac{c_{sf}}{m_{uf}} \quad 0 \quad \frac{c_{sr}}{m_{ur}} \quad 1 \ 1 \quad \left(-\frac{c_{sf}}{m_s} - \frac{c_{sr}}{m_s}\right) \quad \left(\frac{L_f c_{sf}}{I_{sL}} - \frac{L_r c_{sr}}{I_{sL}}\right) \right]^T,$$

$$\mathbf{a_8} = \left[0 \quad -\frac{L_f c_{sf}}{m_{uf}} \quad 0 \quad \frac{L_r c_{sr}}{m_{ur}} \quad -L_f \ L_r \quad \left(\frac{L_f c_{sf}}{m_s} - \frac{L_r c_{sr}}{m_s}\right) \quad \left(-\frac{L_f^2 c_{sf}}{I_{sL}} - \frac{L_r^2 c_{sr}}{I_{sL}}\right) \right]^T,$$

$$\mathbf{B_F} = [\mathbf{b_1} \quad \mathbf{b_2}],$$

where

$$\mathbf{b_1} = \left[0 \quad -\frac{\sin(\alpha_{mr,f})}{m_{uf}} \quad 0 \ 0 \ 0 \ 0 \quad \frac{\sin(\alpha_{mr,f})}{m_s} \quad -L_f \frac{\sin(\alpha_{mr,f})}{I_{sL}} \right]^T, \tag{15}$$

$$\mathbf{b_2} = \left[0 \ 0 \ 0 \quad -\frac{\sin(\alpha_{mr,r})}{m_{ur}} \quad 0 \ 0 \quad \frac{\sin(\alpha_{mr,r})}{m_s} \quad L_r \frac{\sin(\alpha_{mr,r})}{I_{sL}} \right]^T.$$

They are used for evaluation of the LQ time-infinite problem with the solution of the continuous time algebraic Riccati equation:

$$\mathbf{A}^T \mathbf{P} + \mathbf{PA} + \mathbf{PB_F} \mathbf{R}^{-1} \mathbf{B_F}^T \mathbf{P} + \mathbf{Q} = \mathbf{0}. \tag{16}$$

where $\mathbf{u_r} = [z_{rf}, v_{rf}, z_{rr}, v_{rr}]^T$ denotes vector of disturbances of the system, that is, road-induced displacement and velocity of the front and rear vehicle part, respectively. Matrices $\mathbf{A}$ and $\mathbf{B_F}$ are expressed as follows:

$$\mathbf{A} = [\mathbf{a_1} \ \mathbf{a_2} \ \mathbf{a_3} \ \mathbf{a_4} \ \mathbf{a_5} \ \mathbf{a_6} \ \mathbf{a_7} \ \mathbf{a_8}], \tag{13}$$

where

The solution $\mathbf{P}$ of (16) is applied for evaluation of the control gain as follows:

$$\mathbf{G} = \mathbf{R}^{-1} \mathbf{B}^T \mathbf{P}. \tag{17}$$

In order to retain the same experimental conditions for all algorithms, the clipped LQ control was limited to only two measurable state variables, heave and pitch velocities, resulting in the following control force for the front or rear suspension MR damper:

$$F_{\text{alg},(f/r)}(n) = -g_{v_s,(f/r)} \cdot v_s(n) - g_{\omega_s,(f/r)} \cdot \omega_s(n), \tag{18}$$

where gains $g_{v_s,(f/r)}$ and $g_{\omega_s,(f/r)}$ are selected elements of the control gain matrix $\mathbf{G}$ evaluated according to (17).

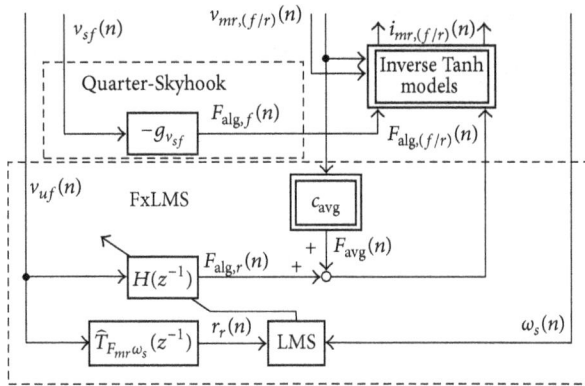

FIGURE 5: Block diagram of the mixed Skyhook and FxLMS control.

3.3. Mixed Skyhook and FxLMS Control. The FxLMS (Filtered-x LMS) algorithm is an example of adaptive feedforward control commonly applied in active noise control [33]. The idea of using the modified FxLMS algorithm for vibration control in semiactive suspension was introduced in [16]. It was assumed that the experimental vehicle is equipped with an additional device scanning the road profile and the algorithm takes advantage of the reference signal generated by this device. Here, in order to simplify the control system, it is assumed that instead of the reference signal taken from the scanner, we can use a signal obtained from the sensor measuring the motion of the front vehicle axle v_{uf} as the reference signal for the feedforward control of the rear damper. Thus, the front MR damper should be controlled by the classical Skyhook scheme related to the quarter-car model (8), whereas the rear MR damper can be controlled by the modified FxLMS algorithm, which results in a control scheme presented in Figure 5. Here, relative motion of the MR damper pistons and motion of other vehicle parts, particularly the vehicle body and axles, are available at the output of the vehicle dynamics simulator.

Similar to the previously referred Skyhook control, the goal of the adaptive algorithm related to the rear vehicle part is to minimize pitch vibrations of the vehicle body. Parameters of the adaptive FIR (Finite Impulse Response) filter $H(z^{-1})$ are updated according to the following expression:

$$\mathbf{h}(n+1) = \gamma \mathbf{h}(n) - \mu \cdot \omega_s(n) \cdot \frac{\mathbf{r}(n)}{\mathbf{r}^T(n) \cdot \mathbf{r}(n) + \zeta}, \quad (19)$$

where $\mu = 0.007$ is the adaptation step of the algorithm. Leakage of the FxLMS algorithm is controlled by γ equal to 0.997. For semiactive devices with the significant force saturation, the leakage parameter is introduced to ensure stability of the algorithm. Parameter $\zeta = 10^{-15}$ is used to avoid division by zero in the case of zeroing $\mathbf{r}(n)$. The parameters mentioned above were adjusted experimentally.

The reference signal $r(n)$ is obtained by filtering the vertical velocity of the front vehicle axle z_{uf} by the appropriate model of the secondary path $\hat{T}_{F_{mr}\omega_s}(z^{-1})$. This signal path was estimated starting from the force generated by the rear MR damper and ending with the pitch velocity assuming

that the additional viscous damper was added to the vehicle suspension. This fictitious damper equivalent to the averaged velocity-force characteristics is used for modification of the dissipative domain of the MR damper, as it was shown in [16]. Operator z^{-1} presented in Figure 5 corresponds to a one-sample delay. Consecutive M samples of $r(n)$, where $M = 128$ is the selected experimentally length of the filter $H(z^{-1})$, create the vector $\mathbf{r}$.

Control signal $F_{\text{alg},r}(n)$ is generated as a result of filtering v_{uf} using the adaptive filter $H(z^{-1})$. Additionally, F_{avg} generated by the fictitious passive damper c_{avg} according to $v_{mr,r}$ related to the rear part of the vehicle is added to the $F_{\text{alg},r}(n)$. Finally, the desired control current $i_{mr,(f/r)}$ is calculated using the inverse MR damper model (7).

4. Results

Simulation environment consists of two time domains related to the simulation of the vehicle dynamics and to vibration control. The differential equations related to the half-car model (1) and (2) and the Bouc-Wen model (3) and (4) were numerically solved using the Runge-Kutta method with variable integration step and interpolated for the sampling interval $T_{s,\text{model}} = 1$ ms, whereas the control algorithm was executed with the sampling interval $T_{s,\text{control}} = 2$ ms. It reflects the real experimental setup with a digital controller.

In the first stage the analysis was carried out in time domain based on heave and pitch acceleration presented for in-phase and antiphase road-induced excitation, respectively. The first resonant frequency of the analysed half-car model is close to 2 Hz. Thus, time diagrams of heave acceleration are presented for road-induced excitation of such frequency in order to clearly show differences in damping the heave resonant peak for all algorithms (see Figure 6). In the case of pitch acceleration the greatest differences in control performance were found for excitation frequency equal to 4.8 Hz. It can be noticed in time diagrams that 2-dimensional heave or pitch Skyhook exhibits good performance for only one in-phase or antiphase case of excitation. Control quality offered by quarter-Skyhook and mixed Skyhook-FxLMS algorithm is comparable, where the latter one is slightly better in mitigation of pitch vibration.

4.1. Transmissibility Characteristics and Control Quality Assessment. Analysis of vibration control was performed within the frequency range from 1.0 to 10 Hz, as the most of vibration modes of the vehicle dynamics occur within this range. In-phase and antiphase road-induced excitation of the front and rear wheels are critical for the different heave and pitch vibration modes, respectively. Thus, the heave transmissibility characteristics and the related quality index were evaluated for the in-phase excitation of both wheels, whereas the pitch dynamics was analysed for the antiphase excitation, that is, when the peak at the front wheel meets with the valley at the rear wheel, or vice versa. The evaluation procedure for the algorithms was carried out in two stages. Generally, the quality of vibration control is evaluated using

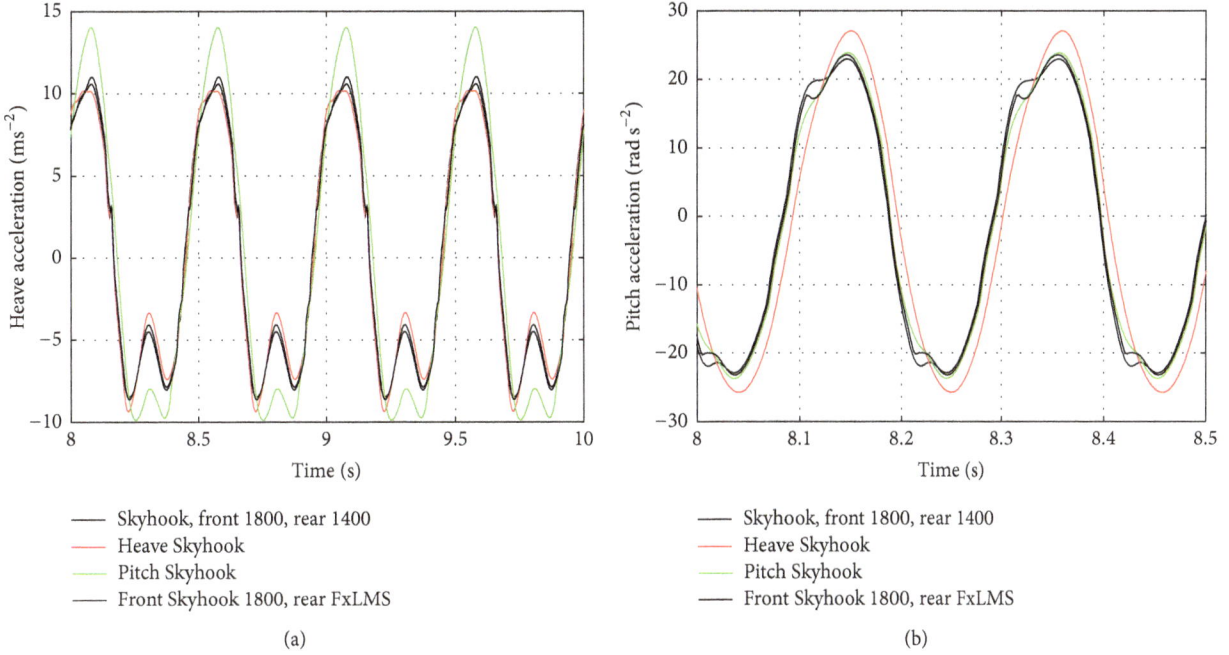

Skyhook, front 1800, rear 1400
Heave Skyhook
Pitch Skyhook
Front Skyhook 1800, rear FxLMS

(a)

Skyhook, front 1800, rear 1400
Heave Skyhook
Pitch Skyhook
Front Skyhook 1800, rear FxLMS

(b)

FIGURE 6: Comparison of algorithms in time domain: (a) heave acceleration for in-phase road-induced excitation of 2 Hz and (b) pitch acceleration for antiphase road-induced excitation of 4.8 Hz.

the velocity or acceleration transmissibility characteristics. The velocity transmissibility is defined for the heave v_s as

$$T_{v_r,v_s}(f) = \sqrt{\frac{\sum_{n=1}^N v_s^2(n)}{\sum_{n=1}^N v_r^2(n)}}\Bigg|_f \quad (20)$$

and for the pitch ω_s as

$$T_{v_r,\omega_s}(f) = \sqrt{\frac{\sum_{n=1}^N \omega_s^2(n)}{\sum_{n=1}^N v_r^2(n)}}\Bigg|_f . \quad (21)$$

The acceleration transmissibility is defined for the heave a_s as

$$T_{a_r,a_s}(f) = \sqrt{\frac{\sum_{n=1}^N a_s^2(n)}{\sum_{n=1}^N a_r^2(n)}}\Bigg|_f \quad (22)$$

and for the pitch ε_s as

$$T_{v_r,\varepsilon_s}(f) = \sqrt{\frac{\sum_{n=1}^N \varepsilon_s^2(n)}{\sum_{n=1}^N a_r^2(n)}}\Bigg|_f . \quad (23)$$

These characteristics are calculated assuming the vehicle is excited with a series of the road bumps with varying frequency f and the velocity amplitude $A = 0.3\,\text{ms}^{-1}$, invariant over the whole frequency range.

Besides, the algorithms were compared using the following normalized quality indices based on the heave velocity transmissibility characteristics:

$$J_{v_s} = \frac{\sum_{1\,\text{Hz}}^{10\,\text{Hz}} T_{v_r,v_s}(f)\,\Delta f}{\sum_{1\,\text{Hz}}^{10\,\text{Hz}} T_{0,v_r,v_s}(f)\,\Delta f} \quad (24)$$

and on the pitch velocity transmissibility characteristics:

$$J_{\omega_s} = \frac{\sum_{1\,\text{Hz}}^{10\,\text{Hz}} T_{v_r,\omega_s}(f)\,\Delta f}{\sum_{1\,\text{Hz}}^{10\,\text{Hz}} T_{0,v_r,\omega_s}(f)\,\Delta f}, \quad (25)$$

where $T_{0,v_r,v_s}(f)$ and $T_{0,v_r,\omega_s}(f)$ denote the reference heave and pitch velocity transmissibility characteristics evaluated for the soft passive suspension, that is, for control currents equal to zero. Furthermore, the following normalized quality indices based on acceleration transmissibility characteristics were also analysed:

$$J_{a_s} = \frac{\sum_{1\,\text{Hz}}^{10\,\text{Hz}} T_{a_r,a_s}(f)\,\Delta f}{\sum_{1\,\text{Hz}}^{10\,\text{Hz}} T_{0,a_r,a_s}(f)\,\Delta f} \quad (26)$$

and on the pitch velocity transmissibility characteristics:

$$J_{\varepsilon_s} = \frac{\sum_{1\,\text{Hz}}^{10\,\text{Hz}} T_{a_r,\varepsilon_s}(f)\,\Delta f}{\sum_{1\,\text{Hz}}^{10\,\text{Hz}} T_{0,a_r,\varepsilon_s}(f)\,\Delta f}. \quad (27)$$

The quality indices are evaluated based on the heave and pitch transmissibilities interpolated with frequency resolution denoted as $\Delta f = 0.1\,\text{Hz}$.

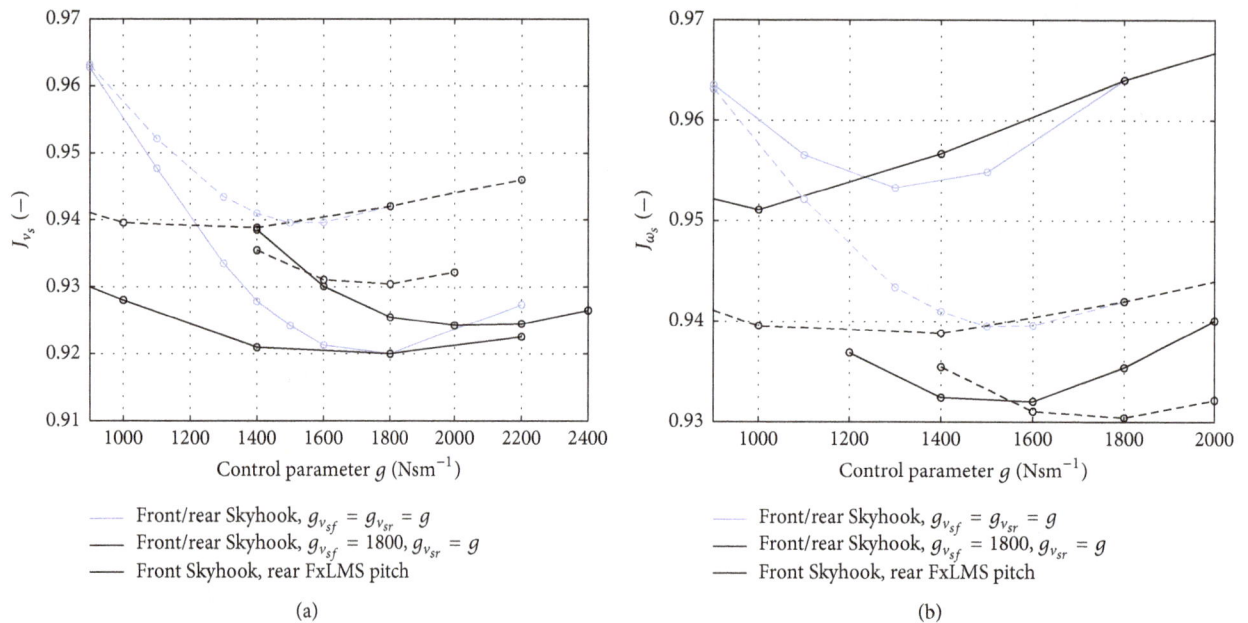

FIGURE 7: Comparison of vibration control quality for the front/rear Skyhook and mixed Skyhook-FxLMS algorithms: (a) the heave index for the front and rear excitation in phase and (b) the pitch index for the front and rear excitation in antiphase. Dashed lines—the average of J_{v_s} and J_{ω_s}.

4.2. Optimization and Validation of Vibration Control Algorithms.

In order to properly indicate advantages of the proposed mixed Skyhook-FxLMS algorithm in comparison to the well-known quarter- and two-dimensional Skyhook control, the latter algorithms need to be initially optimized. The optimization is not a trivial task since the Skyhook control is nonadaptive whereas the vehicle model is highly nonlinear. Thus, the optimal gain in the algorithm was determined by a trial-and-error method, which requires a series of experiments to be performed. The proper value of the gain depends mainly on the suspension parameters. The values corresponding to the damping coefficients presented in Parameters of the Simulated 4 DOFs Half-Car Model in Section 2.1 can be used as a starting point for the search procedure. On the other hand, the gain depends weakly on the amplitude of the excitation or generally on the road class, as it was shown in [34]. However, it was also observed that the choice of the gain is not crucial for the efficiency of vibration mitigation as it varies little over a wide range of gain values (see Figures 7 and 9).

All presented algorithms were compared using vibration control quality indices defined by (24)–(27) and evaluated for the different quarter-Skyhook parameters $g_{v_{s(f/r)}}$ as well as the different two-dimensional Skyhook parameters $g_{v_{s(f/r)}}$ and $g_{\omega_{s(f/r)}}$. Herein, results are presented only for the quality indices which are based on velocity quantities. The control parameters were being adjusted for each algorithm in a wide range in order to find an optimized configuration for each of them. An example of a set of heave velocity transmissibility characteristics is presented in Figure 8, where results were obtained for different values of s_{v_s} parameter of heave Skyhook algorithm which was synthesized focusing on

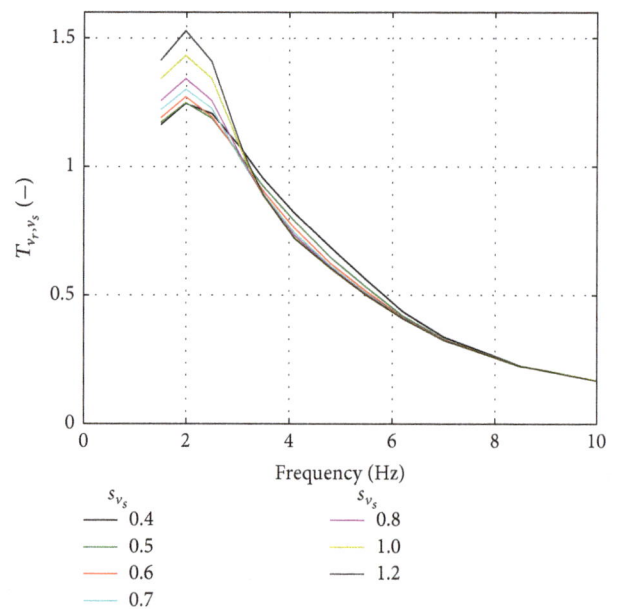

FIGURE 8: Comparison of heave Skyhook algorithm in frequency domain for different values of control parameters and for in-phase road-induced excitation.

mitigation of heave vibration mode. Thus, changing s_{v_s} can lead to heave Skyhook algorithm which is underactuated, optimized or overactuated particularly for the 2 Hz resonant frequency related to the heave mode.

The quarter-Skyhook was validated in two stages with respect to the averaged quality index. Firstly, control gains were assumed to be the same for the front and rear vehicle

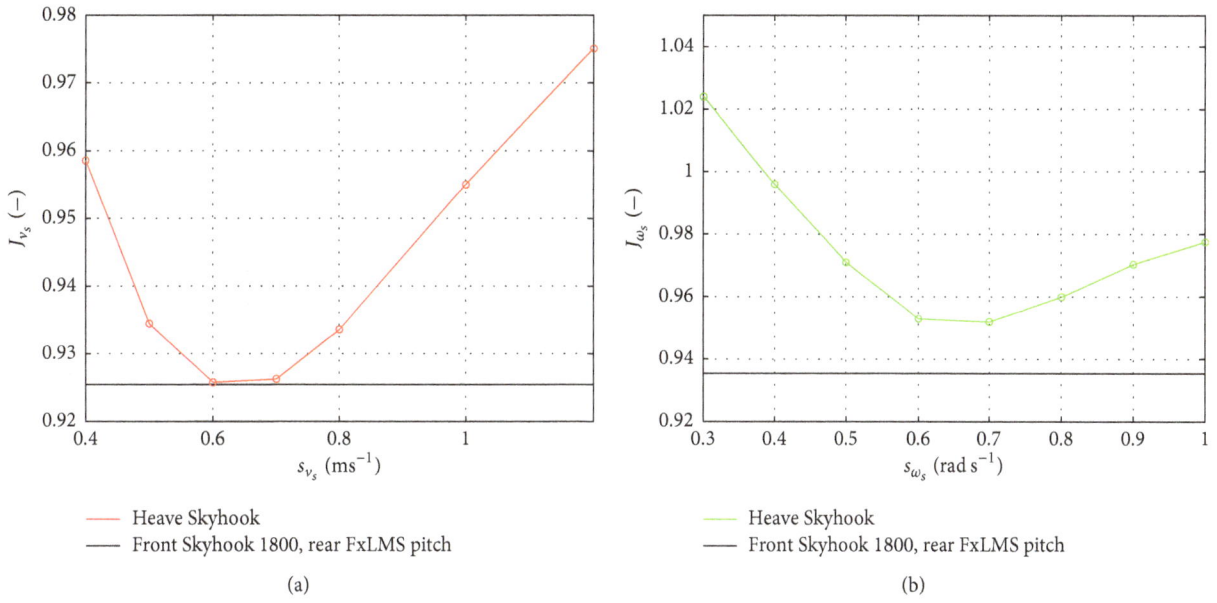

FIGURE 9: Mixed Skyhook/FxLMS versus 2-dimensional Skyhook optimized with respect to (a) heave mitigation or (b) pitch mitigation.

suspension which gave an optimized control quality for $g_{v_{sf}}$ and $g_{v_{sr}}$ equal to 1800. Secondly, neighbourhood of the solution space of such parameters, changing both of them, was examined which gave improved Skyhook control for $g_{v_{sf}}$ = 1800 and $g_{v_{sr}}$ = 1400.

Figure 7 shows a comparison of the quarter-Skyhook and the mixed Skyhook-FxLMS control approaches. Both cases of Skyhook control parameters were presented, that is, the same values of $g_{v_{sf}}$ and $g_{v_{sr}}$ as well as $g_{v_{sf}}$ equal to 1800 and varying $g_{v_{sr}}$. The characteristics were evaluated for the road-induced excitation of the front and rear vehicle in phase and antiphase. It can be seen that vibration control for both algorithms is comparable for the in-phase excitation; however, the difference is clearly visible for the antiphase excitation. Besides, the averaged value of both quality indices was shown for the Skyhook algorithm which confirms that mixed Skyhook-FxLMS is recommended to be used when both heave and pitch vibrations need to be mitigated.

Next, the two-dimensional heave and pitch Skyhook are treated as a reference algorithm separately for heave and pitch vibrations. Optimization results for both algorithms are presented in Figure 9. The heave and pitch Skyhook were optimized assuming that the weights included in the matrix Q, related to the heave or pitch vibration, respectively, are nonzero. Thus, both the heave or pitch Skyhook are fully focused on mitigation of the selected vehicle vibration modes. It can be noticed that, for the in-phase excitation, the proposed mixed algorithm is slightly better than the Skyhook optimized with respect to the heave, and for the antiphase excitation it is significantly better than the Skyhook optimized with respect to the pitch.

Finally, the heave and pitch velocity and acceleration transmissibilities defined by (20)–(23) obtained for

mixed Skyhook/FxLMS and quarter-Skyhook and for two-dimensional Skyhook optimized for either heave or pitch mitigation are presented in Figures 10 and 11. It is easy to notice that optimization of the heave Skyhook leads to deterioration of the pitch mitigation and vice versa. Similar conclusions can be drawn based on either velocity or acceleration transmissibilities. The quarter-Skyhook gives good results for heave vibration; however its efficiency is worse in case of mitigation of pitch vibration. On the contrary, mixed Skyhook-FxLMS performs well for both cases. Moreover, all types of Skyhook control, contrary to FxLMS component, require preliminary optimization, which is not an easy task in the case of real road experiments, as it was shown before.

5. Conclusions

The paper deals with vibration control related to the model of the experimental vehicle equipped with the automotive MR dampers. The analysis is carried out based on the 4-DOFs half-car model, which includes the Bouc-Wen model of the MR dampers. The mixed Skyhook/FxLMS control approach was proposed and compared with quarter-Skyhook and clipped LQ control simplified to the two-dimensional Skyhook algorithm optimized with respect to heave or pitch. Here, the FxLMS algorithm implemented for control of the rear MR damper uses the vertical velocity signal obtained from the front axle as a reference signal.

The quality of vibration control was analysed based on the heave and pitch velocity transmissibility characteristics as well as on the quality indices related to the vibration control of heave and pitch. Initially, all presented types of Skyhook control were numerically optimized to be properly

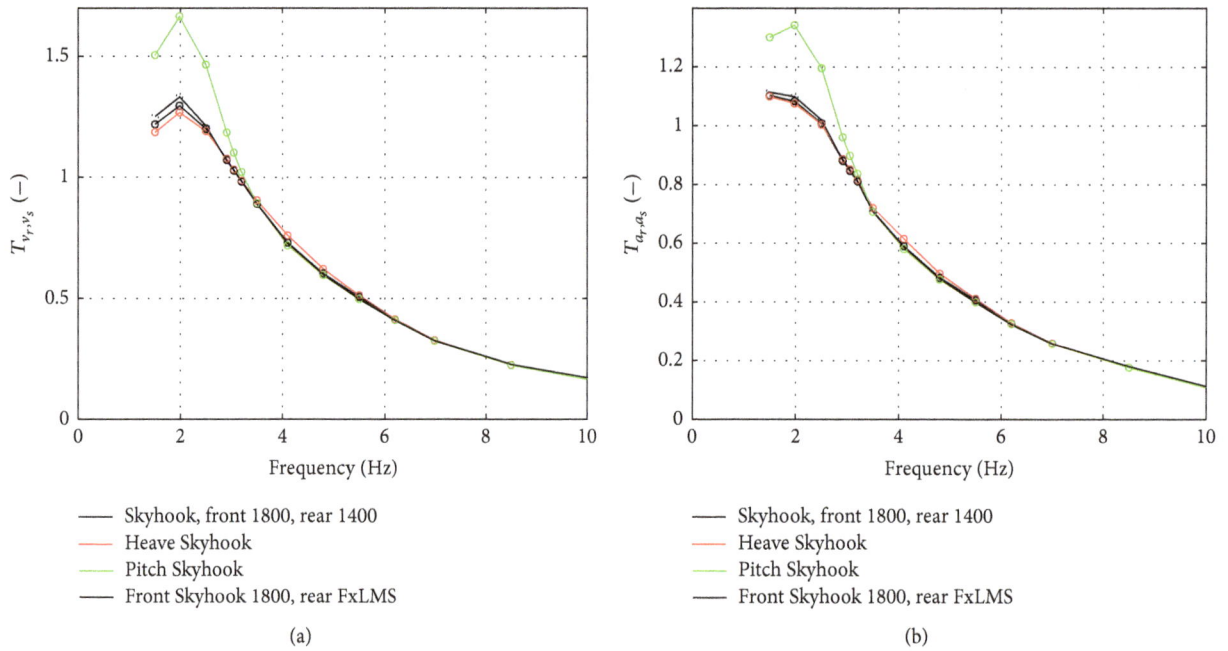

FIGURE 10: Comparison of algorithms in frequency domain for in-phase road-induced excitation based on transmissibilities of (a) heave velocity and (b) heave acceleration.

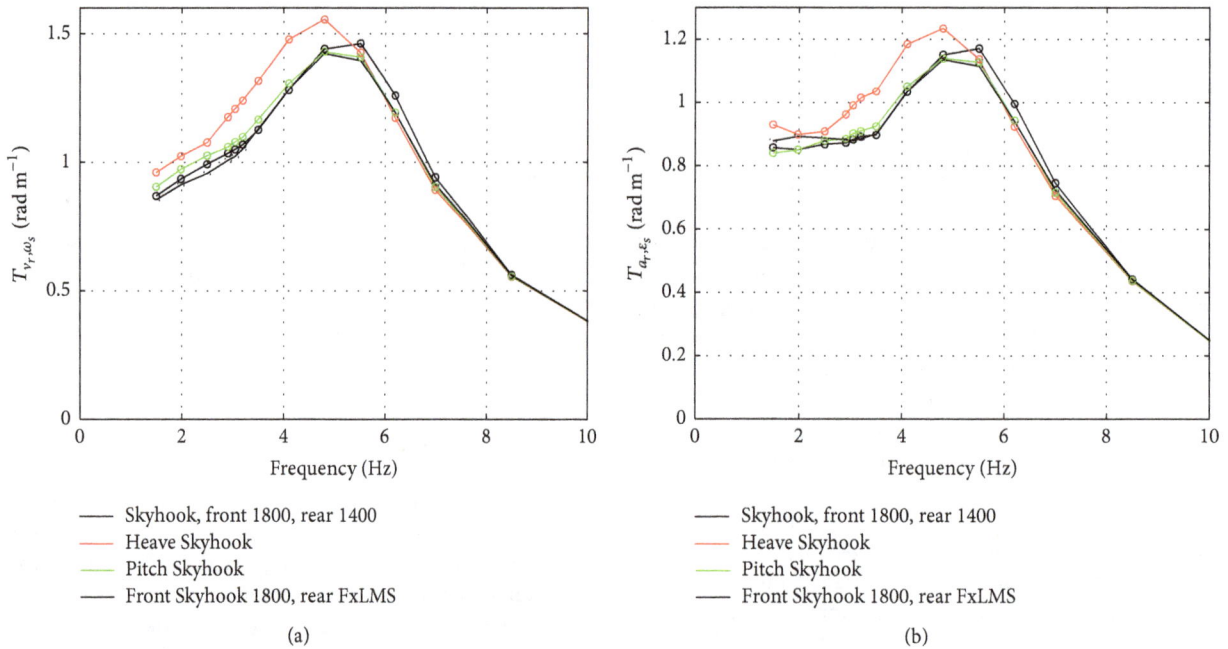

FIGURE 11: Comparison of algorithms in frequency domain for antiphase road-induced excitation based on transmissibilities of (a) pitch velocity and (b) pitch acceleration.

compared with the mixed Skyhook/FxLMS algorithm. Both validation approaches confirmed that the proposed solution should be applied in the case of mitigation of both heave and pitch vibrations. Moreover, this approach is favoured for its adaptability in comparison to the classical Skyhook algorithms.

Competing Interests

The authors confirm that the received funding does not lead to any conflict of interests regarding the publication of this manuscript. Furthermore, the authors declare there is no other possible conflict of interests in the manuscript.

Acknowledgments

The work reported in this paper has been partially financed by the Polish Ministry of Science and Higher Education.

References

[1] D. Karnopp, M. J. Crosby, and R. A. Harwood, "vibration control using semi-active force generators," *Journal of Engineering for Industry, Transactions of ASME*, vol. 96, no. 2, pp. 619–626, 1974.

[2] B. F. Spencer Jr., S. J. Dyke, M. K. Sain, and J. D. Carlson, "Phenomenological model for magnetorheological dampers," *ASCE Journal of Engineering Mechanics*, vol. 123, no. 3, pp. 230–238, 1997.

[3] B. Sapiński, *Magnetorheological Dampers in Vibration Control*, AGH University of Science and Technology Press, Cracow, Poland, 2006.

[4] X. M. Dong, M. Yui, Z. Li, C. Liao, and W. Chen, "A comparison of suitable control methods for full vehicle with four MR dampers, part I: Formulation of control schemes and numerical simulation," *Journal of Intelligent Material Systems and Structures*, vol. 20, no. 7, pp. 771–786, 2009.

[5] P. Krauze, "Comparison of control strategies in a semi-active suspension system of the experimental ATV," *Journal of Low Frequency Noise, Vibration and Active Control*, vol. 32, no. 1-2, pp. 67–80, 2013.

[6] S. Guo, S. Li, and S. Yang, "Improvement of measurement results based on scattered data in cases where averaging is ineffective," in *Proceedings of the IEEE International Conference on Vehicular Electronics and Safety (ICVES '06)*, vol. 13, no. 10, pp. 403–406, Beijing, China, December 2006.

[7] I. Maciejewski, "Modelling and control of semi-active seat suspension with magnetorheological damper," in *Proceedings of the 24th Symposium on Vibrations in Physical Systems*, vol. 13, no. 10, pp. 1–6, Poznan-Bedlewo, Poland, May 2010.

[8] K. Plaza, *Modelling and control for semi-active vibration damping [Ph.D. dissertation]*, Silesian University of Technology, Gliwice, Poland, 2008.

[9] B. Sapiński, *Real-Time Control of Magnetorheological Dampers in Mechanical Systems*, AGH University of Science and Technology Press, Cracow, Poland, 2008.

[10] J.-P. Hyvarinen, *The improvement of full vehicle semi-active suspension through kinematical model [Ph.D. thesis]*, University of Oulu, Oulu, Finnland, 2004.

[11] Y. Cho, B. S. Song, and K. Yi, "Improvement of measurement results based on scattered data in cases where averaging is ineffective," *KSME International Journal*, vol. 13, no. 10, pp. 667–676, 1999.

[12] A. Hać and I. Youn, "Optimal semi-active suspension with preview based on a quarter car model," *Journal of Vibration and Acoustics*, vol. 114, no. 1, pp. 84–92, 1992.

[13] A. Hać and I. Youn, "Optimal design of active and semi-active suspensions including time delays and preview," *Journal of Vibration and Acoustics*, vol. 115, no. 4, pp. 498–508, 1993.

[14] J. Marzbanrad, G. Ahmadi, H. Zohoor, and Y. Hojjat, "Stochastic optimal preview control of a vehicle suspension," *Journal of Sound and Vibration*, vol. 275, no. 3–5, pp. 973–990, 2004.

[15] R. S. Prabakar, C. Sujatha, and S. Narayanan, "Optimal semi-active preview control response of a half car vehicle model with magnetorheological damper," *Journal of Sound and Vibration*, vol. 326, no. 3–5, pp. 400–420, 2009.

[16] P. Krauze and J. Kasprzyk, "Vibration control in quarter-car model with magnetorheological dampers using FxLMS algorithm with preview," in *Proceedings of the 13th European Control Conference (ECC '14)*, pp. 1005–1010, Strasbourg, France, June 2014.

[17] P. Krauze and J. Kasprzyk, "FxLMS algorithm with preview for vibration control of a half-car model with magnetorheological dampers," in *Proceedings of the IEEE/ASME International Conference on Advanced Intelligent Mechatronics (AIM '14)*, pp. 518–523, Besancon, France, July 2014.

[18] S. Budzan and J. Kasprzyk, "Fusion of 3D laser scanner and depth images for obstacle recognition in mobile applications," *Optics and Lasers in Engineering*, vol. 77, pp. 230–240, 2016.

[19] J. Wiora, "Improvement of measurement results based on scattered data in cases where averaging is ineffective," *Sensors and Actuators, B: Chemical*, vol. 201, pp. 475–481, 2014.

[20] H. A. Asl and G. Rideout, "Using lead vehicle response to generate preview functions for active suspension of convoy vehicles," in *Proceedings of the American Control Conference (ACC '10)*, vol. 8671 of *Lecture Notes in Computer Science*, pp. 4594–4600, Baltimore, Md, USA, July 2010.

[21] P. Krauze and J. Kasprzyk, "Vibration control for an experimental off-road vehicle using magnetorheological dampers," in *Proceedings of the International Conference Vibroengineering*, pp. 306–312, Katowice, Poland, October 2014.

[22] D. Hrovat, "Survey of advanced suspension developments and related optimal control applications," *Automatica*, vol. 33, no. 10, pp. 1781–1817, 1997.

[23] X. Song, M. Ahmadian, and S. C. Southward, "Modeling magnetorheological dampers with application of nonparametric approach," *Journal of Intelligent Material Systems and Structures*, vol. 16, no. 5, pp. 421–432, 2005.

[24] X. Q. Ma, S. Rakheja, and C.-Y. Su, "Development and relative assessments of models for characterizing the current dependent hysteresis properties of magnetorheological fluid dampers," *Journal of Intelligent Material Systems and Structures*, vol. 18, no. 5, pp. 487–502, 2007.

[25] B. Sapiński, "Parametric identification of mr linear automotive size damper," *Journal of Theoretical and Applied Mechanics*, vol. 40, no. 3, pp. 703–722, 2002.

[26] E. R. Wang, X. Q. Ma, S. Rakhela, and C. Y. Su, "Modelling the hysteretic characteristics of a magnetorheological fluid damper," *Proceedings of the Institution of Mechanical Engineers, Part D: Journal of Automobile Engineering*, vol. 217, no. 7, pp. 537–550, 2003.

[27] E. R. Wang, X. Q. Ma, S. Rakheja, and C. Y. Su, "Force tracking control of vehicle vibration with mr-dampers," in *Proceedings of the 2005 IEEE International Symposium on Intelligent Control*, pp. 995–1000, Limassol, Cyprus, June 2005.

[28] G. Yang, B. F. Spencer Jr., J. D. Carlson, and M. K. Sain, "Large-scale MR fluid dampers: modeling and dynamic performance considerations," *Engineering Structures*, vol. 24, no. 3, pp. 309–323, 2002.

[29] J.-H. Koo, F. D. Goncalves, and M. Ahmadian, "A comprehensive analysis of the response time of MR dampers," *Smart Materials and Structures*, vol. 15, no. 2, pp. 351–358, 2006.

[30] J. Kasprzyk, J. Wyrwał, and P. Krauze, "Automotive MR damper modeling for semi-active vibration control," in *Proceedings of the IEEE/ASME International Conference on Advanced Intelligent Mechatronics (AIM '14)*, pp. 500–505, Besancon, France, July 2014.

[31] M. Sibielak, W. Rączka, and J. Konieczny, "Modified clipped-LQR method for semi-active vibration reduction systems with hysteresis," *Journal of Solid State Phenomena*, vol. 177, pp. 10–22, 2011.

[32] A. E. J. Bryson and Y.-C. Ho, *Applied Optimal Control: Optimization, Estimation and Control*, Hemisphere Publishing Corporation, 1975.

[33] S. M. Kuo and D. R. Morgan, *Active Noise Control Systems: Algorithms and DSP Implementations*, Wiley Series in Telecommunication and Signal Processing, Wiley-Interscience, 1996.

[34] P. Krauze, *Control of semiactive vehicle suspension system using magnetorheological dampers [Ph.D. dissertation]*, Silesian University of Technology, Gliwice, Poland, 2015.

10

On the Development of Focused Ultrasound Liquid Atomizers

Ahmed M. Al-Jumaily and Ata Meshkinzar

Institute of Biomedical Technologies, Auckland University of Technology, Private Bag 92006, Auckland 1142, New Zealand

Correspondence should be addressed to Ahmed M. Al-Jumaily; ahmed.al-jumaily@aut.ac.nz

Academic Editor: Marc Asselineau

This paper reviews the evolution of focused ultrasonic transducers of various kinds for fluid atomization and vaporization. Ultrasonic transducers used for atomization purposes in biomedical, pharmaceutical, or industrial applications, such as surface acoustic wave (SAW) transducers, array of micromachined nozzles, and Fourier horn micromachined nozzles with or without a central channel, are all presented and compared. For simplicity of manufacturing and low cost, we focus on plates and curved and corrugated structures for biomedical humidification.

1. Introduction

Ultrasonic transducers have been receiving a great deal of attention during the past few decades because of their wide variety of applications as sensors or actuators in flaw detection, thickness gaging, materials research and medical diagnostics, and sonar. In many biomedical, pharmaceutical, and industrial applications, ultrasonic transducers can be used as atomizers. Humidifiers, micro/nanoelectronics, nanoparticles synthesis, spray coating, drug delivery, drug preparation for inhalation, and others are among the most common applications of atomization for which ultrasonic transducers can play an important role and can be extremely beneficial. Droplets <10 μm in diameter are highly desirable for various medical purposes. For pulmonary microcirculation-related applications, the droplets must be smaller than 7 μm in diameter to safely pass through the microvessels of the lung without causing obstruction [1]. For delivering drugs to the respiratory system, droplets of 3 to 5 μm are ideal depending on the disease and its site [2]. For delivery of medications to the alveolar capillary bed, 1–3 μm droplets (optimal at 2 μm) are ideal [2, 3]. However, delivery for humidification, such as in lung therapy, requires humidified air where water particles are much smaller than 1 μm.

Inhalation is an attractive route for noninvasive delivery of drugs [4–7] especially peptides and proteins that are easily broken down by enzymes in the stomach when taken orally [4, 6]. The wide variety of potential applications of ultrasonic transducers justifies their in-depth analysis and investigation. The shape and geometry of the transducer as well as its modes of vibration play the most significant roles in its performance. Various types of transducer geometries and shapes such as flat, curved and corrugated plates, and cylindrical and spherical shells have been implemented for ultrasound generation. The driving mechanism for this generation is the vibration characteristics of the transducer. To quantify the power and strength of the signal, it is essential to understand those characteristics. Analytical, numerical, and experimental investigations have been conducted on various transducers and their vibration characteristics were examined in the existing literature. Different approaches were followed by authors and various designs were devised for the transducers making them applicable to suit specific requirements.

Atomization is an important application of ultrasonic transducers, in particular in biomedical applications. Therefore, some researchers have focused on the mechanism of the atomization itself which is definitely of great help and importance to further improve the performance of the transducers. Two mechanisms were considered by various researchers for atomization, breaking up of capillary waves at the liquid surface and cavitation. However, further research showed that a combination of these two mechanisms is the source of atomization [8, 9]. Most of these studies focus on the performance of such devices for atomization rather than evaporation. However, some investigations have dealt with methods to evaporate water droplets. These methods include resonant excitation and disintegration in air, acoustic squeezing and disintegration, and finally solid surface excitation. Droplets of water can be acoustically levitated by means of an acoustic levitator resulting in the oscillation and disintegration of the water droplets [10, 11]. A similar approach was attempted to acoustically squeeze a droplet to the point where it flattens out and self-disintegrates [12, 13]. The interaction of water droplets with solid surfaces vibrating at ultrasonic frequencies was also investigated [14–16]. Current commercial biomedical devices (Misty-Neb, Aero Eclipse, Omron, Pari eFlow, and Philips I-neb) producing droplets or aerosols by compressed air, a vibrating piezoelectric plate together with a metallic mesh or a vibrating mesh, suffer from broad droplet size distributions and low throughput, which makes it difficult to deliver sufficient dosages of drugs precisely and rapidly to the targeted sites [17, 18]. Furthermore, the eFlow and I-neb which utilize vibrating mesh technology [19] and are considered the most advanced commercial devices suffer from clogging of the mesh orifices [20]. Heating can also be used to atomize and produce droplets. However, ultrasonic atomization provides significant energy and space savings compared to conventional heated humidifiers [21]. Therefore, further improvement of available devices or devising new compact atomizers capable of producing water vapor rather than water droplets is of practical importance, in particular in medical therapy devices such as those used for lung therapy. It is worth noting that an inseparable part to further improve existing vaporizing devices and technologies for atomization is to thoroughly investigate and flourish the vibration characteristics of the transducers. Consequently, an important brief but essential review will also be presented in Section 2 regarding flat plate and shell type transducers followed by another section on atomization methods.

Since the information on the area is widespread and the research field is of great significance, the authors decided to merge many of such topics into a single document which will serve as a helpful resource and guideline for many of those working in the field.

2. Ultrasound Generation

2.1. Vibrating Plate Radiators. Achieving ultrasonic frequencies for gases utilizing conventional ultrasonic transducers has been difficult with low efficiency. Further, electromagnetic, magnetostrictive, and piezoelectric ultrasound sources seem incapable of generating high power ultrasounds with good efficiency and directivity as they cannot undergo large longitudinal vibration amplitude. The problems lying behind ultrasound generation in gases are attributed to low specific acoustic impedance and high absorption of the medium. Therefore, in order to have an efficient energy transmission, good impedance matching between the transducer and gas medium, large amplitudes of vibration and highly directional radiation are needed [22]. Vibrating plates can be used as ultrasonic radiators in fluids; however, the directivity of the generated acoustic waves is a matter of great importance.

The radiations from different parts of a simple flat plate with constant thickness are in counter phase leading to phase cancellation and, therefore, poor directivity. On the contrary, if one considers a plate with some steps on the surface raised half a wavelength of the radiated sound, the radiations become in-phase and directivity increases. Several experimental and analytical approaches have been attempted to determine the improvements achieved in delivering stronger ultrasound using various plate geometries and shapes. Circular aluminum plate transducers with steps were designed and their vibration characteristics were investigated experimentally and analytically. Using uniform plate solution and the energy method, an analytical solution was obtained for the frequency equation by matching boundary conditions at the step interface and free edge condition at the external boundaries [23, 24]. Good agreement was reported between the experimental and the analytical approaches and better directivity. The structure of the transducer and the radiating plate consists of three concentric copper rings with small holes to spray water jets to the nodal circles to cool them. High intensity ultrasound was reported to have been generated in gases with high efficiency (approx. 80%) and directivity. Analytical vibration analysis of stepped plates without any restriction on the number of steps or on the order of the vibration mode, unlike the earlier works in the literature which were restricted to one step and/or first vibration mode, was conducted by San Emeterio et al. [25]. The analytical results agreed well with those of experiments in thick areas of the plate, while discrepancies were observed in thin regions. Therefore, in their analytical model for the deformation function, they proposed a correction factor taking the effect of the base plate and step thicknesses ratio into account and obtained good agreement between their analytical and experimental results. The influence of Poisson's ratio on the natural frequencies of stepped-thickness circular plate has been also dealt with [26]. In that paper, Al-Jumaily and Jameel determined the natural frequencies of the simply supported and clamped stepped plates using classical plate solutions with exact continuity conditions at the step. They concluded that larger Poisson's ratio indicates stiffer plates and larger natural frequencies. Also as the step size increases, the natural frequencies are expected to be larger. They also stated that Poisson's ratio should not be disregarded in the continuity equation in particular for the fundamental frequency.

To have directional underwater radiation, two design methods were introduced by Montero de Espinosa and Gallego-Juárez [27]. In one method, they implemented helical waveguides and in the other one, a delaying liquid was used with the same acoustic impedance as water, but half of the sound velocity as in water. This way, they made in-phase the regions of the plate which had been previously radiating in counter phase resulting in good directivity for the frequency range of 20–27 kHz.

High power generators are needed to make use of sonic and ultrasonic energy in industrial applications. However, their use has been restricted by the inadequacy of such generators. To this aim, Gallego-Juárez et al. worked on the design and development of macrosonic generators for industrial applications taking into account good impedance matching between the transducer and the medium for efficient transmission of energy, high directional or focused radiation for energy concentration and high amplitude of the operating mode for intense acoustic radiation, fatigue failure of the material, the distribution and location of the stresses, and the isolation of the operating vibration mode from other close nontuned modes [28]. Finally, double-stepped rectangular titanium plates were devised achieving high power and good performance in air (7.6 kHz with the applied power of around 2 kW had the efficiency of 67% and dimensions of 1.8 × 0.9 m²). Gallego-Juárez et al. have conducted further research to come up with a solution for large industrial applications of high intensity transducers for air where numerous practical problems are encountered due to the required high-order vibration modes [29]. Another solution was proposed for low sonic applications where stepped-plate transducers seem unsuitable as the height of the steps, which has to be half a wavelength of the radiation, becomes too high and it makes the transducer construction impractical. For the industrial case, they used an array of five circular stepped-plate transducers (each plate 48 cm diameter, 21 kHz, applied power 350 W and 75% efficiency). For the case of low frequency applications in air, a plate transducer with reflectors was used making the radiation in-phase leading to high directivity.

The above literature reveals that for many applications from low sonic transducers to ultrasonic radiation in gas or fluid medium including underwater sonar, manipulations on the geometry of flat plates of different kinds, that is, rectangular or circular, led into a solution suited for that specific application. The common theme for all the cases is that the radiation was made focused and in-phase by geometrical manipulations. Therefore, similar ideas can be employed for smaller biomedical humidifier applications.

2.2. Shell Transducers. In order to gain a better understanding of an ultrasonic transducer, it is essential to investigate its vibration characteristics for various design configurations. In order to increase the focus and directivity, curved structures can be a suitable candidate. Therefore, for airborne ultrasonic ranging measurement, a partially cylindrical (curved) PVDF transducer with silver electrode on both sides was reported to effectively couple ultrasound into the air and generate strong sound pressure [30]. The transducer was observed to have two resonances as length extensional mode and flexural bending mode. Transient surface displacement measurements revealed that vibration peaks were in-phase for the length extensional mode and out of phase for the flexural bending mode generating a stronger ultrasound wave for the length extensional mode. The resonance frequencies and vibration amplitudes of the two modes strongly depend on the structure parameters as well as the material properties. Controlling the thickness of electrode is important as well since it affects the resonance characteristic of the transducer. The resonance frequencies of the two modes should be separated as far as possible from each other to minimize the influence of bending vibration. The effect of variable curvature on the transducer performance was not known. Therefore, Toda and Tosima investigated the vibration modes of a curved, clamped, piezoelectric multilayer film with uniform and

nonuniform film curvature and it was concluded that the nonuniform curvature generates much higher output pressure in air than a uniform one [31]. They did not mention the highest achievable sound pressure level in decibel using their proposed piezoelectric multilayer film. However, according to Figures 9 and 10 in their work where the output acoustic pressure is depicted for various frequencies, the highest sound pressure is approximately 33 Pa which corresponds to 124 dB using the unanimous formula of sound pressure level (dB) = 20 log p + 94, where p is the sound pressure in Pascal.

Since the curvature proved to be influential in increasing the focus and improving the directivity, the next attempt was to implement a row of curvatures placed next to each other forming a corrugated structure. A corrugated PVDF film air transducer was scrutinized to achieve a high power output and a sharp beam angle [32]. The vibration phase of convex section is shifted 180 degrees from the concave section. These waves add constructively to form a strong acoustic beam when the corrugation height is a little larger than one-half of the wavelength since the vibration is distributed over all of the points on the film. The vibration characteristics of axially symmetrical annular corrugated shell piezoelectric transducers were also analyzed [33–35]; however, the motion for such a transducer with fixed edges was reported to be complicated and therefore a global analysis was necessary in order to increase the efficiency of such transducers in operating modes. Blum et al. devised and investigated a two-dimensional, air-coupled array for the noncontact generation of ultrasound [36]. The overall design objective was to position 20 electrostatic transducers in such a way that the signal amplitude at the focal line of the array gets maximized. Then, to identify the most critical parameters affecting the predicted behavior of the array, a sensitivity study was performed and it was concluded that changes in the spatial position of the transmitting transducers have a noticeable effect on the signal at the focal line. The sound pressure level of 142.70 dB SPL was achieved in air. Measurements showed that the amplitude of the ultrasonic waves generated with the air-coupled array is in the order of five times lower than that of a piezoelectric contact transducer.

In therapy devices, in particular those used for lungs, we always look for pure water vapor, not droplets, in the air. In spite of the fact that all of the above devices have shown some ultrasound improvement in directivity and power generation, nevertheless none of them are suited to deliver this objective. However, with the aim of producing an ultrasonic field for drying foodstuff, an aluminum cylindrical chamber was designed which was driven by a piezoelectric transducer [37]. A high intensity acoustic field was obtained inside the tube (155 dB of SPL) with relatively low industrial power applied (75 W). Although this is the only paper which shows a complete water evaporation process, nonetheless it requires very high power; it is large in size and not suited for biomedical humidification such as in lung therapy. In fact, for therapy humidification in particular for the lung, a smaller size transducer with less power and a shape which fits in the air delivery line would be impeccable.

This section clarifies the effect of various shell type geometries on the directivity and intensity of the acoustic field generated. Although none of the references dealt with the atomization or direct biomedical application, the results seem promising since geometrical manipulations on shell type structures created focused radiation. Thus, such an idea can be converted and well suited to the specific requirements of any biomedical application as atomization, humidification, drug delivery, and others.

3. Ultrasonic Transducers as Atomizers

Ultrasonic transducers have been extensively used for droplet generation and atomization and, as already mentioned, their application varies from ultrasonic humidification, micro/nanoelectronics, nanoparticles synthesis, spray coating to drug delivery and drug preparation for inhalation. In this section, their application for atomization purposes is summarized. In the end, a comparison among the available existing methods will be performed in the form of a table to discuss and illustrate the advantages and disadvantages of each method.

Lass et al. presented a paper on the current vibrating membrane nebulizer technology for drug delivery and dealt with current devices in the market which can be of interest [19]. Compressed air and vibrating piezoelectric plates together with a metallic mesh or a vibrating mesh are common technologies for producing aerosols or droplets in commercial devices (Misty-Neb, AeroEclipse, Omron, Pari eFlow, and Philips I-neb). A disadvantage is their broad droplet size distributions and low throughput making it difficult to deliver sufficient dosages of drugs precisely and rapidly to the targeted sites [17, 18]. In addition, blockage of mesh orifices in the vibrating mesh membrane of eFlow and I-neb which are considered to be the most advanced commercial devices is a pitfall [20]. Recent in vivo studies have indicated that in both adults and children, when inhaling typical aerosols from current commercial devices, the upper airways [48], ventilator, and endotracheal tubes are significant barriers to lung deposition [49]. As a direct result of poly disperse droplet size distributions, drugs are delivered to nontargeted sites, resulting in harmful side effects in the pharynx and losses in the ventilator/endotracheal tubes. Therefore, investigation and development of ultrasonic transducers as atomizers in order to improve their performance are a matter of great importance.

In 1988, Elrod et al. worked on droplet formation using tone bursts of focused acoustic energy [50]. In order to generate a spherically converging acoustic beam, a focusing element (acoustic microscope lens) was used without any nozzle. The liquid surface was adjusted to be at the focal plane, where the beam was concentrated. Droplets of 300 to 5 μm were generated within the frequency range of 5 to 300 MHz.

3.1. Vibrating Plate Atomizers. Vibrating plate transducers are commonly being used for humidifiers and other applications since they possess the best combination of performing specifications compared to other types of devices. In these atomizers, an electrical signal is converted to mechanical oscillation using a piezoelectric material immersed in a reservoir of water. The ultrasonic waves created by the mechanical vibration of the plate are directed towards the water surface creating a mist of water droplets. These atomizers are normally available in a high frequency range of 1.65–3 MHz which are capable of producing droplets within the range of 1–5 μm in diameter using 2 up to 30 W of power. The flow rate varies within the range of 5–400 mL/h. They are in contact with water and the water level on their top affects their performance and should be taken into account for the optimum performance of the device. They are available in various plate area dimensions and thicknesses. Examples of such atomizers can be referred to in references [44–47]. Their flow rate also depends on many external factors such as input power, droplet size, frequency and liquid quality, temperature, and level (depth). The good point about them is that they can produce various sizes of droplets and flow rates which makes them suitable for some applications. As already mentioned, flat plates do not produce focused ultrasonic waves. Therefore, curved transducers have been used in some of the aforementioned atomizers providing a little bit of better focus and performance compared to the flat ones as evident in [44–47]. However, depending on the application, a combination of high flow rate, small droplet size, and low power demand may be required which has not been achieved yet. Therefore, seeking alternative ways to create more focused ultrasonic waves seems essential and can lead to the better performance of the device. Going to higher frequencies to obtain more focused waves resulting in better performance of the device can lead to overheating and depolarization of the material [51] and also fatigue failure [28]. Therefore, it is a restricting factor and alternative ways should be sought. One alternative is using stepped-thickness plates. Although, they have been investigated and proven to be useful and practical as discussed elaborately in Section 2.1, to the authors' knowledge they have not been implemented in the current vibrating plate atomizers available in the market. To overcome overheating, Lozano et al. designed an electronic system to excite piezo-ceramic disks for ultrasonic atomization [51]. It was designed based on a DC-AC converter with H-bridge topology switching ZVS (zero voltage switching) mode optimized for R-L-C type loads. In order to reduce ceramic overheating, the system operated in burst mode sending a predetermined number of pulses. There was no sign of atomization for voltages below 10 V. Excitation bursts of 80,000 pulses at a repetition rate of 1 kHz were applied to the transducers. No atomization was initiated at bursts formed by less than 10,000 pulses and only a small proportion of the power consumption led to mechanical vibration, while a large part was dissipated as heat. The resulting droplet size distribution showed two main peaks at 3.5 and 5.5 μm. Droplet diameter was observed to be independent of the excitation amplitude (voltage), while the atomization rate increased with voltage.

3.2. Surface Acoustic Wave Atomizers. Kurosawa et al. proposed a novel way to produce dry fog using a surface acoustic wave (SAW) transducer of LiNbO$_3$ piezoelectric substrate as in Figure 1 [38, 39]. The atomizer consists of a vibrator which has an interdigital aluminum transducer (IDT) consisting of 20 pairs of electrodes supplied with RF (radio frequency) power amplifier at 48 MHz frequency. The surface wave called

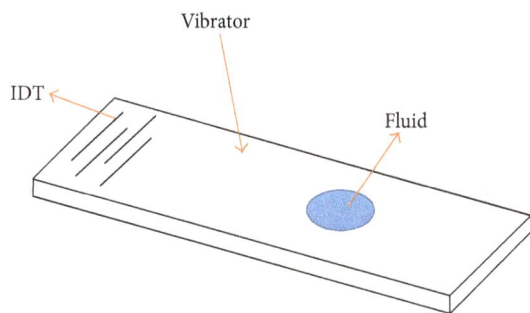

FIGURE 1: Surface acoustic wave atomizer.

capillary wave generated by the radiation of acoustic wave from the SAW device surface was capable of atomization when the liquid layer on the surface was half a millimeter or less. The atomizing mechanism was reported not to be vapor but spray from crests of the surface wave in a fluid. The mean diameter of the mist was about 5 μm. The atomizing rate was 170 μL/min at 2.3 W input power (36 V). Applying 100 MHz was reported to be capable of reducing the size of the device to an extent suiting it for medical applications such as spraying a liquid medicine to the diseased target directly with the atomizer on an endoscope.

A low driving power SAW atomizer was developed consisting of a unidirectional interdigital transducer, a horn, and a waveguide fabricated on a LiNbO$_3$ substrate to increase the focus [40]. For continuous atomization of water to get droplets of 1.5 μm and 40 μL/min flow rate, 1 W of driving power was required at a frequency of about 78 MHz. Low driving power and small droplets make the atomizer attractive for practical applications in chemistry, biology, and medicine.

Ju et al. constructed and tested high frequency SAW atomizers, with resonance frequencies in the range 50–95 MHz [52]. A liquid sample was charged by a high-voltage (about 5 kV) conductive wire located just above the SAW device. Mean diameters of original droplets formed by atomization were estimated to be 5.7 (50 MHz), 4.4 (75 MHz), and 2.7 μm (95 MHz), respectively. Based on the test, the minimum power required for atomization was approximately 4 W (50 MHz), 11 W (75 MHz), and 24 W (95 MHz) and atomization speed was 0.06 μL/s (50 MHz), 0.04 μL/s (75 MHz), and 0.01 μL/s (95 MHz). The high-voltage needed made the use of such atomizer restricted.

3.3. Ultrasonic Horns. High ultimate particle velocity, low acoustic loss, and high thermal conductivity make silicon ideal for high amplitude ultrasonic applications superior to titanium alloys [53]. Therefore, high amplitude ultrasound was generated by a micromachined silicon resonant transducer to atomize liquids. Lead-zirconate-titanate (PZT) plates were used to drive the needle-shaped device which was made by bonding two silicon horns. Water was atomized continuously at a flow rate of 2.4 mL/min into a mist with mean drop size of 25 μm at 72 kHz resonance frequency.

In order to develop a portable device for internal body therapeutic applications, Li et al. devised a 421 kHz miniaturized piezoelectric ultrasonic transducer of one-wavelength design based on the longitudinal vibration mode with a stepped horn to focus the energy [54]. The horn effect was to magnify the vibration amplitude noticeably. The PZT material selected was PZT8 which is widely used for high power applications and brass electrodes were used. Finite element analysis and equivalent circuit models were applied for theoretical analysis. The device was immersed by a quarter of the wavelength and the directivity pattern was omnidirectional within the range from 30 to 150 degrees. The potential applications of the transducers are sonodynamic therapy, drug delivery, and microfluidic pumping. It was also noted that the efficiency decreases as the voltage increases (input power increases more rapidly compared to output power; however, the intensity increases monotonically with increasing voltage). The possible reason was mentioned to be increasing losses at the high-voltage range.

Tsai and others investigated, built, and tested micro-electro-mechanical system- (MEMS-) based miniaturized silicon ultrasonic droplet generators of a new and simple nozzle architecture with multiple-Fourier horns in resonance with and without a central channel [17, 18, 43, 55, 56]. A Schematic is depicted in Figure 2. The silicon resonator is made of a multiple-Fourier horn section where each horn is half-wavelength long with a longitudinal vibration amplitude magnification of 2. The drive section includes a piezoelectric plate transducer bonded to the rectangular silicon base using silver paste. When PZT transducer plates are excited at the nozzle resonance frequency, a standing acoustic wave is created through the nozzle with maximum longitudinal vibration (displacement) at the nozzle tip. As a result of the vibration, standing capillary waves are formed on the free surface of the liquid as it issues from the nozzle tip. Atomization occurs by the breakup of these standing capillary waves. It is worth noting that when the ultrasonic drive frequency deviated from the nozzle resonance frequency by more than 1.5 kHz, a large liquid drop with a diameter greater than the tip width was reported to be formed at the nozzle tip without any atomization taking place.

Tsai et al. produced monodisperse ethanol droplets of 2.4 μm and water droplets of 4.5 μm in diameter in ultrasonic atomization using 1.5 and 1.0 MHz MEMS-based silicon nozzles, respectively, each consisting of 3 Fourier horns in resonance (with a central channel), required electrical drive power as low as 0.25 W and supply flow rates as high as 350 μL/min (21 mL/h) [43]. At the resonance frequency, the measured longitudinal vibration amplitude at the nozzle tip increases as the number of Fourier horns (n) increases in good agreement with the theoretical values of 2^n. Using this design allows for very high vibration amplitude gain at the nozzle tip resulting in no reduction in the tip cross-sectional area for contact of liquid to be atomized. This leads to a noticeable reduction in the electric drive power which in turn decreases the possibility of transducer failure during atomization. They also extended their work to externally liquid fed ultrasonic nozzles without a central channel [17]. Droplets with a diameter range 2.2–4.6 μm for alcohol (2.9–4.6 for

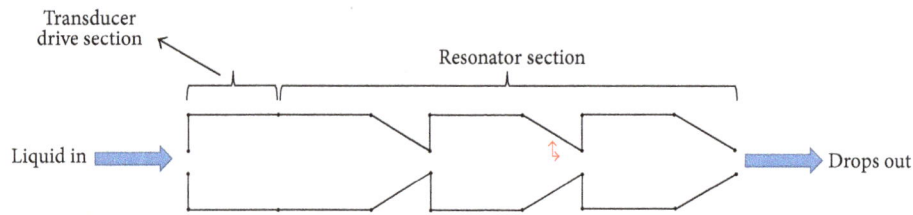

FIGURE 2: Schematic of a multiple-Fourier horn ultrasonic nozzle.

water) were produced at high throughput of 420 μL/min and very low electrical drive power of 80 mW. The electrical loss for the lossy PZT transducer was reported to increase with its thickness. Moreover, the nozzles with a central channel require higher drive power for atomization which is due to the fact that they need a pair of PZT transducers. The range of drive power measured was extremely lower than that required in conventional ultrasonic atomization using MHz disk transducers which was attributable to the new nozzle architecture requiring only a single basic nozzle without a central channel and, thus, a single PZT transducer for activation. The optimum number of Fourier horns was found to be 3 or 4 with electrical drive power significantly below 0.1 W at throughput of 100 μL/min, while the 2-Fourier horn nozzle required a drive power of 0.6 W to initiate atomization. In contrast, no atomization could be initiated with the 1-Fourier horn nozzle even at a drive power as high as 0.8 W. Batch fabrication of nozzles with similar or different design specifications on a common silicon wafer reduces the costs according to the authors. Utilizing the same approach as above, Tsai et al. designed and built a small nozzle requiring low drive power with a pocket-size ultrasonic nebulizer [18]. Various pulmonary drugs were nebulized utilizing the pocket-size unit with different aerosol sizes and output rates. The obtained results demonstrated the capability of the silicon-based MHz MFHUNs (multiple-Fourier horn ultrasonic nozzle) for production of monodisperse droplets of desirable size (2 to 5 μm) and moderate output (up to 0.2 mL/min) at low electrical drive power (sub-Watt). At the typical output rate of 0.15 mL/min with 3.5 μm diameter droplets, the required electrical drive power was 0.27 W. It was reported that a higher output rate can be accomplished readily by using an array of identical ultrasonic nozzles. With 2 MHz of frequency, droplets of 3.1 μm were produced and the output was 350 μL/min.

3.4. Active and Passive Mesh Membranes. Mesh membranes have also been used for atomization purposes. In some of the cases, the mesh membrane vibrates itself and expels the liquid out of the mesh orifices, whereas in some other cases the liquid is between the mesh membrane and another membrane which is vibrating [19]. The first is called active and the latter is called passive mesh membrane. A new type of piezoelectric array microjet was introduced for drug delivery consisting of a fluid chamber, which was formed by a piezoelectric actuator bonded to a silicon chip with nozzles (passive mesh membrane) [41, 42]. Droplets of 5 to 10 μm diameter resulted using piezoelectric transducers

operating at around 36 kHz to actuate arrays of 5 μm diameter micromachined nozzles. Following the same approach as in Figure 3, Meacham focused on a piezoelectrically driven, micromachined atomizer concept that utilizes fluid cavity resonances in the 0.5–5 MHz range combined with acoustic wave focusing for droplet generation or jet ejection [57]. This simple technique capable of producing droplets of sub-5 μm diameter (D50 was 4.9 μm) had low-temperature and low power operation. It also had low cost fabrication with the capacity to scale throughput up or down by using an array. It was reported that when the piezoelectric transducer is driven at the fundamental cavity (fluid reservoir) resonance frequency or any of the higher cavity modes, a standing acoustic wave develops and constructive interference in the pyramidal nozzle focuses the wave so that the peak pressure gradient occurs near the tip of the nozzle and causes the liquid to be ejected through the nozzles as droplets. The geometry of the chamber can be readily modified to increase or decrease the driving frequency of operation (for example by increasing or decreasing the height of the cavity) according to the desired specifications. Results were presented for various affecting parameters with various orifice diameters at various AC voltages (10, 30, and 36 V) having different input power. Experiments performed at different resonance frequencies revealed that the diameter of the ejected droplets decreases with increasing frequency. Doubling the thickness of the piezoelectric transducer reduces the longitudinal resonance of the transducer to half of the original frequency. The power transfer from the piezoelectric transducer to the fluid was more efficient at lower frequencies near the fundamental cavity resonant mode.

Jeng et al. devised a PZT-driven atomizer consisting of a flexible membrane and a micromachined trumpet-shaped nozzle array [58]. The atomizer employed a PZT bimorph plate attached to a liquid-proof HDPE (high density polyethylene) membrane with a low Young's modulus to generate a pressure wave in the liquid reservoir. The experimental results showed that the atomizer can generate droplets of alcohol with a Sauter mean diameter (SMD) of 4.6 μm at a flow rate of 2.5 g/min. The PZT-driven atomizer was tested under an operating voltage of 22 V and current of 180 mA, the frequency was from 12 to 24 kHz, and the power needed to start up was 2 W. During operation, a voltage was applied to the PZT bimorph plate causing the HDPE diaphragm to deflect continuously backward and forward. The volume change in the reservoir due to the diaphragm deflection in the forward direction caused an acoustic pressure wave to propagate towards the nozzles located at the front of the microatomizer

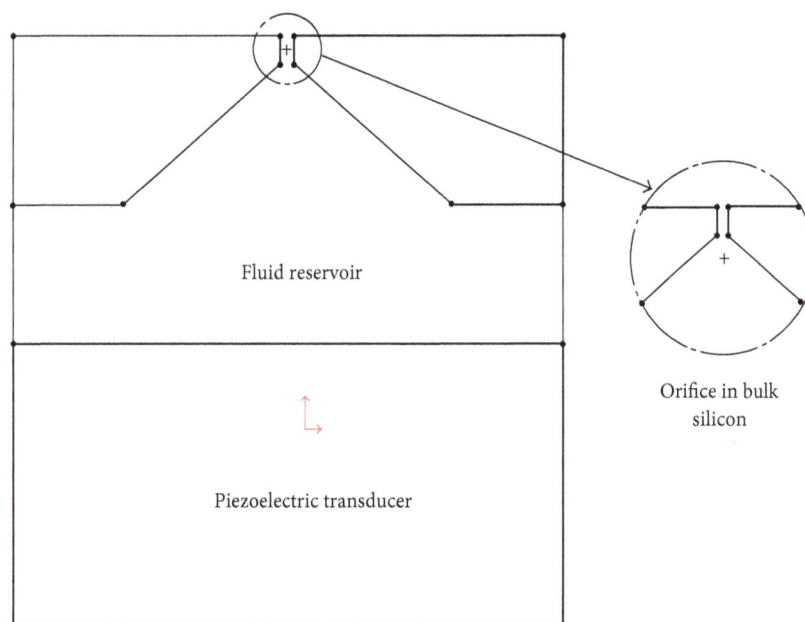

FIGURE 3: Schematic of a piezoelectrically driven, micromachined atomizer with passive mesh membrane.

expelling the droplets through the nozzle plate. Rearward deflection of the diaphragm drew liquid into the reservoir through the replenishment channel. By adjusting the driving waveform, voltage amplitude and frequency, different flow rates, and particle size distributions could be obtained.

4. Discussion

After going through various geometries and design methods, it can be easily understood that the geometry and the vibration mode play important roles in the ultrasound generation and power demands. Whereas uniform thickness plates are not capable of producing in-phase and focused waves, the stepped-thickness plates proved to be useful [22–24, 27]. Directivity patterns were compared in [23] for uniform and stepped-thickness plates and raising the steps to half a wave length of the radiated sound in the medium affected the directivity and strength of the ultrasound generation. However, using stepped-thickness plates for low frequencies is not practical due to the large height of the steps which in specific applications (i.e., small and thin transducers) may be a restriction specially for radiation in liquids and, therefore, not feasible. The operating frequency, size, and shape of the structure and the intended application clarify whether the steps are practically possible or not. Further, the acoustic medium is extremely influential to decide on the applicability of introducing steps in the thickness. Assuming water to be the acoustic medium with the sound velocity of approximately four times that of air, the thickness of the steps at the same frequency will be much higher than that of air and, as a consequence, it may not be viable to have steps. For many types of liquid medicine, the same problem exists. Going to higher frequencies in order to reduce the height of the steps may not be possible since higher frequencies have

some barriers as overheating of the transducer as reported by Lozano et al. [51]. Therefore, alternatives should be sought to achieve highly directional and focused ultrasonic waves. One alternative can be implementing curved transducers which have shown to be good at generating strong ultrasound in air [30, 31]. However, they have not been thoroughly investigated when in contact with liquids suiting them for drug delivery or atomization. Another alternative can be using cylindrical shells to increase the directivity and focus of the generated ultrasonic waves. García-Pérez et al. employed a shell vibrating at one of its resonance frequencies to produce a high level of focused acoustic intensity in the air inside the tube to dry the fruits [37]. They also noted that considering the air flow rate inside the chamber is an important factor affecting the acoustic field extensively. The high level of acoustic intensity achieved in this paper acknowledged the influence of geometry and vibration mode on the ultrasound generation. However, the power consumption of 75 W and electric drives needed may be suitable for industrial purposes, while they are not suitable for biomedical applications. According to this reference and their achievement, cylindrical shells seem a promising alternative to be scrutinized.

In conclusion, improving current devices for biomedical applications such as atomization, drug delivery, humidification, and lung therapy seems crucial considering the drawbacks of the current commercial devices as already mentioned. Broad droplet size distributions, low throughput, clogging of the mesh orifices for drug delivery equipment, and high power consumption and large size of the lung supportive devices should be eliminated. According to the literature and discussions made in this paper, the future investigations on such biomedical devices which could have industrial applications as well should be concentrated and targeted on the geometrical manipulation of the transducer

TABLE 1: Summary of the available methods for atomization.

Method	Schematic and properties	Droplet size and Flow rate	Advantages	Disadvantages
SAW [38, 39]	2.3 W; 48 MHz, 36 V	5 μm 10.2 mL/h	Small size & low power	IDT fabrication process & very high frequency & big droplets
SAW driving structure consisting of a unidirectional interdigital transducer, a horn, and a waveguide [40]	1 W; 78 MHz	1.5 μm 2.4 mL/h	Small size & smaller droplets & low power	Very low flow rate & IDT fabrication & high frequency
Arrays of 5 μm diameter micromachined nozzles [41, 42]	36 kHz; 70, 80 V	5–10 μm, According to [19] for typical frequencies of around 100 kHz, it is 4 μm	Small size & low frequency	Fabrication & big droplets
MEM Fourier-horn ultrasonic nozzle with central channel [43]	0.25 W; 971 kHz	4.5 μm 21 mL/h	Small size & excellent low power	Fabrication & big droplets & average flow rate
MEM Fourier-horn ultrasonic nozzle without central channel (externally fed) [17]	80 mW; 2 MHz	2.9–4.6 μm 25 mL/h	Small size & excellent low power	Fabrication & average droplets & average flow rate
MEM Fourier-horn ultrasonic nozzle without central channel (externally fed) [18]	0.27 W; 2 MHz	2–5 μm & Typically = 3.5 μm Max 21 mL/h & Typically = 9 mL/h	Pocket Size & battery operated & excellent low power	Fabrication & average droplets & average or low flow rate
Current vibrating plate transducers [44–47]	2–30 W; 1.65–3 MHz	1–5 μm 5–400 mL/h	Various droplet size and flow rate	Low-power devices with small droplets have low flow rate

to increase the acoustic intensity and its directivity which could in turn reduce the size of the transducer itself as well as corresponding electrical drives. The corresponding mode of vibration should also be carefully taken into account since it extensively affects the acoustic radiation. Therefore, extensive investigation on the proper mode of vibration for the considered geometry is essential. Such manipulation is anticipated to decrease the power demands since Gallego-Juárez et al. reported 147 dB of acoustic intensity with 350 W of power using five circular stepped-plate transducers covering a radiating area of about 1.64×1.64 m^2 [29]. Gallego-Juárez et al. managed to get approximately 140 dB with just 1 W and over 160 dB with 200 W (around 160 dB with 100 W) for aluminum and titanium stepped plates of 200 mm diameter which were excited by means of a prestressed composite half-wavelength cylindrical transducer consisting of two piezoelectric ceramic rings [22]. García-Pérez et al. reported 154 dB with only 75 W employing an aluminum cylindrical shell of 100 mm diameter and 310 mm length which was driven by a piezoelectric composite transducer consisting of an extensional piezoelectric sandwich element together with a mechanical amplifier [37]. Although the size and/or power consumption of the abovementioned examples may not suit biomedical applications, the intensity and directivity of the generated ultrasonic field achieved by the geometrical manipulations can verify the great role of the geometry. Moreover, the focused transducer with high intensity may result in the elimination of mesh orifices in the new design approaches

solving the blockage problem. All these should encourage the future investigations to be concentrated on the transducer geometry, its suitable vibration mode, and the strength and directivity of the generated acoustic field.

5. Conclusion

Due to a great deal of research on ultrasonic transducers and their widespread use in many different applications, they have been reviewed in this paper. Moreover, focus has been on reviewing the ultrasonic transducers as atomizers, which have a lot of biomedical and industrial applications. Various methods for generating focused ultrasound by means of flat plate transducers and shell configurations were presented. In the end, transducers for atomization were introduced. Table 1 summarizes the available approaches for atomization together with the positive and negative points of each method. As evident in the Table, vibrating plate atomizers possess the best combination of performing specifications, so that they are very common atomizers. None of the above methods with the exception of reference [37] produce complete ultrasound humidification such as that used for lung supportive devices. Biomedical humidification in lung therapies requires complete water vapor in the air rather than water droplets. For this humidification, it is preferable to have an in-line humidifier which can fit in a breathing tube of about 2 cm diameter such as those used in continuous positive airway pressure (CPAP) devices [59]. This paper can serve as a compact reference and guideline for various investigations on focused ultrasound generation as well as atomization using ultrasonic transducers.

Competing Interests

The authors declare that there is no conflict of interests in the submitted manuscript.

References

[1] K. Hettiarachchi, E. Talu, M. L. Longo, P. A. Dayton, and A. P. Lee, "On-chip generation of microbubbles as a practical technology for manufacturing contrast agents for ultrasonic imaging," *Lab on a Chip*, vol. 7, no. 4, pp. 463–468, 2007.

[2] J. S. Patton and P. R. Byron, "Inhaling medicines: delivering drugs to the body through the lungs," *Nature Reviews Drug Discovery*, vol. 6, no. 1, pp. 67–74, 2007.

[3] J. Heyder, "Deposition of inhaled particles in the human respiratory tract and consequences for regional targeting in respiratory drug delivery," *Proceedings of the American Thoracic Society*, vol. 1, no. 4, pp. 315–320, 2004.

[4] D. A. Edwards, J. Hanes, G. Caponetti et al., "Large porous particles for pulmonary drug delivery," *Science*, vol. 276, no. 5320, pp. 1868–1871, 1997.

[5] R. Langer, "Drugs on target," *Science*, vol. 293, no. 5527, pp. 58–59, 2001.

[6] R. Langer and N. A. Peppas, "Advances in biomaterials, drug delivery, and bionanotechnology," *AIChE Journal*, vol. 49, no. 12, pp. 2990–3006, 2003.

[7] D. A. LaVan, T. McGuire, and R. Langer, "Small-scale systems for in vivo drug delivery," *Nature Biotechnology*, vol. 21, no. 10, pp. 1184–1191, 2003.

[8] M. N. Topp, "Ultrasonic atomization-a photographic study of the mechanism of disintegration," *Journal of Aerosol Science*, vol. 4, no. 1, pp. 17–25, 1973.

[9] B. Avvaru, M. N. Patil, P. R. Gogate, and A. B. Pandit, "Ultrasonic atomization: effect of liquid phase properties," *Ultrasonics*, vol. 44, no. 2, pp. 146–158, 2006.

[10] A. L. Yarin, D. A. Weiss, G. Brenn, and D. Rensink, "Acoustically levitated drops: drop oscillation and break-up driven by ultrasound modulation," *International Journal of Multiphase Flow*, vol. 28, no. 6, pp. 887–910, 2002.

[11] E. H. Trinh, "Compact acoustic levitation device for studies in fluid dynamics and material science in the laboratory and microgravity," *Review of Scientific Instruments*, vol. 56, no. 11, pp. 2059–2065, 1985.

[12] A. V. Anilkumar, C. P. Lee, and T. G. Wang, "Stability of an acoustically levitated and flattened drop: an experimental study," *Physics of Fluids*, vol. 5, no. 11, pp. 2763–2774, 1993.

[13] W. Tao Shi, R. E. Apfel, and R. Glynn Holt, "Instability of a deformed liquid drop in an acoustic field," *Physics of Fluids*, vol. 7, no. 11, pp. 2601–2607, 1995.

[14] A. L. Yarin, "Drop impact dynamics: splashing, spreading, receding, bouncing," *Annual Review of Fluid Mechanics*, vol. 38, pp. 159–192, 2006.

[15] M. Rein, "Phenomena of liquid drop impact on solid and liquid surfaces," *Fluid Dynamics Research*, vol. 12, no. 2, pp. 61–93, 1993.

[16] R. Rioboo, C. Tropea, and M. Marengo, "Outcomes from a drop impact on solid surfaces," *Atomization and Sprays*, vol. 11, no. 2, pp. 155–165, 2001.

[17] C. S. Tsai, R. W. Mao, S. K. Lin, N. Wang, and S. C. Tsai, "Miniaturized multiple fourier-horn ultrasonic droplet generators for biomedical applications," *Lab on a Chip*, vol. 10, no. 20, pp. 2733–2740, 2010.

[18] C. Tsai, R. Mao, S. Lin, Y. Zhu, and S. Tsai, "Faraday instability-based micro droplet ejection for inhalation drug delivery," *Technology*, vol. 2, no. 1, pp. 75–81, 2014.

[19] J. S. Lass, A. Sant, and M. Knoch, "New advances in aerosolised drug delivery: vibrating membrane nebuliser technology," *Expert Opinion on Drug Delivery*, vol. 3, no. 5, pp. 693–702, 2006.

[20] B. L. Rottier, C. J. P. van Erp, T. S. Sluyter, H. G. M. Heijerman, H. W. Frijlink, and A. H. de Boer, "Changes in performance of the pari eflow® rapid and pari LC plus™ during 6 months use by CF patients," *Journal of Aerosol Medicine and Pulmonary Drug Delivery*, vol. 22, no. 3, pp. 263–269, 2009.

[21] B. Mortimer, G. Saban, A. Samboer, and B. Verveckken, "The development of an ultrasonic humidifier for domestic applications," in *Proceedings of the Domestic Use of Electrical Energy Conference*, pp. 138–140, 1999.

[22] J. A. Gallego-Juárez, G. Rodriguez-Corral, and L. Gaete-Garreton, "An ultrasonic transducer for high power applications in gases," *Ultrasonics*, vol. 16, no. 6, pp. 267–271, 1978.

[23] A. Barone and J. A. G. Juarez, "Flexural vibrating free-edge plates with stepped thicknesses for generating high directional ultrasonic radiation," *The Journal of the Acoustical Society of America*, vol. 51, no. 3B, pp. 953–959, 1972.

[24] J. A. Gallego Juárez, "Axisymmetric vibrations of circular plates with stepped thickness," *Journal of Sound and Vibration*, vol. 26, no. 3, pp. 411–416, 1973.

[25] J. L. San Emeterio, J. A. Gallego-Juárez, and G. Rodriguez-Corral, "IGH axisymmetric modes of vibration of stepped circular plates," *Journal of Sound and Vibration*, vol. 114, no. 3, pp. 495–505, 1987.

[26] A. M. Al-Jumaily and K. Jameel, "Influence of the Poisson ratio on the natural frequencies of stepped-thickness circular plate," *Journal of Sound and Vibration*, vol. 234, no. 5, pp. 881–894, 2000.

[27] F. Montero de Espinosa and J. A. Gallego-Juárez, "A directional single-element underwater acoustic projector," *Ultrasonics*, vol. 24, no. 2, pp. 100–104, 1986.

[28] J. A. Gallego-Juárez, G. Rodríguez-Corral, E. Riera-Franco de Sarabia, F. Vázquez-Martínez, V. M. Acosta-Aparicio, and C. Campos-Pozuelo, "Development of industrial models of high-power stepped-plate sonic and ultrasonic transducers for use in fluids," in *Proceedings of the Ultrasonics Symposium*, vol. 2, pp. 571–578, Atlanta, Ga, USA, October 2001.

[29] J. A. Gallego-Juárez, G. Rodríguez-Corral, E. Riera-Franco De Sarabia, F. Vázquez-Martínez, C. Campos-Pozuelo, and V. M. Acosta-Aparicio, "Recent developments in vibrating-plate macrosonic transducers," *Ultrasonics*, vol. 40, no. 1–8, pp. 889–893, 2002.

[30] H. Wang and M. Toda, "Curved PVDF airborne transducer," *IEEE Transactions on Ultrasonics, Ferroelectrics, and Frequency Control*, vol. 46, no. 6, pp. 1375–1386, 1999.

[31] M. Toda and S. Tosima, "Theory of curved, clamped, piezoelectric film, air-borne transducers," *IEEE Transactions on Ultrasonics, Ferroelectrics, and Frequency Control*, vol. 47, no. 6, pp. 1421–1431, 2000.

[32] M. Toda, "Phase-matched air ultrasonic transducers using corrugated PVDF film with half wavelength depth," *IEEE Transactions on Ultrasonics, Ferroelectrics, and Frequency Control*, vol. 48, no. 6, pp. 1568–1574, 2001.

[33] H. Du, L. Xu, H. Hu et al., "High-frequency vibrations of corrugated cylindrical piezoelectric shells," *Acta Mechanica Solida Sinica*, vol. 21, no. 6, pp. 564–572, 2008.

[34] L. Xu, M. Chen, H. Du et al., "Vibration characteristics of a corrugated cylindrical shell piezoelectric transducer," *IEEE Transactions on Ultrasonics, Ferroelectrics, and Frequency Control*, vol. 55, no. 11, pp. 2502–2508, 2008.

[35] H. Li, F. Yang, H. Du et al., "Dynamic characteristics of axially-symmetrical annular corrugated shell piezoelectric transducers," *Acta Mechanica Solida Sinica*, vol. 22, no. 5, pp. 499–509, 2009.

[36] F. Blum, J. Jarzynski, and L. J. Jacobs, "A focused two-dimensional air-coupled ultrasonic array for non-contact generation," *NDT and E International*, vol. 38, no. 8, pp. 634–642, 2005.

[37] J. V. García-Pérez, J. A. Cárcel, S. de la Fuente-Blanco, and E. Riera-Franco de Sarabia, "Ultrasonic drying of foodstuff in a fluidized bed: parametric study," *Ultrasonics*, vol. 44, pp. e539–e543, 2006.

[38] M. Kurosawa, T. Watanabe, A. Futami, and T. Higuchi, "Surface acoustic wave atomizer," *Sensors and Actuators: A. Physical*, vol. 50, no. 1-2, pp. 69–74, 1995.

[39] M. Kurosawa, A. Futami, and T. Higuchi, "Characteristics of liquids atomization using surface acoustic wave," in *Proceedings of the International Conference on Solid State Sensors and Actuators (TRANSDUCERS '97)*, IEEE, Chicago, Ill, USA, June 1997.

[40] W. Soluch and T. Wrobel, "Low driving power SAW atomiser," *Electronics Letters*, vol. 42, no. 24, Article ID 1432, 2006.

[41] S. Yuan, Z. Zhou, and G. Wang, "Experimental research on piezoelectric array microjet," *Sensors and Actuators A: Physical*, vol. 108, no. 1–3, pp. 182–186, 2003.

[42] S. Yuan, Z. Zhou, G. Wang, and C. Liu, "MEMS-based piezoelectric array microjet," *Microelectronic Engineering*, vol. 66, no. 1-4, pp. 767–772, 2003.

[43] S. C. Tsai, C. H. Cheng, N. Wang, Y. L. Song, C. T. Lee, and C. S. Tsai, "Silicon-based megahertz ultrasonic nozzles for production of monodisperse micrometer-sized droplets," *IEEE Transactions on Ultrasonics, Ferroelectrics, and Frequency Control*, vol. 56, no. 9, pp. 1968–1979, 2009.

[44] L. APC International, Ed., *1.65 mhz Nebulizer Boards Catalog*, 2014.

[45] Siansonic general atomizing kit catalog, S.T.C. Ltd.

[46] Siansonic mini nebulizer kit catalog, S.T.C. Ltd.

[47] Siansonic high efficiency atomizer transducer catalog, S.T.C. Ltd.

[48] S. Sangwan, R. Condos, and G. C. Smaldone, "Lung deposition and respirable mass during wet nebulization," *Journal of Aerosol Medicine: Deposition, Clearance, and Effects in the Lung*, vol. 16, no. 4, pp. 379–386, 2003.

[49] T. G. O'Riordan, L. I. Kleinman, K. Hughes, and G. C. Smaldone, "Predicting aerosol deposition during neonatal ventilation: feasibility of bench testing," *Respiratory Care*, vol. 39, no. 12, pp. 1162–1168, 1994.

[50] S. A. Elrod, B. Hadimioglu, B. T. Khuri-Yakub, E. G. Rawson, and C. F. Quate, "Focused acoustic beams for nozzleless droplet formation," in *Proceedings of the Ultrasonics Symposium*, IEEE, Chicago, Ill, USA, 1988.

[51] A. Lozano, H. Amaveda, F. Barreras, X. Jordà, and M. Lozano, "High-frequency ultrasonic atomization with pulsed excitation," *Journal of Fluids Engineering, Transactions of the ASME*, vol. 125, no. 6, pp. 941–945, 2003.

[52] J. Ju, Y. Yamagata, H. Ohmori, and T. Higuchi, "High-frequency surface acoustic wave atomizer," *Sensors and Actuators, A: Physical*, vol. 145-146, no. 1-2, pp. 437–441, 2008.

[53] A. Lal and R. M. White, "Micromachined silicon ultrasonic atomizer," in *Proceedings of the 1996 IEEE Ultrasonics Symposium*, pp. 339–342, IEEE, San Antonio, Tex, USA, November 1996.

[54] T. Li, Y. Chen, and J. Ma, "Development of a miniaturized piezoelectric ultrasonic transducer," *IEEE Transactions on Ultrasonics, Ferroelectrics, and Frequency Control*, vol. 56, no. 3, pp. 649–659, 2009.

[55] S. C. Tsai, Y. L. Song, T. K. Tseng, Y. F. Chou, W. J. Chen, and C. S. Tsai, "High-frequency, silicon-based ultrasonic nozzles using multiple fourier horns," *IEEE Transactions on Ultrasonics, Ferroelectrics, and Frequency Control*, vol. 51, no. 3, pp. 277–285, 2004.

[56] C. S. Tsai, R. W. Mao, S. K. Lin, E. Chien, and S. C. Tsai, "MEMS-based multiple fourier-horn silicon ultrasonic atomizer for inhalation drug delivery," in *Proceedings of the IEEE International Ultrasonics Symposium (IUS '11)*, pp. 1119–1122, Orlando, Fla, USA, October 2011.

[57] J. M. Meacham, *A Micromachined Ultrasonic Droplet Generator: Design, Fabrication, Visualization, and Modeling*, School of Mechanical Engineering, Georgia Institute of Technology, 2006.

[58] Y.-R. Jeng, C.-C. Su, G.-H. Feng, Y.-Y. Peng, and G.-P. Chien, "A PZT-driven atomizer based on a vibrating flexible membrane and a micro-machined trumpet-shaped nozzle array," *Microsystem Technologies*, vol. 15, no. 6, pp. 865–873, 2009.

[59] A. M. Al-Jumaily and P. Reddy, *Medical Devices for Respiratory Dysfunction: Principles and Modeling of Continuous Airway Pressure (CPAP)*, ASME, New York, NY, USA, 2012.

11

A Note on an Analytic Solution for an Incompressible Fluid-Conveying Pipeline System

Vincent O. S. Olunloyo,[1] Charles A. Osheku,[2] and Patrick S. Olayiwola[3]

[1]*Department of Systems Engineering, Faculty of Engineering, University of Lagos, Akoka-Yaba, Lagos 23401, Nigeria*
[2]*Centre for Space Transport and Propulsion, National Space Research and Development Agency,
Federal Ministry of Science and Technology, FCT, PMB 437, Abuja, Nigeria*
[3]*Department of Mechanical & Biomedical Engineering, College of Engineering, Bells University of Technology, Ota 234037,
Ogun State, Nigeria*

Correspondence should be addressed to Charles A. Osheku; charlesosheku2002@yahoo.com

Academic Editor: Mohammad Tawfik

This paper presents an integral transform analytic solution to the equations governing a fluid-conveying pipeline segment where a gyroscopic or Coriolis force effect is taken into consideration. The mathematical model idealizes a segment of the pipeline as an elastic beam conveying an incompressible fluid. It is clearly shown that when such a system is supported at both ends and in a free motion, the Coriolis force dissipates no energy (or simply does not work) as it generates conjugate complex vibratory components for all flow velocities. It is demonstrated that the modal natural frequencies can be computed from the algebraic products of the complex frequency pairs. Clearly, the patterns of the characteristics of the system's natural frequencies are seen partly when the real and imaginary components are plotted, as widely seen in the literature. Nonetheless, results from this study revealed that a continuity profile exists to connect the subcritical, critical, and postcritical vibratory behaviours when the absolute values are plotted for any velocity. In the meantime, the efficacy and versatility of this method against the usual assumed spatial or temporal modal solutions are demonstrated by confirming the predictions and validity of results of earlier workers such as Paidoussis, Ziegler, and others where pre- and postdivergence behaviours are exhibited.

1. Introduction

Fluid-conveying pipes are parts of the most common engineering examples of slender systems interacting with axial flows; another good example is the deployment of flexible conduits in the oil and gas exploration and production industry. A compendium of other examples can be found in Paidoussis [1] work spanning over the last 50 years. The list of examples is not limited to the field of engineering but cuts across other areas of human endeavor such as the study of pulmonary and urinary tract systems or even haemodynamics within human physiology.

These problems have generated a lot of research interests over the years partly because some served as models for studying the stability of certain classes of dynamical systems leading to the development of novel numerical and analytical methods for solving such problems. It has also turned out, over the years, that several of these techniques have found wider applications in other areas of research that otherwise appeared unrelated and have in fact occasionally led to the development of unanticipated practical applications and devices. Thus for linear dynamics for axial flows along slender structures, the pipe conveying fluid is regarded as the main paradigm. It also serves as model for classical problems involving axial momentum transport or axially moving continua such as high speed magnetic and paper tapes [1].

However, a careful study of the development of this area of research revealed that most of the research interests were curiosity-driven as some of the interesting phenomena observed occurred at velocities and conditions outside practical engineering and operational working limits as at that time [2]. This however gradually changed with the study of high velocity flow within light-gauge piping used in rocket engines and stability problems experienced by oil pipelines at modest conveyance speeds. Nowadays, people are looking at areas of direct applications, for example, in the behaviour of aspirating pipes for ocean mining and LNG in-situ production. This concept is to be utilized for the proposed offshore mining of methane liquid-crystal deposits and carbon sequestration. Here, the interest is in flow-induced vibrations and instabilities that can arise at high flow rates.

Furthermore, with the advent of High Pressure-High Temperature (HP-HT) oil and gas exploration and production, the lengths of the flexible risers deployed surely qualify them as hoses or pipe strings whose vibration behaviour is of interest in the exploration field.

In the modeling of the mechanics of fluid-conveying pipes, one of the terms that has received considerable attention over the years is the Coriolis force that was assigned the role of energy absorption that counters the centrifugal effect that normally arose in free motions. Broadly speaking, such an energy absorption affects the stability in conservative and nonconservative systems, as was shown by Section 7 of Elishakoff's work (2005) [2]; Öz and Boyaci (2000) [3]; Szmidt and Przybyłowicz (2013) [4]; Askarian et al. (2014) [5]; Kuiper and Metrikine (2004) [6]; Chellapilla and Simha (2008) [7]. Other contributors that attempted to investigate and explain the behaviour included Leklong et al. (2008) [8]; Al-Hilli and Ntayeesh (2013) [9]; Guo et al. (2006) [10]; Zhang et al. (2000) [11]; Modarres-Sadeghi and Païdoussis (2009) [12] as well as Ibrahim (2010) [13].

Principal among these findings is that while the centrifugal force imparts energy to the system, the Coriolis force absorbs energy from the system, such that the balance between the two, in the absence of dissipation, gives rise to flutter. However, when confronted with a nonconservative system the effect of the Coriolis force can lead to destabilization. These conclusions were arrived at partly on experimental work as can be found in [14, 15]. However, in the 1950s, proofs were claimed of researchers and mathematicians' findings showing that the Coriolis has negligible effects even on the molecular scale interaction.

Part of the interest in this present study rests on the fact that previous explanations in literature as to the effect of the Coriolis force on the natural frequency of the system derived from ad hoc heuristic arguments accompanied sometimes by authoritative and masterly analyses and interpretation of results of numerical and experiments. What is however missing is adequate proof based on the results of blind solutions of the unabridged governing differential equations for the linear problems as simple examples or justifications of such rationalizations.

A separate but issue related is that although the gyroscopic (Coriolis) forces do no work in the course of free motions, they nonetheless exert important influences on the overall dynamical behaviour of a pipeline system as pointed out in [1]. It would therefore be useful to know exactly what their influences are in such cases.

Another issue that has arisen over the years is that of the efficacy of the methodology. To be sure, several methods have been used to tackle the class of problems associated with the dynamics of fluid-conveying pipes but prominent amongst these is the original work of Gregory and Paidoussis and the sequel as presented in [1] where the use of an eigenfunction expansion in a modified Galerkin scheme was introduced. Part of the initial challenge was the absence of computers and the availability of validation modules to check the results of numerical work. With the development of the Finite element method more emphasis was placed on numerical schemes and laboratory experiments were framed up to confirm the predictions of these studies. Some of the other methods used for analyses over the years included the spectral method, for example, deployed by Lee and Park (2006) [16] or the differential transformation method recently applied by Qiao et al. (2006) [17].

Other recent works are those of Dodds Jr. and Runyan (1965) [18], who used flow visualization and velocity measurement to experimentally clarify the mechanism underlying the fluid-induced vibration in double T-junction of pipeline systems; Yamaguchi et al. (2016) [19], who used a solution method based on the Frobenius power series on a derived asymptotic model from the solutions of the Pridmore-Brown equation for the Fourier transform of the vibrational fluid pressure; Kutin and Bajsić (2014) [20], who use smart materials; and Jweeg and Ntayeesh (2015) [21] who made use of application of method of multiple scales to analyse approximately the gyroscopic system for a nonlinear fluid-conveying pipeline.

This paper further establishes the method of complex integral transforms (where the cosine and sine transforms are special cases) as one other effective method that can be used to tackle such problems within the context of linear theory for a start and is organized as follows. Section 1 introduces the problems under investigation. In Section 2, the analysis of the pipeline conveying an incompressible fluid with the governing partial differential equation and appropriate boundary conditions are presented. In Section 3, the complex and natural frequencies of the system are computed together with the relationship of the system's natural frequency with the flow critical velocity. Section 4 analyses the dynamic responses for purely elastic pipe in three cases, namely, simple supports at both ends, cantilever pipe, and a clamped-pinned ends pipe. In Section 5, results are analysed and discussed and Section 6 concludes the paper, whilst references are listed in the final section and at the end found the Appendices A and B.

2. Analysis of the Pipeline System Conveying an Incompressible Fluid

The homogeneous Partial Differential Equation (PDE) governing the flow-induced vibration of a pipeline conveying an incompressible fluid is given by

$$\mathrm{EI}\frac{\partial^4 w}{\partial x^4} + \left(\rho_s + \rho_f\right)\frac{\partial^2 w}{\partial t^2} + \rho_f U^2 \frac{\partial^2 w}{\partial x^2} + 2\rho_f U \frac{\partial^2 w}{\partial t \partial x} = 0, \tag{1}$$

where EI is the flexural rigidity of the pipe, ρ_s and ρ_f are the mass per unit length of pipe and fluid, respectively, flowing with a steady flow velocity U, and w is the lateral deflection of the pipe. The parameters x and t are the axial coordinate and time variables, respectively.

In literature of fluid structure interaction mechanics, where fluid-induced vibrations are studied, (1) is often nondimensionalized as

$$\frac{\partial^4 \overline{W}}{\partial \xi^4} + \overline{U}^2 \frac{\partial^2 \overline{W}}{\partial \xi^2} + 2\sqrt{\beta}\overline{U}\frac{\partial^2 \overline{W}}{\partial \xi \partial \tau} + \frac{\partial^2 \overline{W}}{\partial \tau^2} = 0, \quad (2)$$

where

$$\overline{U} = UL\sqrt{\frac{\rho_f}{\text{EI}}},$$

$$\beta = \frac{\rho_f}{\rho_s + \rho_f},$$

$$\tau = \frac{t}{L^2}\sqrt{\frac{\text{EI}}{\rho_s + \rho_f}}, \quad (3)$$

$$\xi = \frac{x}{L},$$

$$\overline{W} = \frac{w}{L},$$

whereas, for the present study, the dimensionless form is given by

$$\frac{\partial^4 \overline{W}}{\partial \overline{x}^4} + \zeta_f \overline{U}^2 \frac{\partial^2 \overline{W}}{\partial \overline{x}^2} + 2\zeta_f \overline{U}\frac{\partial^2 \overline{W}}{\partial \overline{x} \partial \overline{t}} + \left(1 + \zeta_f\right)\frac{\partial^2 \overline{W}}{\partial \overline{t}^2} = 0, \quad (4)$$

where

$$\overline{U} = UL\sqrt{\frac{\rho_f}{\text{EI}}},$$

$$\zeta_f = \frac{\rho_f}{\rho_s},$$

$$\overline{t} = \frac{t}{\tau}, \quad (5)$$

$$\tau = L^2\sqrt{\frac{\rho_s}{\text{EI}}},$$

$$\overline{x} = \frac{x}{L},$$

$$\overline{W} = \frac{w}{L},$$

and the components involved are, respectively, the restoring, inertia, bending/centrifugal, and Coriolis force terms.

For comparison purposes, we present the following.

Case 1. We expect that if the pipeline segment is conveying a fluid and an eventual situation calls for the valve to be closed or the pump/compressor shut-off, an entrained fluid,

be it hot, cold, pressurized, or otherwise, will be trapped in the pipe. When the flow velocity $\overline{U} = 0$ (2) based on the nondimensionalizing method in the literature becomes

$$\lim_{\overline{U}\to 0}\left(\frac{\partial^4 \overline{W}}{\partial \xi^4} + \overline{U}^2 \frac{\partial^2 \overline{W}}{\partial \xi^2} + 2\sqrt{\beta_f}\overline{U}\frac{\partial^2 \overline{W}}{\partial \xi \partial \tau} + \frac{\partial^2 \overline{W}}{\partial \tau^2}\right)$$
$$= \frac{\partial^4 \overline{W}}{\partial \xi^4} + \frac{\partial^2 \overline{W}}{\partial \tau^2} = 0 \quad (6)$$

leading to

$$\Omega_n = \pm\lambda^2 = \pm n^2\pi^2 \quad (7)$$

which shows that there is no fluid inside the pipe if the flow velocity is zero. *This is not to be so.* However, considering (4), based on this paper's dimensionless method, with $\overline{U} = 0$, it becomes

$$\lim_{\overline{U}\to 0}\left(\frac{\partial^4 \overline{W}}{\partial \overline{x}^4} + \zeta_f \overline{U}^2 \frac{\partial^2 \overline{W}}{\partial \overline{x}^2} + 2\zeta_f \overline{U}\frac{\partial^2 \overline{W}}{\partial \overline{x} \partial \overline{t}}\right.$$
$$\left. + \left(1 + \zeta_f\right)\frac{\partial^2 \overline{W}}{\partial \overline{t}^2}\right) = \frac{\partial^4 \overline{W}}{\partial \overline{x}^4} + \left(1 + \zeta_f\right)\frac{\partial^2 \overline{W}}{\partial \overline{t}^2} = 0 \quad (8)$$

leading to

$$\Omega_n = \pm\frac{\lambda^2}{\sqrt{\left(1 + \zeta_f\right)}} = \pm\frac{n^2\pi^2}{\sqrt{\left(1 + \zeta_f\right)}}. \quad (9)$$

Revealing that, the system's vibration, configuration, and response are strongly affected by the entrainment fluid in the pipe.

Case 2. If there is no fluid in the pipe, that is, $\zeta_f = \beta = 0$, (2) and (4), respectively, become

$$\lim_{\zeta_f\to 0}\left(\frac{\partial^4 \overline{W}}{\partial \overline{x}^4} + \zeta_f \overline{U}^2 \frac{\partial^2 \overline{W}}{\partial \overline{x}^2} + 2\zeta_f \overline{U}\frac{\partial^2 \overline{W}}{\partial \overline{x} \partial \overline{t}}\right.$$
$$\left. + \left(1 + \zeta_f\right)\frac{\partial^2 \overline{W}}{\partial \overline{t}^2}\right) = 0 \quad \text{that is} \quad \frac{\partial^4 \overline{W}}{\partial \overline{x}^4} + \frac{\partial^2 \overline{W}}{\partial \overline{t}^2} = 0 \quad (10)$$

leading to

$$\Omega_n = \pm\lambda\sqrt{\lambda^2 - \overline{U}^2} = \pm n\pi\sqrt{n^2\pi^2 - \overline{U}^2}, \quad (11)$$

$$\lim_{\zeta_f \to 0} \left(\frac{\partial^4 \overline{W}}{\partial \overline{x}^4} + \zeta_f \overline{U}^2 \frac{\partial^2 \overline{W}}{\partial \overline{x}^2} + 2\zeta_f \overline{U} \frac{\partial^2 \overline{W}}{\partial \overline{x} \partial \overline{t}} \right.$$

$$\left. + \left(1 + \zeta_f\right) \frac{\partial^2 \overline{W}}{\partial \overline{t}^2} \right) = 0 \qquad (12)$$

that is $\dfrac{\partial^4 \overline{W}}{\partial \overline{x}^4} + \dfrac{\partial^2 \overline{W}}{\partial \overline{t}^2} = 0$.

The literature method (10) is showing that when no fluid is present in the pipe, there is still a flow velocity. This is not possible and is likely a fundamental error in physics. However, (12) shows that the restoring and the inertial accelerations in dimensionless form are the balancing vectors.

Case 3. We now examine the governing equation (1) in its original form without nondimensionalizing, such that, when $U = 0$, it becomes

$$EI \frac{\partial^4 w}{\partial x^4} + \left(\rho_s + \rho_f\right) \frac{\partial^2 w}{\partial t^2} = 0 \qquad (13)$$

leading to

$$\Omega_n = \pm \frac{EI}{\rho_s \left(1 + \zeta_f\right)} \left(\frac{\lambda}{L}\right)^2 = \pm \frac{EI}{\left(1 + \zeta_f\right)} \frac{n^2 \pi^2}{L^2}. \qquad (14)$$

In the absence of fluid in the pipe, that is, $m_f = 0$.

$$EI \frac{\partial^4 w}{\partial x^4} + \rho_s \frac{\partial^2 w}{\partial t^2} = 0 \qquad (15)$$

leading to

$$\Omega_n = \pm \frac{EI}{\rho_s} \left(\frac{\lambda}{L}\right)^2 = \pm \frac{EI}{\rho_s} \frac{n^2 \pi^2}{L^2}. \qquad (16)$$

It is confirmed that (13) is similar to (8), demonstrating that the natural frequency is dependent on the mass or density of fluid flowing in the pipeline.

Now, the following definitions hold for the Fourier complex integral transforms Wrede and Spiegel [22], Olayiwola [23], and Jeffrey [24]; namely,

$$F\left\{\overline{W}\left(\overline{x}, \overline{t}\right)\right\} = \frac{1}{\sqrt{2\pi}} \int_{-\infty}^{\infty} e^{-i\lambda x} \overline{W}\left(\overline{x}, \overline{t}\right) dx$$

$$= \overline{W}^F \left(\lambda, \overline{t}\right);$$

$$F^{-1}\left\{\overline{W}^F \left(\lambda, \overline{t}\right)\right\} = \frac{1}{\sqrt{2\pi}} \int_{-\infty}^{\infty} e^{i\lambda x} \overline{W}^F \left(\lambda, \overline{t}\right) d\lambda \qquad (17)$$

$$= W\left(\overline{x}, \overline{t}\right)$$

such that in this case

$$F\left(\overline{x}, \overline{t}\right) = \begin{cases} 0 & \text{when } -\infty \le \overline{x} < 0 \\ \overline{W}\left(\overline{x}, t\right) & \text{when } 0 \le \overline{x} \le 1 \\ 0 & \text{when } 1 < \overline{x} \le \infty. \end{cases} \qquad (18)$$

Using (18) on (4), the following governing equation ensues in the transforms plane:

$$\frac{d^2 \overline{W}^F}{d\overline{t}^2} + i2 \frac{\lambda \zeta_f \overline{U}}{\left(1 + \zeta_f\right)} \frac{d\overline{W}^F}{d\overline{t}} + \frac{\left(\lambda^4 - \zeta_f \overline{U}^2 \lambda^2\right)}{\left(1 + \zeta_f\right)} \overline{W}^F$$

$$= \frac{1}{\left(1 + \zeta_f\right)} \left(-\overline{W}_{\overline{xxx}} e^{-i\lambda \overline{x}}\Big|_0^1 - i\lambda \overline{W}_{\overline{xx}} e^{-i\lambda \overline{x}}\Big|_0^1 \right.$$

$$+ \left(\lambda^2 - \zeta_f \overline{U}^2\right) \overline{W}_{\overline{x}} e^{-i\lambda \overline{x}}\Big|_0^1 \qquad (19)$$

$$\left. + \left(i\lambda^3 - \zeta_f \overline{U}^2\right) \overline{W} e^{-i\lambda \overline{x}}\Big|_0^1 - \zeta_f \overline{U} \frac{d}{d\overline{t}} \left(\overline{W} e^{-i\lambda \overline{x}}\Big|_0^1\right) \right),$$

subject to simply supported conditions at both ends; namely,

$$\overline{W}\left(0, t\right) = \overline{W}_{xx}\left(0, t\right) = 0,$$

$$\overline{W}\left(1, t\right) = \overline{W}_{xx}\left(1, t\right) = 0. \qquad (20)$$

In conjunction with the following conditions:

$$\frac{d\overline{W}\left(0, t\right)}{d\overline{t}} = \frac{d\overline{W}\left(1, t\right)}{d\overline{t}} = 0. \qquad (21)$$

Substituting (20) into (19) leads to the following ordinary differential equation (ode) in the transform plane:

$$\left(1 + \zeta_f\right) \frac{d^2 \overline{W}^F}{d\overline{t}^2} + i2\lambda \zeta_f \overline{U} \frac{d\overline{W}^F}{d\overline{t}} + \left(\lambda^4 - \zeta_f \overline{U}^2 \lambda^2\right) \overline{W}^F$$

$$= \left(-\overline{W}_{\overline{xxx}} e^{-i\lambda \overline{x}}\Big|_0^1 + \left(\lambda^2 - \zeta_f \overline{U}^2\right) \overline{W}_{\overline{x}} e^{i\lambda \overline{x}}\Big|_0^1 \right) = 0. \qquad (22)$$

This is a nonhomogeneous second-order ordinary differential equation in time domain. We can now proceed to solve for the frequency and displacement responses.

3. Complex and Natural Frequencies

In order to find the natural frequencies of the system we seek to solve the complimentary equation of the system in its Fourier complex transform plane by using the trial solution, namely,

$$\overline{W}^F \left(\lambda, \overline{t}\right) = Ae^{s\overline{t}}, \qquad (23)$$

to obtain the following characteristic equation:

$$\left(\left(1 + \zeta_f\right) s^2 + i2\zeta_f \lambda \overline{U} s + \left(\lambda^4 - \zeta_f \overline{U}^2 \lambda^2\right)\right) Ae^{s\overline{t}} = 0. \qquad (24)$$

The preceding equation can now be solved for the roots of s to obtain complex conjugate pairs of the forms:

$$s_1 = \left(-i\frac{\zeta_f \overline{U}\lambda}{(1+\zeta_f)} \right.$$

$$\left. + \sqrt{\left(i\frac{\zeta_f \overline{U}\lambda}{(1+\zeta_f)} \right)^2 - \left(\frac{(\lambda^4 - \zeta_f \overline{U}^2\lambda^2)}{(1+\zeta_f)} \right)} \right),$$

$$(25)$$

$$s_2 = \left(-i\frac{\zeta_f \overline{U}\lambda}{(1+\zeta_f)} \right.$$

$$\left. - \sqrt{\left(i\frac{\zeta_f \overline{U}\lambda}{(1+\zeta_f)} \right)^2 - \left(\frac{(\lambda^4 - \zeta_f \overline{U}^2\lambda^2)}{(1+\zeta_f)} \right)} \right).$$

In order to isolate the effect of the Coriolis force, we introduce the expression

$$\Omega_{\text{Cor.}} = i\frac{\zeta_f \overline{U}\lambda}{(1+\zeta_f)} \qquad (26)$$

into (25). On comparing these equations with the natural frequency, the relative frequencies are related as follows:

$$\Omega_1 = -\Omega_n\left(\xi_n + \sqrt{(1+\xi_n^2)} \right),$$

$$\Omega_2 = -\Omega_n\left(\xi_n - \sqrt{(1+\xi_n^2)} \right), \qquad (27)$$

where

$$\xi_n = \frac{\Omega_{\text{cor}}}{\Omega_n}. \qquad (28)$$

Moreover, the product of the conjugate pairs gives the natural frequencies of the system; namely,

$$\Omega_n^2 = \Omega_1 \times \Omega_2; \qquad (29)$$

that is,

$$\Omega_n^2 = \left(\frac{(\lambda^4 - \zeta_f \overline{U}^2\lambda^2)}{(1+\zeta_f)} \right). \qquad (30)$$

Although, from mathematical physics, the complex notation "i" indicates that $\Omega_{\text{Cor.}}$ is acting perpendicularly to the natural frequency Ω_n, nonetheless, an algebraic functional relation can be deduced as follows.

Substituting a square of (26) into (30) gives

$$\Omega_n = \frac{\lambda^2}{\sqrt{(1+\zeta_g)}} \left(1 - \frac{\Omega_{\text{cor}}^2 (1+\zeta_g)^2}{\lambda^4 \zeta_g} \right)^{1/2}. \qquad (31)$$

On expanding, it yields

$$\Omega_n = \frac{\lambda^2}{\sqrt{(1+\zeta_g)}} \left\{ 1 - \frac{\Omega_{\text{cor}}^2 (1+\zeta_g)^2}{2\lambda^4 \zeta_g} \right.$$

$$\left. + \frac{1}{2!}\frac{\Omega_{\text{cor}}^4 (1+\zeta_g)^4}{4\lambda^8 \zeta_g^2} - \frac{1}{3!}\frac{3\Omega_{\text{cor}}^6 (1+\zeta_g)^6}{8\lambda^{12}\zeta_g^3} + \cdots \right\}. \qquad (32a)$$

But for $\overline{U} \to 0$, $\Omega_{\text{cor}} \to 0$ or with $\Omega_{\text{cor}}(1+\zeta_g) \ll \lambda^2\sqrt{\zeta_g}$, (24) tends to a limit; that is,

$$\Omega_n = \frac{\lambda^2}{\sqrt{(1+\zeta_g)}}. \qquad (32b)$$

The first observation to make from relations (18) and (22) is that the natural frequency is not in the same component with the frequency due to the Coriolis force.

Having noted this, it is also useful to conform with the general practice in the literature by expressing the result for the natural frequency, where possible, in a way that relates it to physical parameters or benchmarks associated with the flow. Thus from relation (19), it is straightforward to deduce the following.

(a) The critical axial velocity at which the natural frequency of the system is zero satisfies the relation

$$\overline{U}_{\text{cr}}^2 = \frac{\lambda^2}{\zeta_f} \qquad (33a)$$

and defines the condition for the onset of irregular oscillations.

(b) When there is no axial flow, that is, $\overline{U} = 0$, the natural frequency of the system satisfies the relation

$$\Omega_0^2 = \left(\frac{\lambda^4}{(1+\zeta_f)} \right). \qquad (33b)$$

(c) The general relation for the magnitude of the natural frequency can be rearranged as

$$\Omega_n^2 = \Omega_0^2\left(1 - \frac{\overline{U}^2}{\overline{U}_{\text{cr}}^2} \right). \qquad (34)$$

This is a simple expression that relates the natural frequency to the critical flow velocity. Equation (34) actually proves the existence of such relationship as it could be argued that the results of the experiment of Dodds Jr. and Runyan (1965) [18] provide indirect evidence of the existence of such a relationship for the eigenfrequency mode $n = 1$.

Furthermore, evaluation of the critical velocity and fundamental frequency can be carried out by substituting the appropriate eigenvalues into (24). Thus it can be asserted that, for this case, this method facilitates the derivation of explicit closed form expressions for possible design parameters such as the critical velocity. This method of solution also sets the stage for deriving equivalent results for $\overline{U}_{\text{cr}}$ and Ω_n for the dynamic response when other effects such as damping for example are included.

4. Dynamic Response Analysis for Purely Elastic Pipe

If we consider the dynamic response of the simple system of a purely elastic horizontal pipe with uniformly distributed loads of f Newton per unit length, then (4) is transformed as follows:

$$\frac{d^2\overline{W}^F}{d\bar{t}^2} + 2\Omega_{\text{cor}}\frac{d\overline{W}^F}{d\bar{t}} + \Omega_n^2\overline{W}^F = \frac{J(\lambda)}{(1+\zeta_f)}, \qquad (35)$$

where

$$\Omega_{\text{cor}} = i\frac{\lambda\zeta_f\overline{U}}{(1+\zeta_f)},$$

$$\Omega_n^2 = \frac{\left(\lambda^4 - \zeta_f\overline{U}^2\lambda^2\right)}{(1+\zeta_f)}, \qquad (36)$$

$$J(\lambda) = \left(-\overline{W}_{\overline{xxx}}(\overline{x},0)\,e^{-i\lambda\overline{x}}\Big|_0^1 - i\lambda\overline{W}_{\overline{xx}}(\overline{x},0)\,e^{-i\lambda\overline{x}}\Big|_0^1 \right.$$

$$+ \left(\lambda^2 - \zeta_f\overline{U}^2\right)\overline{W}_{\overline{x}}(\overline{x},0)\,e^{-i\lambda\overline{x}}\Big|_0^1$$

$$+ \left(i\lambda^3 - \zeta_f\overline{U}^2\right)\overline{W}(\overline{x},0)\,e^{-i\lambda\overline{x}}\Big|_0^1$$

$$\left. - \zeta_f\overline{U}\frac{d}{d\bar{t}}\left(\overline{W}(\overline{x},0)\,e^{-i\lambda\overline{x}}\Big|_0^1\right)\right). \qquad (37)$$

The general solution for the deflection response of the system is hence given as

$$\overline{W}^F(\lambda,\bar{t}) = A(\lambda)\,e^{-i\Omega_1\bar{t}} + B(\lambda)\,e^{-i\Omega_2\bar{t}} + \frac{J(\lambda)}{\Omega_n^2(1+\zeta_f)}, \qquad (38)$$

where

$$A(\lambda) = -\frac{1}{\Omega_1 - \Omega_2}\left(\Omega_1 H(\lambda,0) - \frac{\Omega_2 J(\lambda)}{\Omega_n^2(1+\zeta_f)}\right)$$

$$- \frac{2\Omega_{\text{cor}}}{\Omega_1 - \Omega_2}H(\lambda,0),$$

$$B(\lambda) = \frac{2\Omega_{\text{cor}}}{\Omega_1 - \Omega_2}H(\lambda,0) \qquad (39)$$

$$+ \frac{1}{\Omega_1 - \Omega_2}\left(\Omega_2 H(\lambda,0) - \frac{\Omega_1 J(\lambda)}{\Omega_n^2(1+\zeta_f)}\right).$$

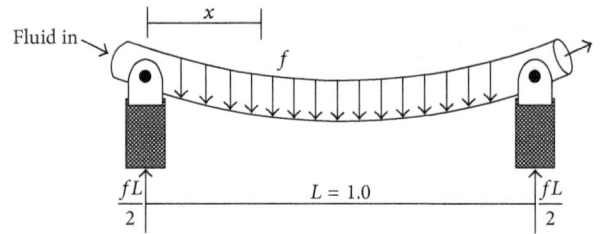

FIGURE 1: Simply supported beam with distributed load.

Therefore, in the Fourier plane, the dynamic response is obtained as

$$\overline{W}^F(\lambda,\bar{t})$$

$$= -\frac{\Omega_2 e^{-i\Omega_2\bar{t}} - \Omega_1 e^{-i\Omega_1\bar{t}} - 2\Omega_{\text{cor}}\left(e^{-i\Omega_1\bar{t}} - e^{-i\Omega_2\bar{t}}\right)}{\Omega_1 - \Omega_2}$$

$$\cdot H(\lambda,0) - \frac{\left(\Omega_1 e^{-i\Omega_2\bar{t}} - \Omega_2 e^{-i\Omega_1\bar{t}}\right) - (\Omega_1 - \Omega_2)}{(\Omega_1 - \Omega_2)(1+\zeta_f)} \qquad (40)$$

$$\cdot \frac{J(\lambda)}{\Omega_n^2},$$

where

$$H(\lambda,0) = \Im\left\{\overline{W}(\overline{x},0)\right\}. \qquad (41)$$

We now examine the deflection responses for three cases of horizontal pipeline system with regard to solution (40) above.

Case 1.

Pipe with Simple Supports at Both Ends. In this case, as shown in Figure 1, the initial configuration of the pipe before it is dynamically excited shows that deflection is symmetric about the middle of the beam (or the pipe).

For this case, when the system is at the static state, time $\bar{t} = 0$, the flow velocity is also zero, though the pipe might have trapped some fluid within it. (See Appendix for further analysis.)

The deflection function is given by Nash [25]:

$$\overline{W} = \overline{w}\left(\frac{\overline{x}^3}{12} - \frac{\overline{x}^4}{24} - \frac{\overline{x}}{24}\right), \qquad (42)$$

where $\overline{w} = \beta_f(1 + \zeta_f)$ and $\beta_f = m_p L/2E\widehat{I}$.

Now, (37) reduces to

$$J_{ss}(\lambda) = \left(-\overline{W}_{\overline{xxx}}(\overline{x},0)\,e^{-i\lambda\overline{x}}\Big|_0^1 \right.$$

$$\left. + \left(\lambda^2 - \zeta_f\overline{U}^2\right)\overline{W}_{\overline{x}}(\overline{x},0)\,e^{-i\lambda\overline{x}}\Big|_0^1\right) \qquad (43)$$

so that

$$J_{ss}(\lambda) = \frac{\beta_f}{12}\left(1+\zeta_f\right)\left\{\left(\lambda^2 - \zeta_f\overline{U}^2 + 12\right)\left(e^{-i\lambda}-1\right)\right\}, \quad (44)$$

$$H(\lambda, 0) = \mathfrak{S}\left\{\overline{W}(\overline{x}, 0)\right\}$$

$$= \beta_f\left(1+\zeta_f\right)\int_0^1 \left(\frac{\overline{x}^3}{12} - \frac{\overline{x}^4}{24} - \frac{\overline{x}}{24}\right)e^{-i\lambda\overline{x}}d\overline{x}. \quad (45)$$

The inverse Fourier transform is given as

$$\overline{W}(\overline{x}, \overline{t}) = \frac{1}{\sqrt{2\pi}}\int_{-\infty}^{\infty}\overline{W}(\lambda, \overline{t})\,e^{i\lambda\overline{x}}d\lambda. \quad (*a)$$

With $\overline{t} = 0$ and using (45) and (40) the response becomes

$$\overline{W}(\overline{x}, \overline{t}) = \left(-\int_0^\infty \frac{\Omega_2 e^{-i\Omega_2\overline{t}} - \Omega_1 e^{-i\Omega_1\overline{t}} - 2\Omega_{\text{cor}}\left(e^{-i\Omega_1\overline{t}} - e^{-i\Omega_2\overline{t}}\right)}{\Omega_1 - \Omega_2}H(\lambda, 0)\,e^{i\lambda\overline{x}}d\lambda \right.$$
$$\left. -\int_0^\infty \frac{\left(\Omega_1\left(e^{-i\Omega_2\overline{t}}-1\right) - \Omega_2\left(e^{-i\Omega_1\overline{t}}-1\right)\right)}{\left(\Omega_1 - \Omega_2\right)}\frac{J_{ss}(\lambda)}{\Omega_n^2}e^{i\lambda\overline{x}}d\lambda \right). \quad (*a')$$

That is,

$$\overline{W}(\overline{x}, \overline{t}) = \overline{w}\left\{\left(\frac{\Omega_2 e^{-i\Omega_2\overline{t}} - \Omega_1 e^{-i\Omega_1\overline{t}} - 2\Omega_{\text{cor}}\left(e^{-i\Omega_1\overline{t}} - e^{-i\Omega_2\overline{t}}\right)}{\Omega_2 - \Omega_1}\right)\left(\frac{\overline{x}^3}{12} - \frac{\overline{x}^4}{24} - \frac{\overline{x}}{24}\right)\right.$$
$$\left. +\int_0^\infty \frac{\left(\Omega_1\left(e^{-i\Omega_2\overline{t}}-1\right) - \Omega_2\left(e^{-i\Omega_1\overline{t}}-1\right)\right)\left(\lambda^2 - \zeta_f\overline{U}^2 + 12\right)\left(e^{-i\lambda}-1\right)}{24\left(\Omega_1 - \Omega_2\right)}\frac{}{\lambda^2\left(\lambda^2 - \zeta_f\overline{U}^2\right)}d\lambda\right\}. \quad (*b)$$

On enforcing the dynamic boundary conditions, namely, $\overline{W}(0, \overline{t}) = \overline{W}(1, \overline{t}) = 0$; $\overline{W}_{\overline{x}\overline{x}}(0, \overline{t}) = \overline{W}_{\overline{x}\overline{x}}(1, \overline{t}) = 0$, the system's dynamic response is given by

$$\overline{W}(\overline{x}, \overline{t})$$
$$= \overline{w}\left\{\left(\frac{\Omega_2 e^{-i\Omega_2\overline{t}} - \Omega_1 e^{-i\Omega_1\overline{t}} - 2\Omega_{\text{cor}}\left(e^{-i\Omega_1\overline{t}} - e^{-i\Omega_2\overline{t}}\right)}{\Omega_2 - \Omega_1}\right)\right.$$
$$\cdot\left(\frac{\overline{x}^3}{12} - \frac{\overline{x}^4}{24} - \frac{\overline{x}}{24}\right) + i$$
$$\cdot\frac{\left(\Omega_1\left(e^{-i\Omega_2\overline{t}}-1\right) - \Omega_2\left(e^{-i\Omega_1\overline{t}}-1\right)\right)}{2n\pi\left(\Omega_1 - \Omega_2\right)}\left\{\sin n\pi(1-\overline{x})\right.$$
$$\left.\left. -\sin n\pi\overline{x}\right\}\right\} \quad (46)$$

or

$$\overline{W}(\overline{x}, \overline{t})$$
$$= \overline{w}\left\{\left(\frac{\Omega_2 e^{-i\Omega_2\overline{t}} - \Omega_1 e^{-i\Omega_1\overline{t}} - 2\Omega_{\text{cor}}\left(e^{-i\Omega_1\overline{t}} - e^{-i\Omega_2\overline{t}}\right)}{\Omega_2 - \Omega_1}\right)\right.$$

$$\cdot\left(\frac{\overline{x}^3}{12} - \frac{\overline{x}^4}{24} - \frac{\overline{x}}{24}\right) + i$$
$$\cdot\frac{\left(\Omega_1\left(e^{-i\Omega_2\overline{t}}-1\right) - \Omega_2\left(e^{-i\Omega_1\overline{t}}-1\right)\right)}{2n\pi\left(\Omega_1 - \Omega_2\right)}\left(1-(-1)^{n+1}\right)$$
$$\left.\cdot\sin n\pi\overline{x}\right\}, \quad (47)$$

where

$$\Omega_1 = -\Omega_{\text{cor}} - \sqrt{\Omega_{\text{cor}}^2 + \Omega_n^2},$$
$$\Omega_2 = -\Omega_{\text{cor}} + \sqrt{\Omega_{\text{cor}}^2 + \Omega_n^2},$$
$$\Omega_{\text{cor}} = \frac{(\Omega_2 + \Omega_1)}{i2} = \frac{\zeta_f\overline{U}\lambda}{(1+\zeta_f)},$$
$$\Omega_n^2 = \left(\frac{\left(\lambda^4 - \zeta_f\overline{U}^2\lambda^2\right)}{(1+\zeta_f)}\right) \quad (48)$$
$$\lambda = n\pi;$$
$$\forall \lambda = \overline{U}_{\text{cr}} = \frac{n\pi}{\sqrt{\zeta_f}} \quad \text{(See Appendix A for expanded analysis).}$$

To enable us understand fully the characteristics of the natural frequencies of such a pipeline system, the three scenarios, namely, critical, subcritical, and postcritical flow points, are examined; that is, comparing (30) and (34), we deduce

$$1 - \Gamma^2 = \frac{\lambda^2}{\Omega_0^2 \left(1 + \zeta_f\right)} \left(\lambda^2 - \zeta_f \overline{U}^2\right), \qquad (49)$$

where

$$\Gamma^2 = \frac{\overline{U}^2}{\overline{U}_{cr}^2}. \qquad (50)$$

The arguments for computing the residues in (40) are $\lambda = \pm \overline{U}\sqrt{\zeta_f}$, when

$$\frac{\lambda^2 \left(\lambda^2 - \zeta_f \overline{U}^2\right)}{\left(1 + \zeta_f\right)} = 0. \qquad (51)$$

Nonetheless, a critical flow point $\Gamma = 1$ is attained for any corresponding velocity, when (47) is zero; that is,

$$\overline{U} = \overline{U}_{cr}. \qquad (52)$$

For a subcritical flow $\Gamma < 1$, this corresponds to the characteristics of the real part of natural frequency against flow velocity as normally seen in literature; that is,

$$\overline{U} < \overline{U}_{cr}, \qquad (53)$$

and for any postcritical flow point, $\Gamma > 1$ which corresponds to the characteristics of the imaginary part of natural frequency against flow velocity; that is,

$$\overline{U} > \overline{U}_{cr}. \qquad (54)$$

The point to note here is that, in actual practice, continuity must exist. As such the natural frequency cannot be zero perpetually for postcritical flow velocity. This necessitated the essence of the plot of absolute characteristics of the natural frequency for all regimes of flow, as demonstrated in this paper. These three scenarios are described graphically in Section 5.

Case 2. A cantilever pipe is described.
     From Figure 2, the initial configuration is described by

$$\overline{W}(\overline{x}, 0) = \frac{\overline{w}}{24} \left(6\overline{x}^2 + 4\overline{x}^3 - \overline{x}^4\right); \qquad (*\,*\,a)$$

thus,

$$J_c(\lambda) = \left(\overline{W}_{\overline{x}\overline{x}\overline{x}}(0,0) - i\lambda \overline{W}_{\overline{x}\overline{x}}(0,0)\right.$$
$$+ \left(\lambda^2 - \zeta_f \overline{U}^2\right) \overline{W}_{\overline{x}}(1,0) e^{-i\lambda} \qquad (*\,*\,a')$$
$$+ \left(i\lambda^3 - \zeta_f \overline{U}^2\right) \overline{W}(1,0) e^{-i\lambda}\right).$$

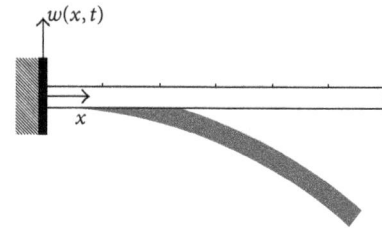

FIGURE 2: A fluid-conveying cantilever beam.

$J_c(\lambda)$ becomes

$$J_c(\lambda)$$
$$= \overline{w}\left\{-12(2 - i\lambda) + \left(4\lambda^2 + i\lambda^3 - 5\overline{U}^2\zeta_f\right)e^{-i\lambda}\right\}, \qquad (*\,*\,b)$$

$$H(\lambda, 0) = \Im\left\{\overline{W}(\overline{x}, 0)\right\}$$
$$= \frac{\overline{w}}{24} \int_0^1 \left(6\overline{x}^2 + 4\overline{x}^3 - \overline{x}^4\right) e^{-i\lambda\overline{x}} d\overline{x}. \qquad (55)$$

The response for the fluid-conveying cantilever beam is therefore given as

$$\overline{W}(\overline{x}, \overline{t})$$
$$= \overline{w}\left\{\frac{1}{24}\left(\frac{\Omega_2 e^{-i\Omega_2\overline{t}} - \Omega_1 e^{-i\Omega_1\overline{t}} - 2\Omega_{cor}\left(e^{-i\Omega_1\overline{t}} - e^{-i\Omega_2\overline{t}}\right)}{\Omega_2 - \Omega_1}\right)\right.$$
$$\cdot \left(6\overline{x}^2 + 4\overline{x}^3 - \overline{x}^4\right) - \frac{\Omega_1\left(e^{-i\Omega_2\overline{t}} - 1\right) - \Omega_2\left(e^{-i\Omega_1\overline{t}} - 1\right)}{\left(1 + \zeta_f\right)\left(\Omega_1 - \Omega_2\right)} \qquad (56)$$
$$\left.\cdot \frac{J_c(\lambda) e^{i\lambda\overline{x}} d\lambda}{\Omega_n^2}\right\};$$

that is,

$$\overline{W}(\overline{x}, \overline{t})$$
$$= \frac{\overline{w}}{24}\left\{\left(\frac{\Omega_2 e^{-i\Omega_2\overline{t}} - \Omega_1 e^{-i\Omega_1\overline{t}} - 2\Omega_{cor}\left(e^{-i\Omega_1\overline{t}} - e^{-i\Omega_2\overline{t}}\right)}{\Omega_2 - \Omega_1}\right)\right.$$
$$\cdot \left(6\overline{x}^2 + 4\overline{x}^3 - \overline{x}^4\right) - i4$$
$$\cdot \frac{\Omega_1\left(e^{-i\Omega_2\overline{t}} - 1\right) - \Omega_2\left(e^{-i\Omega_1\overline{t}} - 1\right)}{2\lambda\left(\Omega_1 - \Omega_2\right)}\left(-24\sin\lambda\overline{x}\right.$$
$$\left.\left. - \lambda^2\sin\lambda(1 - \overline{x}) + \lambda\cos\lambda\overline{x} + \lambda^3\cos\lambda(1 - \overline{x})\right)\right\}, \qquad (57)$$

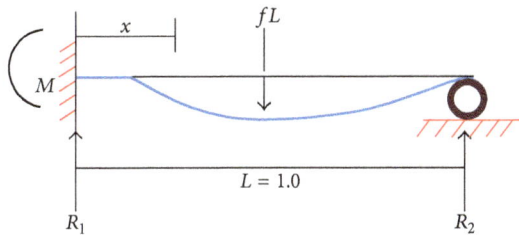

FIGURE 3: A clamped-pinned ended fluid-carrying pipe.

where $\overline{w} = \beta_f(1 + \zeta_f)$

$$\Omega_1 = -\Omega_{\text{cor}} + \sqrt{\Omega_{\text{cor}}^2 - \Omega_n^2},$$

$$\Omega_2 = -\Omega_{\text{cor}} - \sqrt{\Omega_{\text{cor}}^2 - \Omega_n^2},$$

$$\Omega_{\text{cor}} = i\frac{\overline{U}\zeta_f\lambda}{1 + \zeta_f}, \tag{58}$$

$$\Omega_n = \sqrt{\frac{\lambda^4 - \lambda^2\overline{U}^2\zeta_f}{1 + \zeta_f}},$$

for which the characteristic equation at $\overline{x} = 0, 1$ is given by

$$\left(2 - \lambda^2\right)\tan\lambda - 3\lambda = 0, \tag{59}$$

Case 3. For clamped and pinned ended supports: see Figure 3.

From Figure 4, the initial configuration is

$$\overline{W}(\overline{x}, 0) = \frac{\overline{w}}{48}\overline{x}^2\left(-3 + 5\overline{x} - 2\overline{x}^2\right). \tag{60}$$

Dynamic response of the clamped-pinned pipeline is given as

$$\overline{W}(\overline{x}, \overline{t})$$

$$= \frac{\overline{w}}{48}\left\{\left(\frac{\Omega_2 e^{-i\Omega_2\overline{t}} - \Omega_1 e^{-i\Omega_1\overline{t}} - 2\Omega_{\text{cor}}\left(e^{-i\Omega_1\overline{t}} - e^{-i\Omega_2\overline{t}}\right)}{\Omega_2 - \Omega_1}\right)\right.$$

$$\cdot \overline{x}^2\left(-3 + 5\overline{x} - 2\overline{x}^2\right) - i$$

$$\cdot \frac{1}{\lambda}\left(\frac{\Omega_1\left(e^{-i\Omega_2\overline{t}} - 1\right) - \Omega_2\left(e^{-i\Omega_1\overline{t}} - 1\right)}{(\Omega_1 - \Omega_2)}\right) \tag{61}$$

$$\left. \cdot \left(6\lambda\cos\lambda\overline{x} - 30\sin\lambda\overline{x} - 18\sin\lambda(1 - \overline{x})\right)\right\},$$

where

$$\Omega_1 = -\Omega_{\text{cor}} + \sqrt{\Omega_{\text{cor}}^2 - \Omega_n^2},$$

$$\Omega_2 = -\Omega_{\text{cor}} - \sqrt{\Omega_{\text{cor}}^2 - \Omega_n^2},$$

$$\Omega_{\text{cor}} = i\frac{\overline{U}\zeta_f\lambda}{1 + \zeta_f}, \tag{62}$$

$$\Omega_n = \sqrt{\frac{\lambda^4 - \lambda^2\overline{U}^2\zeta_f}{1 + \zeta_f}}$$

with the characteristic equation for the values of λ at both ends as

$$\tan\lambda - \frac{5}{3}\lambda = 0 \tag{63}$$

and Appendix B gives the summary of the above results.

5. Analysis of Results and Discussion

Although the effect of Coriolis force on the dynamics of a fluid-conveying pipeline has been well known over the decades, early studies were shrouded in ad hoc heuristic arguments on the one hand and elegant interpretation of physical and numerical experiments on the other. The challenge has been to find a simple and straightforward way of demonstrating what has come to be accepted as the general pattern of behaviour of these systems.

In several investigations to date on the general behaviour and conditions of stability associated with unstable oscillation of pipe conveying fluid, researchers were preoccupied with finding answers to two basic questions, namely,

(a) constructing the general pattern of the dynamic response of the system,

(b) establishing the region and or regime where such a solution holds validity.

Within the context of linear theory, the latter question invariably involves determining the critical velocity of internal fluid flow for the system while the former exercise has unraveled curious and unexpected paradoxical patterns of flow.

Most of the difficulties encountered can be traced to the methodology used in tackling these problems wherein reliance was put on solving abridged equations whose solutions were fortified with powerful and masterly interpretation of physical and numerical experimental results.

In this short note we recovered some of the well-known results from a straightforward application of Fourier complex transform to the full linear equations. This affords us the opportunity to

(a) obtain closed form explicit expressions for classical flow variables such as the critical flow velocity and natural frequency of the system,

(b) avoid the tedium of laborious numerical work by being able to locate the eigenvalues of the critical flow velocity in a relatively simple fashion,

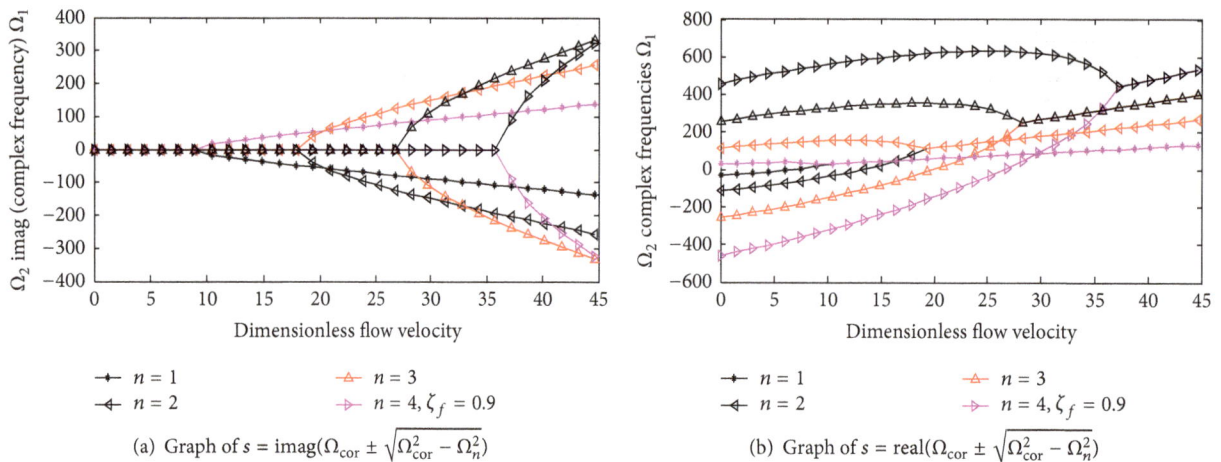

(a) Graph of $s = \text{imag}(\Omega_{\text{cor}} \pm \sqrt{\Omega_{\text{cor}}^2 - \Omega_n^2})$

(b) Graph of $s = \text{real}(\Omega_{\text{cor}} \pm \sqrt{\Omega_{\text{cor}}^2 - \Omega_n^2})$

FIGURE 4: Variation of complex frequency with internal axial flow velocity for different modes.

(c) undertake easy comparative parametric analysis of the variables contained in the solution without heavy reliance on numerical computation,

(d) guarantee, ab initio, that the derived solution complies with the exact or complete configuration and profile of the entire pipe length at onset so that the evolution of the final configuration as a direct result of the initial pipe profile does not constitute an issue,

(e) also provide confirmation of predictions of earlier workers and facilitates a better understanding, organization, and interpretation of some of the phenomena reported to date in the literature.

In respect of the case of a simply supported beam hanging above the ground, our investigation clearly shows from (26) that the frequency associated with the Coriolis force is partly driven by the mass ratio of the internal fluid and increases linearly with the flow velocity. The critical flow velocity is found to be dependent on the fluid-to-pipe mass ratio ζ_f, thereby showing that it is a function of the mass of the fluid flowing in the pipe. Equations (47), (57), and (61) also demonstrated that the system's dynamic responses are dependent on the initial configurations of the pipeline system and the boundaries' forces and moments.

The profiles illustrated (Figures 4–6) confirmed the mirror imaging or characteristics for the real, imaginary, and absolute values of the complex frequencies. At the critical points, flow bifurcations are demonstrated. As for the absolute values of the complex frequency pairs, the critical values and the two regimes of the beam divergence are clearly represented in the figures. We have equally noted that, for a given mass ratio, critical flow velocities are ordered in consonant with the number of modes.

The family of curves for the natural frequency of the system are illustrated (Figures 7–9) as a function of the flow velocity and fluid-pipe mass ratio. With (34) and (52), the subcritical velocity points corresponding to the predivergence values of the system are shown by the curves representing the imaginary parts of Ω_n against the flow velocity $\overline{U}$ in Figures 7(a) and 8(a) and against mass fluid

ratio ζ_f in Figure 9(a). Also, as described by (34) and (53) and Figures 7(b), 8(b), and 9(b), the postcritical velocity points corresponding to the postdivergence values are demonstrated for the real part of Ω_n, while the curves for the absolute values of Ω_n as illustrated with (51) show clearly the critical points for both the flow velocity and the fluid mass ratio. The curves for both the imaginary and real parts of Ω_n appear to depict quasi-static configurations from zero flow velocity and mass fluid ratio to the neighbourhood of the critical points and from the same neighbourhood, such quasi-static behaviour is observable as the flow velocity and mass fluid ratio increases beyond the critical points, respectively.

From the foregoing, it does not sound reasonable and possible even from actual experiments for a buckled fluid-conveying pipeline to exhibit a quasi-static geometrical configuration beyond the critical points ad infinitum as shown theoretically in literature. It therefore seems scientifically sensible to study the dynamic response and natural frequency characteristics of a pipeline from the subcritical through critical and postcritical points in order to clearly describe the behavioural divergences of the system.

As shown in Figures 1 and 10, buckling occurs even in the absence of flow for a pipeline segment simply supported at both ends. This condition arises from the deformation of the neutral axis of an originally straight beam, at equilibrium, but is subjected to the action of its weight and the two end reactions. Thus, for a pipeline segment closed at both ends that is carrying an entrained static fluid, the degree of buckling increases as the mass of internal flow increases. However, if dynamic buckling occurs when a fully developed fluid is flowing through the pipe under the same conditions at any flow velocity, the direction of buckling alternates different values of flow velocity as demonstrated in Dodds Jr. and Runyan (1965) [18] for the case of the principal buckling mode $n = 1$ and the behavioural pattern of curves in Figure 10. As demonstrated in this figure, the curves describing the physics of dynamically excited simply supported pipeline systems are parabolas and then show that the downward sagging of the pipe for time $\bar{t} = 0$ and as the pipe is dynamically moving due to the fluid

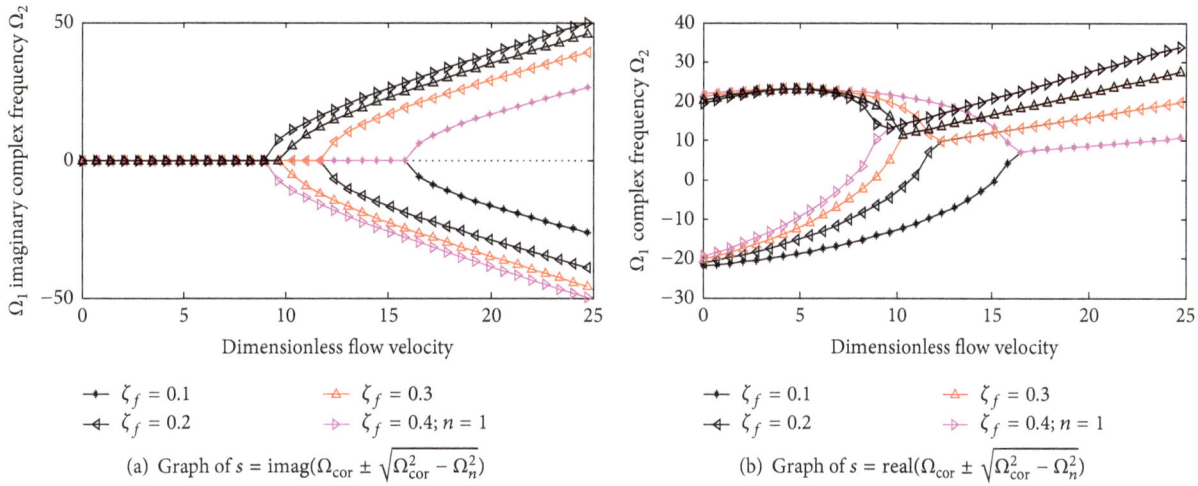

(a) Graph of $s = \text{imag}(\Omega_{\text{cor}} \pm \sqrt{\Omega_{\text{cor}}^2 - \Omega_n^2})$

(b) Graph of $s = \text{real}(\Omega_{\text{cor}} \pm \sqrt{\Omega_{\text{cor}}^2 - \Omega_n^2})$

FIGURE 5: Variation of complex frequency with axial flow velocity for different fluid mass ratios.

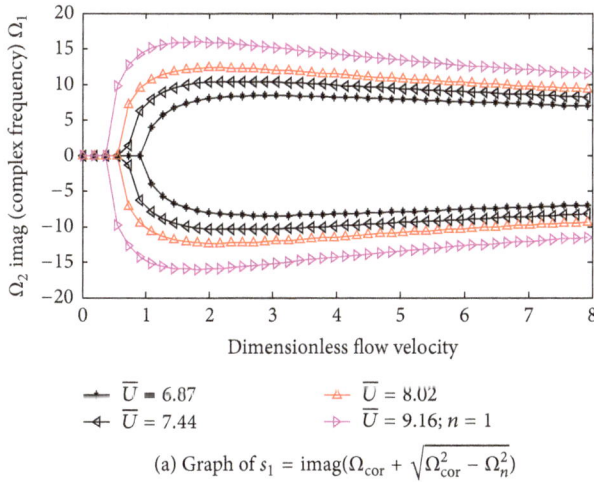

(a) Graph of $s_1 = \text{imag}(\Omega_{\text{cor}} + \sqrt{\Omega_{\text{cor}}^2 - \Omega_n^2})$

FIGURE 6: Movement of critical flow velocity as a function of fluid mass ratio for different axial flow velocities.

flow, depending on its magnitude, the pipe characteristics alternate to give convex and concave curvatures. More so, as interchange of energy develops between the generated Coriolis force and the system, symmetric and antisymmetric oscillations emerge. Also, the slow-moving frequency driven responses enveloping the fast-moving frequency driven ones are demonstrated in Figure 11, thus generating two distinct modes for the system.

As for Case 2 results, in Figures 12–17, the frequency curves for both the complex and natural frequencies are demonstrated to be similar to those for the simply supported pipe, except that their wave numbers are smaller λ values and of complex forms as the transcendental characteristic equation is solved. The fluttering behaviours that are expected of cantilever systems are illustrated in Figures 15 and 16. This naturally explains the results that a pipe with a free-end gives. For these pipeline arrangements, the response-time characteristics shown in Figure 17 explain the envelopments

of the fast-moving frequency responses by their slow-moving counterparts and that as time increases the amplitudes decay. These indeed show that the system performs what is known as beats phenomenon. These features are of paramount significance for design purposes.

Illustrated by Figures 18 and 19 are the results for Case 3, that is, the clamped-pinned pipeline segment. In the figures, the dynamic responses against the pipe lengths are clearly demonstrated as the clamped end reveals the zero gradients due to the moments and reactions at that end as well as zero moments and deflections at the simply supported end.

Case 1 Results. For responses of the pipe simply supported at both ends see Figures 4–11.

Case 2. For results of the cantilever pipe, see Figures 12–17.

Case 3 Results. For responses of the clamped-pinned ends pipe, see Figures 18 and 19.

6. Conclusion

The suitability of Fourier complex exponential transform method for the solution of the homogeneous Coriolis-term dependent mathematical physics equation governing the flow-induced vibration propelled by the conveyance of an incompressible fluid through a pipe segment is presented. The study assumed a linear theory for the fluid structure interaction mechanics where the relevant forces are properly accounted for.

Notably, this fluid structure interaction process as discussed in this paper has the tendency to induce the occurrences of conjugate modal complex and natural frequencies for all flow velocities and other associated parametric variables. This conjugateness of the complex frequencies is the prime sources for the initiation of bifurcation responses. Nevertheless, the modal natural frequencies as inferred in the investigation above can be functionally related to the Coriolis frequency. It is instructive to state that their lines

(a) Graph of imag(Ω_n)

(b) Graph of real(Ω_n)

(c) Graph of absolute(Ω_n)

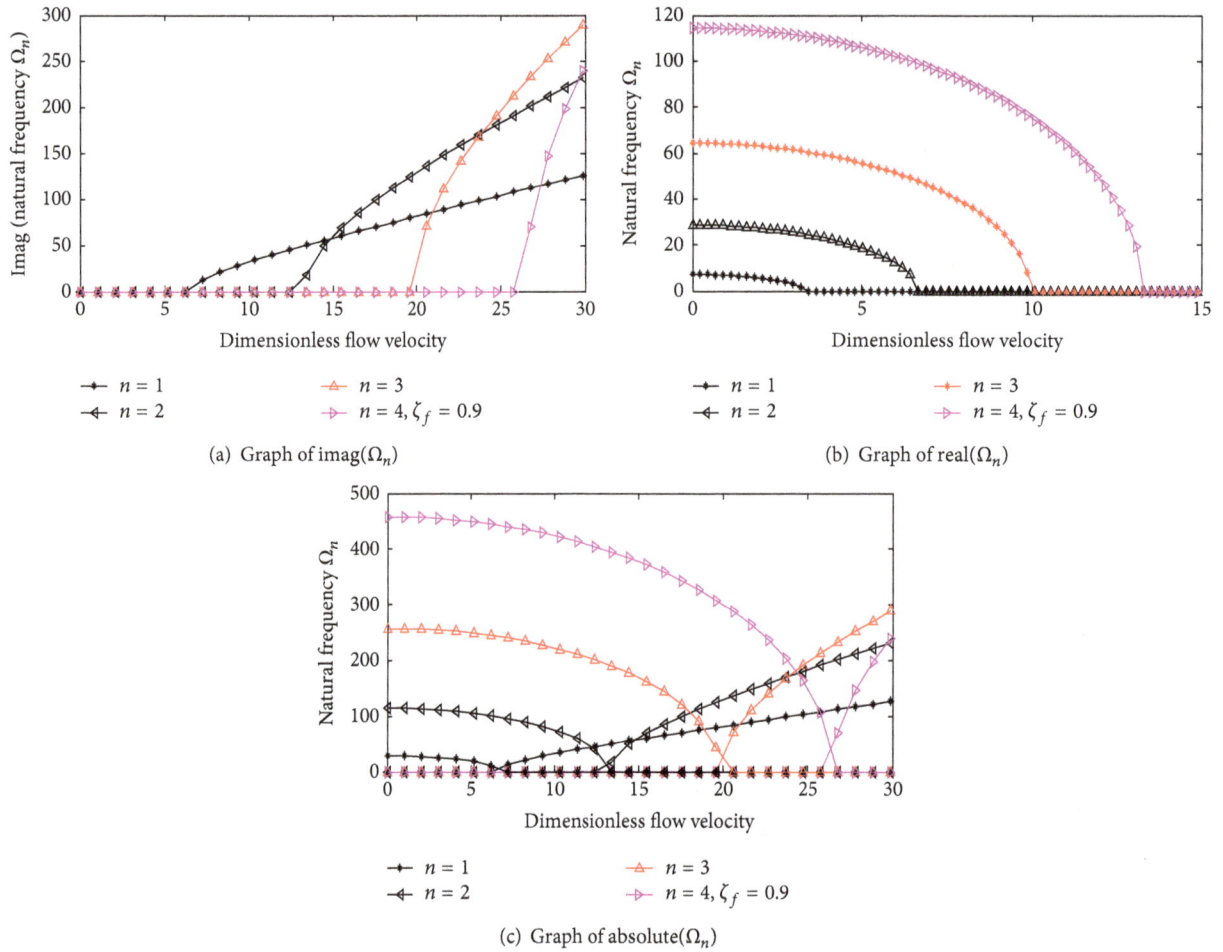

FIGURE 7: Variation of natural frequency with internal axial flow velocity for different modes.

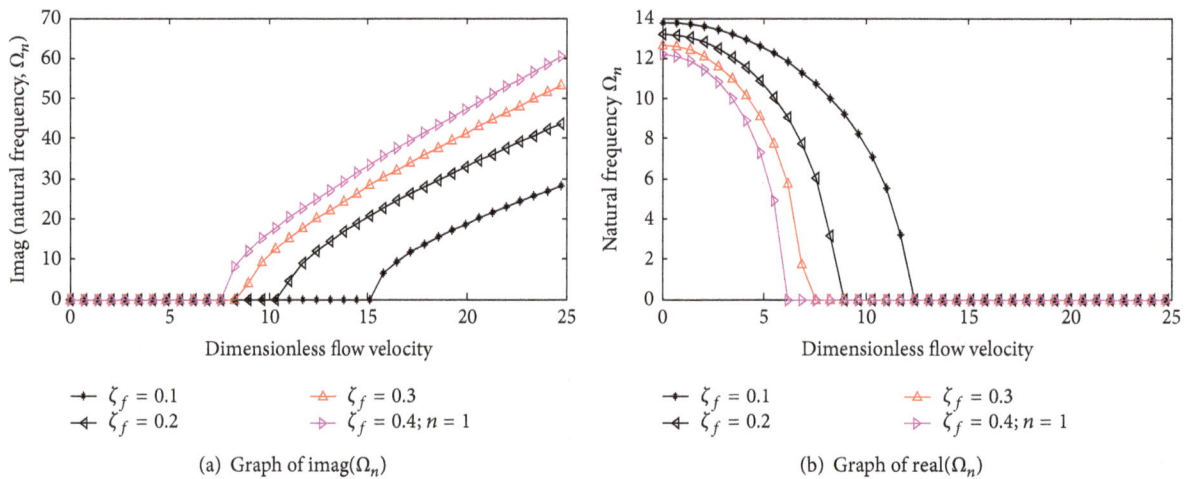

(a) Graph of imag(Ω_n)

(b) Graph of real(Ω_n)

FIGURE 8: Variation of natural frequency with axial flow velocity for different fluid mass ratios.

of action are directed orthogonally to the directrices of the complex frequencies where the Coriolis force is playing a central role. It is also noteworthy to state that whilst the modal natural frequency equation can be alternatively conjured to be independent of Coriolis force as an option, the dynamic responses and complex frequencies are seen to be explicitly tied to the modulating roles of the Coriolis acceleration. Consequently, the use of this transform method can be very apt to conjure approximate closed form solutions for non-linear problems in conjunction with homotopy perturbation

(a) Graph of imag(Ω_n)

(b) Graph of real(Ω_n)

(c) Graph of $\|\Omega_n\|$

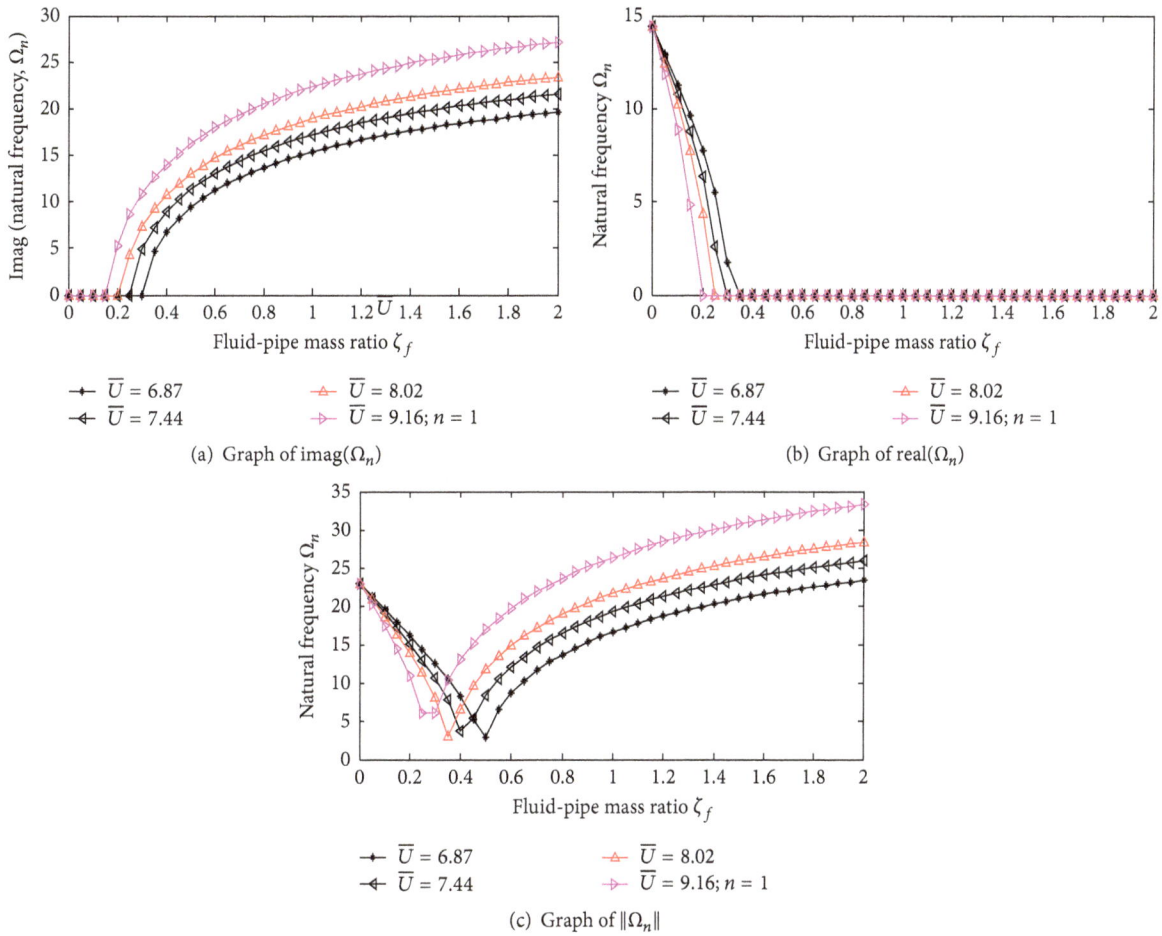

FIGURE 9: Movement of critical flow velocity as a function of fluid mass ratio for different axial flow velocities.

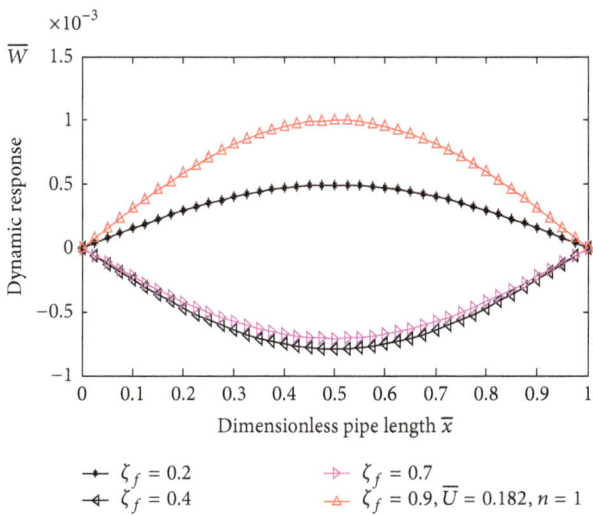

FIGURE 10: Dynamic response of pipe as a function of mass ratio.

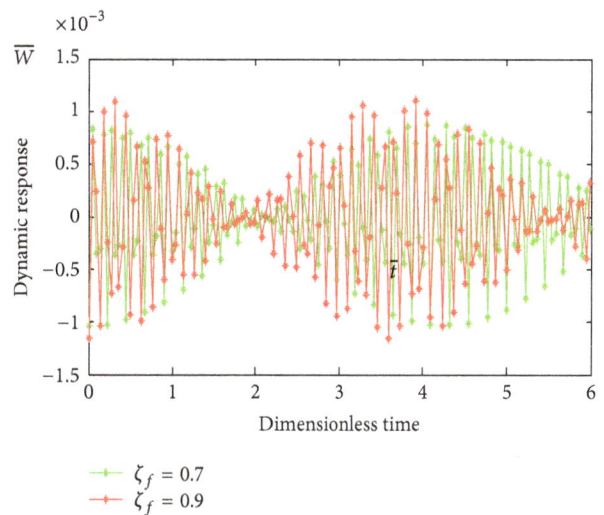

FIGURE 11: Graph of response against dimensionless time for given fluid-pipe mass ratios.

method (HPM), homotopy analysis method, and differential transformation method (DTM), without having recourse to the uses of singular or parameter perturbation method where

the issue of smaller parameters for the perturbation series expansions is an issue.

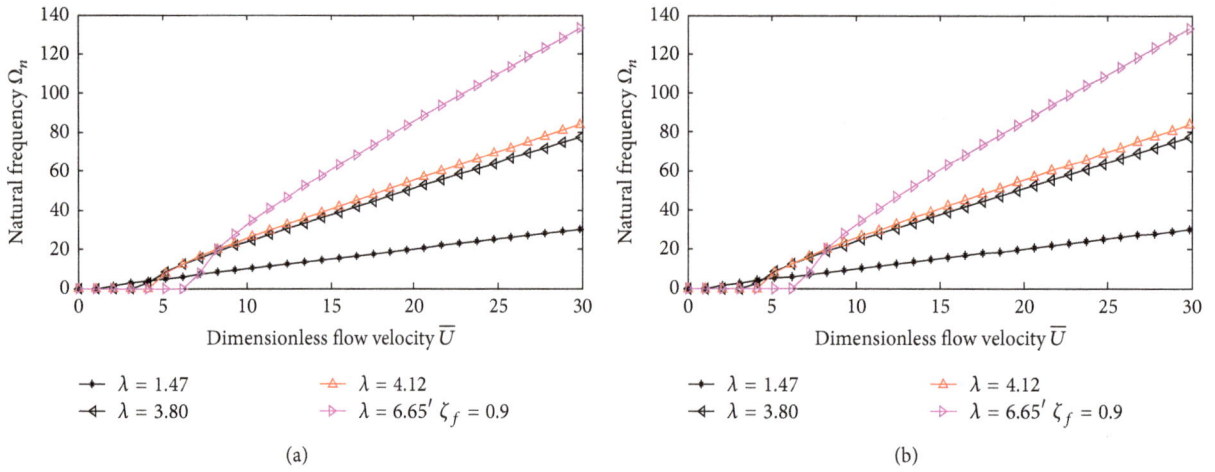

FIGURE 12: (a) Graphs of natural frequency versus flow velocity. (b) Graphs of natural frequency versus flow velocity.

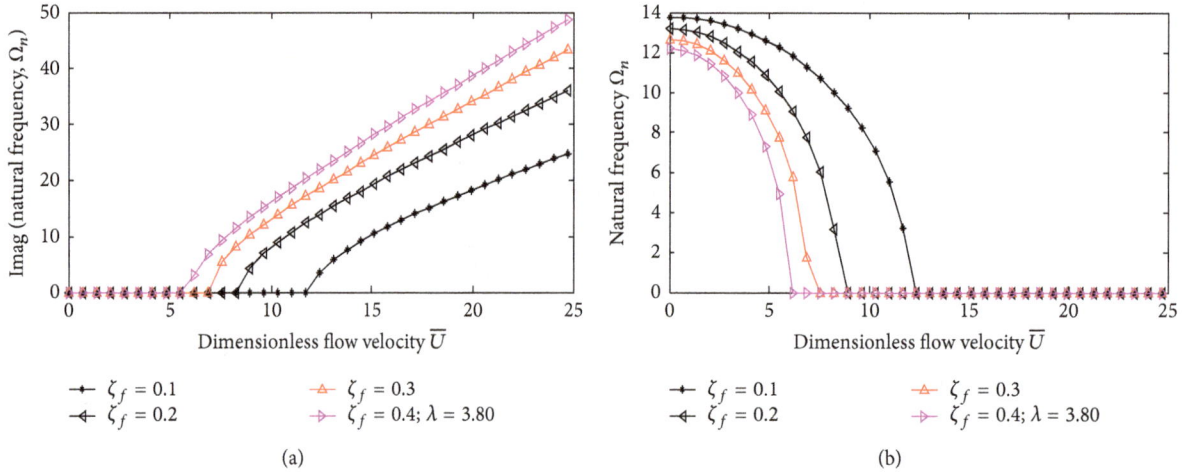

FIGURE 13: (a) Graphs of natural frequency versus flow velocity. (b) Graphs of natural frequency versus flow velocity.

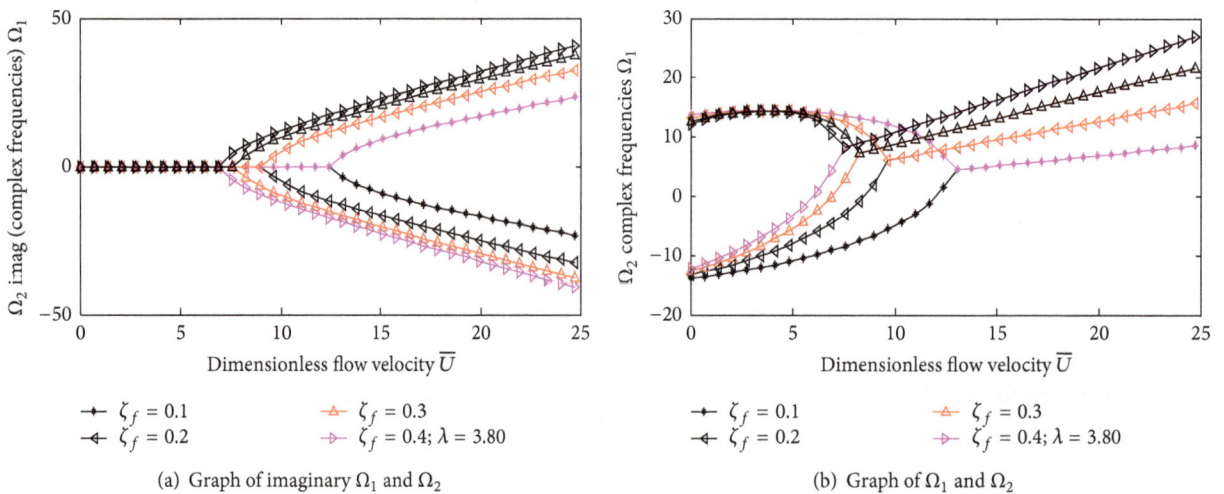

FIGURE 14: Graphs of complex frequency versus flow velocity.

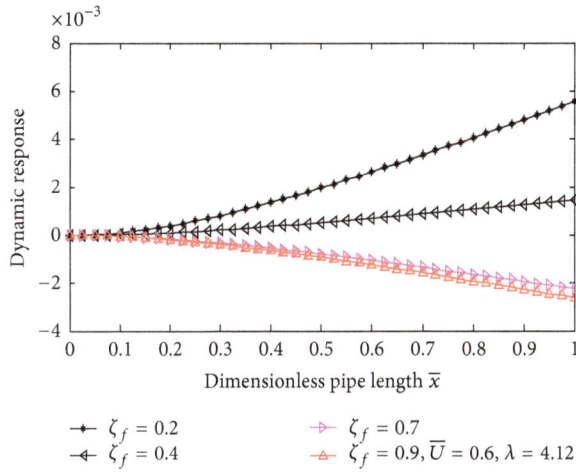

FIGURE 15: Dynamic response of pipe of a given internal flow velocity and different mass ratio.

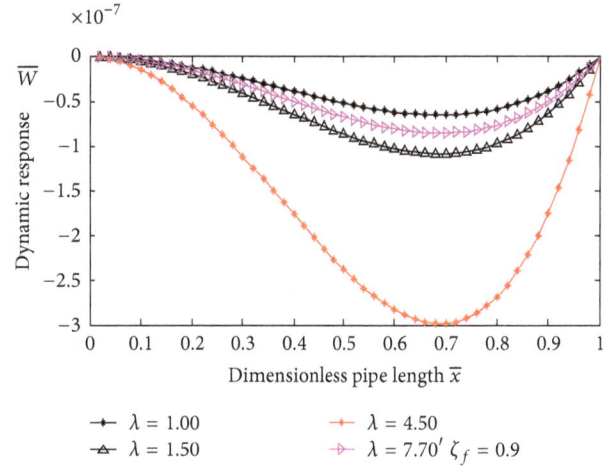

FIGURE 18: Dynamic response of pipe of a given mass ratio for different wave numbers.

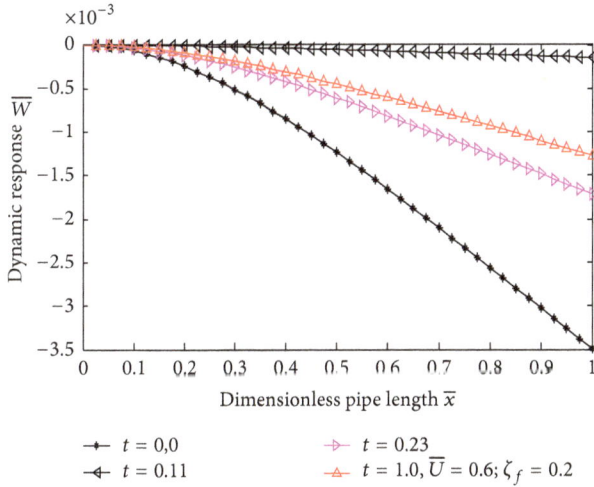

FIGURE 16: Dynamic response of pipe of a given mass ratio for different time.

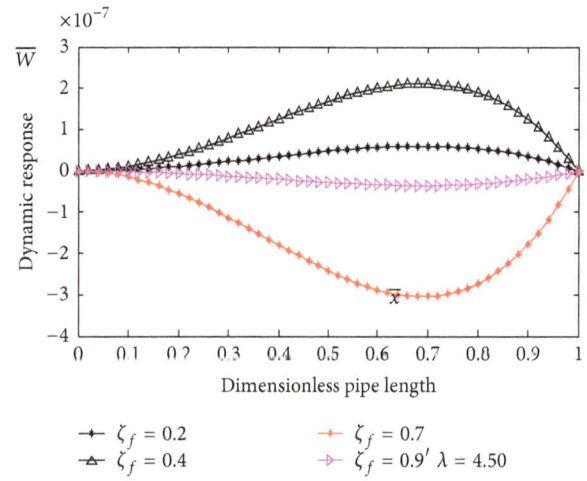

FIGURE 19: Dynamic response of pipe of a given value of λ for different mass ratios.

Appendix

A. Critical Flow Velocity

For Case 1, a pipe simply supported at both ends has physical configuration as shown schematically in Figure 20. Since the pipe and the fluid it is carrying have uniformly distributed downward vertical loads of intensity f on the system, then we consider the shearing force F at a distance $\overline{x}$ from the left end; it is given as shown in Figure 20.

The deflection function is given as

$$\overline{W} = \overline{w}\left(\frac{\overline{x}^3}{12} - \frac{\overline{x}^4}{24} - \frac{\overline{x}}{24} \right), \tag{A.1}$$

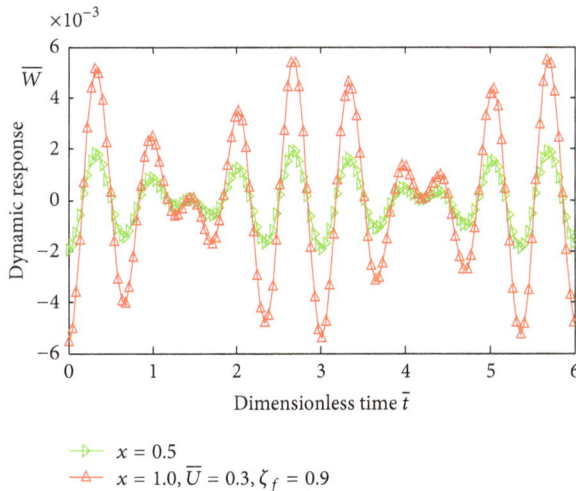

FIGURE 17: Dynamic response of pipe at $x = 0.5$ and 1.0 for a given axial flow velocity and fluid-pipe mass ratio of 0.9.

where $\overline{w} = \beta_f(1 + \zeta_f)$ and $\beta_f = m_p L/2E\widehat{I}$.

TABLE 1: Table of results including other cases.

Cases	Initial configuration $\overline{W}(\overline{x},0)$	$J(\lambda)$	Characteristic equations and $\overline{U}_{cr}$
(1) Pinned-pinned ends	$\overline{W} = \beta_s\left(\dfrac{\overline{x}^3}{6} - \dfrac{\overline{x}^4}{12} - \dfrac{\overline{x}}{12}\right)$	$\left(-\overline{W}_{\overline{x}\overline{x}\overline{x}}(\overline{x},0) + \left(\lambda^2 - \zeta_f\overline{U}^2\right)\overline{W}_{\overline{x}}(\overline{x},0)\right)e^{-i\lambda\overline{x}}\Big|_0^1$	for $\overline{x}=0,1$, $\sin\lambda = 0$; thus $\lambda = n\pi$; $\overline{U}_{cr} = \dfrac{n\pi}{\sqrt{\zeta_f}}$
(2) Cantilever $w(x,t)$	$\overline{W} = \beta_c\{(1-\overline{x})^4 + 4\overline{x} - 1\}$	$\left(\overline{W}_{\overline{x}\overline{x}\overline{x}}(0,0) - i\lambda\overline{W}_{\overline{x}\overline{x}}(0,0) + \left(\lambda^2 - \zeta_f\overline{U}^2\right)\overline{W}_{\overline{x}}(1,0)e^{-i\lambda} + \left(i\lambda^3 - \zeta_f\overline{U}^2\right)\overline{W}(1,0)e^{-i\lambda}\right)$	for $\overline{x}=0,1$, $(2-\lambda^2)\tan\lambda - 3\lambda = 0$; $\overline{U}_{cr} = \dfrac{\lambda}{\sqrt{\zeta_f}}$
(3) Clamped-pinned ends	$\overline{W} = \beta_{cp}\overline{x}^2\left(5 - 2\overline{x}^2 + 3\overline{x}\right)$	$\overline{W}_{\overline{x}\overline{x}\overline{x}}(0,0) - i\lambda\overline{W}_{\overline{x}\overline{x}}(0,0) + \left(\lambda^2 - \zeta_f\overline{U}^2\right)\overline{W}_{\overline{x}}(1,0)e^{-i\lambda} - \overline{W}_{\overline{x}\overline{x}\overline{x}}(1,0)e^{-i\lambda}$	for $\overline{x}=0,1$, $\tan\lambda - \dfrac{3}{5}\lambda = 0$; $\overline{U}_{cr} = \dfrac{\lambda}{\sqrt{\zeta_f}}$
(4) Clamped-clamped ends	$\overline{W}(\overline{x},0) = \dfrac{\overline{w}}{24}\left\{\overline{x}^4 + \overline{x}^2 - 2\overline{x}^3\right\}$	$\left(\overline{W}_{\overline{x}\overline{x}\overline{x}}(\overline{x},0) - i\lambda\overline{W}_{\overline{x}\overline{x}}(\overline{x},0)\right)e^{-i\lambda\overline{x}}\Big|_{\overline{x}=0,1} = J(\lambda)$	for $\overline{x}=0,1$, $\tan\lambda - \dfrac{12\lambda}{36-\lambda^2}$; $\overline{U}_{cr} = \dfrac{\lambda}{\sqrt{\zeta_f}}$

TABLE 1: Continued.

Cases	Initial configuration $\overline{W}(\overline{x},0)$	$I(\lambda)$	Characteristic equations and $\overline{U}_{cr}$
(5) Simply supported viscoelastic foundation $L = 1$, Fluid in, $M + m_f$, R_2, R_1 $p(x,t) = k_o W + \mu \dfrac{dW}{dt}$	$\overline{W} = \beta_f(1+\zeta_f)$ $\times \left(\dfrac{\overline{x}^4}{6} - \dfrac{\overline{x}^5}{4} + \dfrac{2\overline{x}^6}{15} - \dfrac{\overline{x}^7}{42} - \dfrac{11\overline{x}}{420} \right)$	$\left(-\overline{W}_{\overline{x}\overline{x}\overline{x}}(0,0)\, e^{-i\lambda\overline{x}}\Big|_0^1 + \left(\lambda^2 - \zeta_f\overline{U}^2\right)\overline{W}_{\overline{x}}(0,0)\, e^{-i\lambda\overline{x}}\Big|_0^1 \right)$	for $\overline{x} = 0,1$, $\sin\sqrt{\eta} = \sin\lambda = 0$ $\overline{U}_{cr} = \dfrac{n\pi}{\sqrt{\zeta_f}} + \dfrac{k_0}{2n^3\pi^3\sqrt{\zeta_f}}$
(6) Vertical pipe (riser) with end tension R_2, T, z, $A(z,W)$, $T = 1.0$, R_1, T	$\overline{W}(\overline{z},0) = (\overline{w} - 2\pi R\rho_w g)$ $\left(\dfrac{1-\cos\psi}{\psi^4\sin\psi}\left((\psi - \sin\psi)\overline{z} - \psi\overline{z} - \sin\psi\overline{z}\right) - \dfrac{1}{2\psi^4}\left[\begin{array}{l} \psi^2\overline{z} + 2\overline{z}(\cos\psi - 1) \\ -(\psi^2\overline{z}^2 + 2\cos\psi\overline{z} - 2) \end{array} \right] \right)$	$\left(-\overline{W}_{\overline{z}\overline{z}\overline{z}} + \left\{\lambda^2 - \left(\overline{U}^2\zeta_f + \psi^2\right)\right\}\overline{W}_{\overline{z}} \right) e^{-i\lambda\overline{z}}\Big|_0^1$	for $\overline{x} = 0,1, \lambda = n\pi =$ $\sqrt{(\zeta_f\overline{U}^2 + \psi^2)}$ $\overline{U}_{cr} = \dfrac{n\pi}{\sqrt{\zeta_f}}\left(1 - \dfrac{\psi^2}{2n^2\pi^2} \right)$

$\beta_s = \beta_f(1 + \zeta_f); \beta_c = \beta_f(1 + \zeta_f)/24; \beta_{cp} = \beta_f(1 + \zeta_f)/240; \psi = \sqrt{pA(1 - 2v)} + E\alpha\overline{\Delta T} - \overline{\overline{T}}e.$

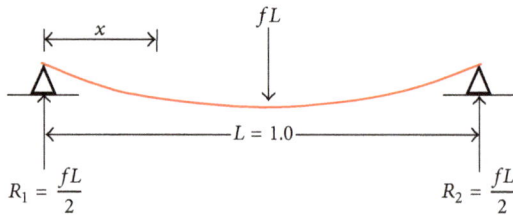

FIGURE 20: Simply supported beam.

In this case, while the shearing force varies linearly, the bending moment varies parabolically along the length of the beam.

$$
\begin{aligned}
\overline{W}(\overline{x},\overline{t}) &= \left\{ \overline{w} \left(\frac{\left(\Omega_2 e^{-i\Omega_1 \overline{t}} - \Omega_1 e^{-i\Omega_2 \overline{t}}\right) - i2\Omega_{\text{cor}}\left(e^{-i\Omega_1 \overline{t}} - e^{-i\Omega_2 \overline{t}}\right)}{(\Omega_2 - \Omega_1)} \right) \right. \\
&\quad \cdot \left(\frac{\overline{x}^4}{24} - \frac{\overline{x}^3}{12} + \frac{\overline{x}}{24} \right) - \frac{\left(\Omega_1\left(e^{-i\Omega_2 \overline{t}} - 1\right) - \Omega_2\left(e^{-i\Omega_1 \overline{t}} - 1\right)\right)}{(\Omega_1 - \Omega_2)} \\
&\quad \left. \cdot \int_0^\infty \frac{J_{ss}(\lambda)}{\Omega_n^2}\, d\lambda \right\};
\end{aligned} \tag{A.2}
$$

$$
-\int_0^\infty \frac{\left(\Omega_1\left(e^{-i\Omega_2 \overline{t}} - 1\right) - \Omega_2\left(e^{-i\Omega_1 \overline{t}} - 1\right)\right)}{12(\Omega_1 - \Omega_2)} \frac{\left\{12\left(2e^{-i\lambda(1-\overline{x})} + e^{i\lambda\overline{x}}\right) + \left(\lambda^2 - \zeta_f \overline{U}^2\right)\left(e^{-i\lambda(1-\overline{x})} + e^{i\lambda\overline{x}}\right)\right\}}{\Omega_n^2}\, d\lambda = 0. \tag{A.5}
$$

That is, for all t,

$$
\int_0^\infty \frac{\left(\Omega_1\left(e^{-i\Omega_2 \overline{t}} - 1\right) - \Omega_2\left(e^{-i\Omega_1 \overline{t}} - 1\right)\right)}{12(\Omega_1 - \Omega_2)} \frac{\left\{12\left(2e^{-i\lambda(1-\overline{x})} + e^{i\lambda\overline{x}}\right) + \left(\lambda^2 - \zeta_f \overline{U}^2\right)\left(e^{-i\lambda(1-\overline{x})} + e^{i\lambda\overline{x}}\right)\right\}}{\left(\lambda^4 - \lambda^2 \overline{U}^2 \zeta_f\right)}\, d\lambda = 0 \tag{A.6}
$$

becoming

$$
\int_0^\infty \frac{\left(\Omega_1\left(e^{-i\Omega_2 \overline{t}} - 1\right) - \Omega_2\left(e^{-i\Omega_1 \overline{t}} - 1\right)\right)}{12(\Omega_1 - \Omega_2)} \frac{\left\{12\left(2e^{-i\lambda(1-\overline{x})} + e^{i\lambda\overline{x}}\right) + \left(\lambda^2 - \zeta_f \overline{U}^2\right)\left(e^{-i\lambda(1-\overline{x})} + e^{i\lambda\overline{x}}\right)\right\}}{\left(\lambda - \sqrt{\eta_1}\right)\left(\lambda + \sqrt{\eta_1}\right)}\, d\lambda = 0, \tag{A.7}
$$

where

$$
\eta_1 = \frac{\overline{U}^2 \zeta_f}{2} \pm \sqrt{\frac{\overline{U}^4 \zeta_f^2}{2}}. \tag{A.8}
$$

that is,

$$
\begin{aligned}
\overline{W}(\overline{x},\overline{t}) &= \overline{w} \left\{ \left(\frac{\left(\Omega_2 e^{-i\Omega_1 \overline{t}} - \Omega_1 e^{-i\Omega_2 \overline{t}}\right) - i2\Omega_{\text{cor}}\left(e^{-i\Omega_1 \overline{t}} - e^{-i\Omega_2 \overline{t}}\right)}{(\Omega_2 - \Omega_1)} \right) \right. \\
&\quad \cdot \left(\frac{\overline{x}^4}{24} - \frac{\overline{x}^3}{12} + \frac{\overline{x}}{24} \right) - \frac{\left(\Omega_1\left(e^{-i\Omega_2 \overline{t}} - 1\right) - \Omega_2\left(e^{-i\Omega_1 \overline{t}} - 1\right)\right)}{12(\Omega_1 - \Omega_2)} \\
&\quad \left. \cdot \int_0^\infty \frac{\left\{\left(\lambda^2 - \zeta_f \overline{U}^2 - 12\right)\left(e^{-i\lambda(1-\overline{x})} + e^{i\lambda\overline{x}}\right)\right\}}{\Omega_n^2}\, d\lambda \right\}.
\end{aligned} \tag{A.3}
$$

But with the boundary conditions, that is, $\overline{W}(0,\overline{t}) = \overline{W}(1,\overline{t}) = 0$, the above equation shows that

$$
\begin{aligned}
\overline{w} &\left(\frac{\left(\Omega_2 e^{-i\Omega_1 \overline{t}} - \Omega_1 e^{-i\Omega_2 \overline{t}}\right) - i2\Omega_{\text{cor}}\left(e^{-i\Omega_1 \overline{t}} - e^{-i\Omega_2 \overline{t}}\right)}{(\Omega_2 - \Omega_1)} \right) \\
&\cdot \left(\frac{\overline{x}^4}{24} - \frac{\overline{x}^3}{12} + \frac{\overline{x}}{24} \right) = 0
\end{aligned} \tag{A.4}
$$

and thus the second term must be zero also; that is,

Using residue theory and contour integration, we deduce that

$$
\lambda = \sqrt{\eta_1}
$$

$$
\text{or } \lambda = \pm\overline{U}\sqrt{\zeta_f}. \tag{A.9}
$$

Adding the residues of (A.7) gives

$$\frac{12}{2\sqrt{\eta_1}}\left\{2\left(e^{-i\sqrt{\eta_1}(1-\bar{x})}-e^{i\sqrt{\eta_1}(1-\bar{x})}\right)\right.$$
$$\left.+\left(e^{i\sqrt{\eta_1}\bar{x}}-e^{-i\sqrt{\eta_1}\bar{x}}\right)\right\}\Big|_{\bar{x}=0,1}=0, \tag{A.10}$$

that is,

$$i\frac{12}{\sqrt{\eta_1}}\left\{-2\sin\sqrt{\eta_1}\left(1-\bar{x}\right)+\sin\sqrt{\eta_1}\bar{x}\right\}\Big|_{\bar{x}=0,1}=0 \tag{A.11}$$

so that when $\bar{x}=0$, at one end, then,

$$\sin\sqrt{\eta_1}=0,\quad\forall\sqrt{\eta_1}=\lambda=n\pi. \tag{A.12}$$

Similarly with $\bar{x}=1$, at the other end, therefore the critical velocity is given as

$$\bar{U}_{\text{critical}}=\frac{n\pi}{\sqrt{\zeta_f}}\quad\forall n=1,2,\ldots, \tag{A.13}$$

and the dynamic response becomes

$$\overline{W}\left(\bar{x},\bar{t}\right)$$
$$=\bar{w}\left\{\left(\frac{\left(\Omega_2 e^{-i\Omega_1\bar{t}}-\Omega_1 e^{-i\Omega_2\bar{t}}\right)-i2\Omega_{\text{cor}}\left(e^{-i\Omega_1\bar{t}}-e^{-i\Omega_2\bar{t}}\right)}{(\Omega_2-\Omega_1)}\right)\right.$$
$$\cdot\left(\frac{\bar{x}^4}{24}-\frac{\bar{x}^3}{12}+\frac{\bar{x}}{24}\right)+i$$
$$\cdot\frac{\left(\Omega_1\left(e^{-i\Omega_2\bar{t}}-1\right)-\Omega_2\left(e^{-i\Omega_1\bar{t}}-1\right)\right)}{n\pi\left(\Omega_1-\Omega_2\right)}\left\{2\sin n\pi\left(1-\bar{x}\right)\right.$$
$$\left.\left.-\sin n\pi\bar{x}\right\}\right\}. \tag{A.14}$$

B. Summary of Derived Results

See Table 1.

Nomenclature

E: Young's modulus
I: Moment of inertia
ρ_s: Mass density of pipe per unit length
ρ_f: Mass density of fluid per unit length
$\overline{U}$:   Internal flow velocity
w: Transverse deflection
$\overline{W}$:   Dimensionless dynamic response
$\overline{W}_{\bar{x}}$:   Pipe's end slope
$\overline{W}_{\overline{xx}}$:   Pipe's end moment
$\overline{W}_{\overline{xxx}}$:   Pipe's end shear force
$\bar{t}$:   Dimensionless time
$\bar{x}$:   Coordinate of pipe's neutral axis
g: Gravitational acceleration
L: Pipe length

λ: Wave number
ζ_f: Fluid-pipe mass ratio
Ω_0:   System frequency at zero flow velocity
Ω_n: Natural frequency
Ω_1:   Complex frequency
Ω_2:   Conjugate complex frequency
$\bar{p}$:   Dimensionless pressure
$\overline{T}$:   Dimensionless temperature
v: Poison ratio
α: Linear expansivity
$\bar{p}$:   Dimensionless pressure
Ω_{cor}:   Frequency due to Coriolis force
$\overline{U}_{\text{cr}}$:   Critical flow velocity.

References

[1] M. P. Paidoussis, *Fluid-Structure Interactions: Slender Structures and Axial Flows*, vol. 1, Academic Press, 2014.

[2] I. Elishakoff, "Controversy associated with the so-called 'follower forces': critical overview," *Applied Mechanics Reviews*, vol. 58, no. 1–6, pp. 117–142, 2005.

[3] H. R. Öz and H. Boyaci, "Transverse vibrations of tensioned pipes conveying fluid with time-dependent velocity," *Journal of Sound and Vibration*, vol. 236, no. 2, pp. 259–276, 2000.

[4] T. Szmidt and P. Przybyłowicz, "Critical flow velocity in a pipe with electromagnetic actuators," *Journal of Theoretical and Applied Mechanics*, vol. 51, no. 2, pp. 487–496, 2013.

[5] A. Askarian, H. Abtahi, and H. Haddaapour, "Dynamic instability of cantilevered composite pipe conveying flow with an end nozzle," in *Proceedings of the International Congress on Sound and Vibration (ICSV '14)*, Beijing, China, 2014.

[6] G. L. Kuiper and A. V. Metrikine, "On stability of a clamped-pinned pipe conveying fluid," *HEBRON*, vol. 49, no. 3, 2004.

[7] K. R. Chellapilla and H. S. Simha, "Vibrations of fluid-conveying pipes resting on two-parameter foundation," *The Open Acoustics Journal*, vol. 1, pp. 24–33, 2008.

[8] J. Leklong, S. Chucheepsakul, and S. Kaewunruen, "Dynamic responses of marine risers/pipes transporting fluid subject to top end excitations," in *Proceedings of the 8th ISOPE Pacific/Asia Offshore Mechanics Symposium*, Bangkok, Thailand, 2008.

[9] A. H. Al-Hilli and T. J. Ntayeesh, "Free vibration characteristics of elastically supported pipe conveying fluid," *Nahrain University College of Engineering Journal*, vol. 16, no. 1, pp. 9–19, 2013.

[10] Q. Guo, L. Zhang, and L. Xiao, "Damage analysis of the vehicle's pipe conveying fluid induced by complex random-shock loads," *Journal of Pressure Equipment and Systems*, vol. 4, pp. 100–103, 2006.

[11] Y. L. Zhang, D. G. Gorman, J. M. Reese, and J. Horacek, "Observations on the vibration of axially tensioned elastomeric pipes conveying fluid," *Proceedings of the Institution of Mechanical Engineers, Part C: Journal of Mechanical Engineering Science*, vol. 214, no. 3, pp. 423–433, 2000.

[12] Y. Modarres-Sadeghi and M. P. Païdoussis, "Nonlinear dynamics of extensible fluid-conveying pipes, supported at both ends," *Journal of Fluids and Structures*, vol. 25, no. 3, pp. 535–543, 2009.

[13] R. A. Ibrahim, "Overview of mechanics of pipes conveying fluids—part I: fundamental studies," *Journal of Pressure Vessel Technology*, vol. 132, no. 3, Article ID 034001, 32 pages, 2010.

[14] M. Murai and M. Yamamoto, "An experimental analysis of the internal flow effects on marine risers," in *Proceedings of the MARTEC, BUET*, pp. 159–165, Dhaka, Bangladesh, 2010.

[15] N. Marakala, K. Appukutttan, and R. Kadoli, "Experimental and theoretical investigation of combined effects of fluid and thermal induced vibration on vertical thin slender tube," *IOSR Journal of Mechanical and Civil Engineering (IOSR-JMCE)*, pp. 63–68, 2014.

[16] U. Lee and J. Park, "Spectral element modelling and analysis of a pipeline conveying internal unsteady fluid," *Journal of Fluids and Structures*, vol. 22, no. 2, pp. 273–292, 2006.

[17] N. Qiao, W. Lin, and Q. Qin, "Bifurcations and chaotic motions of a curved pipe conveying fluid with nonlinear constraints," *Computers and Structures*, vol. 84, no. 10-11, pp. 708–717, 2006.

[18] H. L. Dodds Jr. and H. L. Runyan, "Effect of high velocity fluid flow on the bending vibrations and static divergence of a simply-supported pipe," NASA Technical Note D-2870, 1965.

[19] R. Yamaguchi, G. Tanaka, H. Liu, and T. Hayase, "Fluid vibration induced in T-junction with double side branches," *World Journal of Mechanics*, vol. 6, no. 4, pp. 169–179, 2016.

[20] J. Kutin and I. Bajsić, "Fluid-dynamic loading of pipes conveying fluid with a laminar mean-flow velocity profile," *Journal of Fluids and Structures*, vol. 50, pp. 171–183, 2014.

[21] M. J. Jweeg and T. J. Ntayeesh, "Dynamic analysis of pipes conveying fluid using analytical, numerical and experimental verification with the aid of smart materials," *International Journal of Science and Research*, vol. 4, no. 12, 2015.

[22] R. C. Wrede and M. Spiegel, *Thoery and Problems of Advanced Calculus*, Schaum's Outline Series, McGraw-Hill, New York, NY, USA, 2nd edition, 2002.

[23] P. S. Olayiwola, "Mechanics of a fluid-conveying pipeline system resting on a viscoelastic foundation," *Journal of Multidisciplinary Engineering Science Studies (JMESS)*, vol. 2, no. 3, 2016.

[24] A. Jeffrey, *Advanced Engineering Mathematics*, Harcourt Academic Press, Burlington, Mass, USA, 2002.

[25] W. A. Nash, *Theory and Problems of Strength of Materials*, vol. 83 of *Schaum's Outline Series*, McGraw-Hill, 2nd edition, 1977.

12

Vibrational Interaction of Two Rotors with Friction Coupling

H. Larsson and K. Farhang

Department of Mechanical Engineering and Energy Processes, Southern Illinois University Carbondale, Carbondale, IL 62901, USA

Correspondence should be addressed to K. Farhang; farhang@siu.edu

Academic Editor: Sven Johansson

A lumped parameter model is presented for studying the dynamic interaction between two disks in relative rotational motion and in friction contact. The contact elastic and dissipative characteristics are represented by equivalent stiffness and damping coefficient in the axial as well as torsional direction. The formulation accounts for the coupling between the axial and angular motions by viewing the contact normal force a result of axial behavior of the system. The model is used to investigate stick-slip behavior of a two-disk friction system. In this effort the friction coefficient is represented as an exponentially decaying function of relative angular velocity, varying from its static value at zero relative velocity to its kinetic value at very high velocities. This investigation results in the establishment of critical curve defining two-parameter regions: one in which stick-slip occurs and that in which stick-slip does not occur. Moreover, the onset and termination of stick-slip, when it occurs, are related to the highest component frequency in the system. It is found that stick-slip starts at a period nearly equal to that of the highest component frequency and terminates at a period almost three times that of the highest component frequency.

1. Introduction

It has been known that mechanical systems with friction exhibit a phenomenon known as stick-slip. For example, in a brake system, a stator and a rotor in friction contact are in slip state initially at the start of braking. If the intent is to bring the rotor to a complete stop, then a permanent state of stick (zero relative velocity between the rotor and stator) will exist at the end of braking. In addition to the states of slip and stick, a stick-slip transition region may exist which is generally characterized by high frequency oscillations. In the stick-slip transition region, the system displays a series of transient states, that is, slip to stick, and vice versa. With progression of time, the states corresponding to stick gain increasingly larger duration up to the point at which a steady state of stick is achieved.

The existence of stick-slip gives rise to friction forces which are complex in nature and have been known to be responsible for adverse dynamic characteristics. The effects of stick-slip friction are manifested through the development of excessive vibration and noise. Friction-induced vibration is undesirable due to its detrimental effects on the operation

and acceptable performance of mechanical systems. Frequent vibration leads to accelerated wear of components, surface damage, fatigue, and noise.

Many aspects of friction-induced vibration are found in the literature. For a comprehensive review of friction the reader may refer to Ibrahim [1, 2], Armstrong-Hélouvry et al. [3–6], Crolla and Lang [7], and Tabor [8]. Specifically, Oden and Martins [9] include a review on frictional contact of metallic surfaces and in a more recent publication by Martins et al. [10] some aspects of low speed frictional sliding phenomena are reviewed.

A simple system which has been used to examine the stick-slip phenomenon is that of a mass sliding on a moving belt [11, 12]. Another typical system is the pin-on-disk apparatus that has been utilized to study dynamic instabilities [13]. Analytical studies of a pin-on-disk apparatus [14] show that the coupling between rotational and normal modes of vibration is the primary mechanism of the resulting self-exited oscillations. Soom and Kim [15, 16] as well as Dweib and D'Souza [17] came to similar conclusions, showing that the force oscillations are primarily associated with the normal and tangential contact vibrations. This coupling effect was considered in the steady sliding point characterization of a simple system [18]. A dynamic stability study [19], considering the Schallamach waves (growing surface oscillations which propagate from front to rear at the friction surface), showed that, for large coefficients of friction and sufficiently large Poisson's ratio, steady sliding is unstable.

The most widely accepted cause of stick-slip motion is the speed dependence of kinetic friction [3, 4, 20–22]. The velocity term is often included in analytical models [5, 23]. Bengisu and Akay [24] modeled the friction force as an exponential function of the relative sliding velocity. The friction forces in a model of an automotive disk brake are defined in terms of relative sliding speed, pressure, and temperature at the friction interface [25]. Based on experimental data on stick-slip motion, Bo and Pavelescu [26] developed a friction model consisting of two exponential functions of relative speed. Linear stability theory has shown that there is a critical value of the normal load and the average coefficient of friction, above which the steady state sliding motion becomes unstable [27]. Similar results were found by [28] by studying the stability of a simple model where the friction force was assumed to be steady. It has been observed that the amplitude of the stick-slip motion decreases when the driving velocity, damping coefficient, and spring stiffness increase and the mass of the sliding body decreases [9]. Sherif and Kossa [29] measured and modeled the normal and tangential contact stiffness of two rough surfaces in contact. Sherif [30] analyzed the effect of contact stiffness on the instability condition of frictional vibration. Gao et al. [31] focus on the amplitude of stick-slip motion as a function of humidity, speed, and applied load. It is found that the stick-slip amplitude decreases with increasing substrate speed. In lubricated contacts, the cause of stick-slip motion is attributed to local region of instability where shear stress falls with respect to shear rate [32]. A set of lubricated contact experiments are presented by Polycarpou and Soom [33]. The transitions between sliding and sticking are measured and discussed.

In contrast to the speed dependence of the kinetic friction, Adams [34] presents a model for the dynamic interaction of two sliding surfaces, consisting of a series of moving linear springs at the friction interface. The springs are used to account for the elastic properties of asperities on one of the surfaces. The dynamic analysis indicates that the system is dynamically unstable for any finite speed, even though the coefficient of friction is constant. The idea of representing the friction interaction of two bodies using multiple elastic members such as springs or "bristles" is also presented by [35]. The authors assume that the friction between the two surfaces is caused by a large number of bristles, each contributing a small portion to the total load. The model is numerically inefficient and the authors therefore present a refined model.

In this paper a lumped parameter vibration model for a two-disk friction system is presented. A friction coefficient function that expresses the kinetic coefficient as a function of relative velocity is introduced. The interfacial force between the two plates is calculated based on the dynamic behavior of the disks in the axial direction. This is an important feature in that it allows direct coupling between the axial and rotational

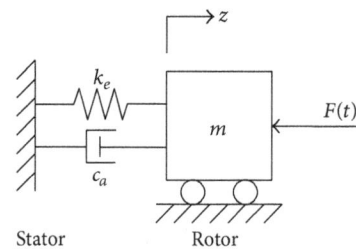

FIGURE 1: Schematic of the model in axial motion.

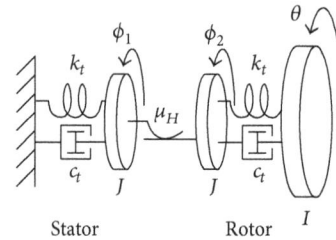

FIGURE 2: Schematic of the model in rotational sliding (slip) state.

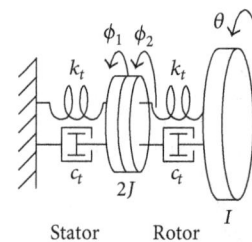

FIGURE 3: Schematic of the model in rotational stick state.

motions. That is, the rotational motion must be influenced by the actuating axial force and material properties of the friction material (stiffness and damping) in compression. Therefore, in addition to the properties of the friction material in torsional direction (shear), the model accounts for the properties in axial direction and how the combination of the mechanical properties can influence the dynamic behavior of the system. This capability enables simulations to accurately demonstrate effects of the mechanical properties on dynamic and vibration behavior of the friction system.

2. The Two-Disk Model

The interaction between two disks in frictional contact is represented using a lumped parameter model shown in Figures 1–3. The two disks are engaged by the actuating force $F(t)$. The compression of the disks depends on $F(t)$, m (mass of the rotor), k_e (equivalent axial stiffness of the disk pair), and c_a (axial damping of the disk pair). The axial behavior of the disks determines the normal interface force between the disk surfaces. Figure 1 shows the schematic of the axial vibration model.

The differential equation of the axial motion for the system can be written as

$$m\ddot{z} + c_a\dot{z} + k_e z = -F(t). \qquad (1)$$

The rotational motion of the system is modeled in two states, that is, sliding friction (slip state) and static friction (stick state). The schematics corresponding to the slip and stick states are shown in Figures 2 and 3, respectively. J, k_t, and c_t are the lumped contact parameters (equivalent inertia, torsional stiffness, and damping coefficient) contributed by each disk. The friction torque at the surface depends on the two nonconstant variables: the normal interface force and the relative angular velocity.

We consider a situation in which the rotor, initially at a nonzero angular velocity, is brought to a complete stop by the application of an axial force and as a result of friction contact with the stator. The differential equation for the angular motion of the rotor can be written as

$$I\ddot{\theta} + c_t\left(\dot{\theta} - \dot{\phi}_2\right) + k_t\left(\theta - \phi_2\right) = 0. \tag{2}$$

The motion of the lumped inertia representing the contact will be governed by two sets of differential equations. When the two moments of inertia are in slip state,

$$J\ddot{\phi}_2 - c_t\left(\dot{\theta} - \dot{\phi}_2\right) - k_t\left(\theta - \phi_2\right) = -\mu_H r F_n,$$
$$J\ddot{\phi}_1 + c_t\dot{\phi}_1 + k_t\phi_1 = \mu_H r F_n. \tag{3}$$

In stick state, the above reduces to a single differential equation

$$2J\ddot{\phi}_2 = c_t\left(\dot{\theta} - \dot{\phi}_2\right) - c_t\dot{\phi}_2 + k_t\left(\theta - \phi_2\right) - k_t\phi_2,$$
$$\phi_1 = \phi_2. \tag{4}$$

The normal surface interface force depends on the axial motion as follows:

$$F_n = \begin{cases} -c_a\dot{z} - k_e z, & -c_a\dot{z} - k_e z > 0, \\ 0, & -c_a\dot{z} - k_e z < 0. \end{cases} \tag{5}$$

The effective radius of tangential force (r) can be estimated if the type of wear is known. Shigley and Mischke [36] show that, after the initial wear has taken place and the disks have worn down to the point where uniform wear is the governing mechanism, the following simple formula can be used to calculate r:

$$r = \frac{1}{4}\left(D + d\right), \tag{6}$$

where d and D are the inner and outer diameters of the contact surface, respectively. When uniform pressure can be assumed over the area of the disk, the following formula may be used [36]:

$$r = \frac{1}{3}\frac{\left(D^3 - d^3\right)}{\left(D^2 - d^2\right)}. \tag{7}$$

3. Stick-Slip Friction

A proposed friction coefficient function for the relative angular velocity ω_r is presented as follows:

$$\mu_H = \text{sgn}\left(\omega_r\right)\left(\mu_d + \left(\mu_s - \mu_d\right)e^{-\alpha|\omega_r|}\right) + \delta\left(\omega_r\right)\mu_0, \tag{8}$$

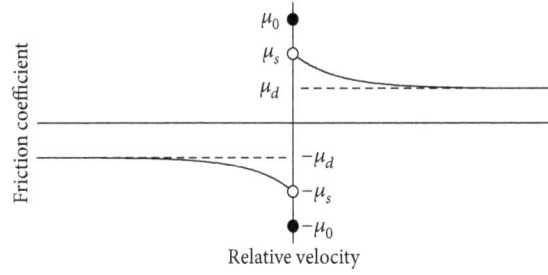

FIGURE 4: Proposed coefficient of friction function, μ_H.

where δ is the Dirac delta function and the function parameters are defined as follows:

$\omega_r = \dot{\phi}_2 - \dot{\phi}_1$ is the relative angular surface velocity.

$\mu_s = \mu(\omega_r \to 0^+)$ is the friction coefficient for low ω_r.

$\mu_d = \mu(\omega_r \to \infty)$ is the friction coefficient for high ω_r.

$\mu_0 = \mu(\omega_r = 0)$ is the friction coefficient for zero ω_r.

α is the friction coefficient decay rate.

Stick-slip motion is highly nonlinear behavior which occurs at the point where the relative velocity between the moving bodies is near zero. At zero relative velocity static friction dictates the behavior. This is a multivalued function implying that the friction force will take any value in the range of $-\mu_0 F_n$ to $\mu_0 F_n$ depending on the internal force within the material. As shown in Figure 4, the breakaway force can be larger than that corresponding to zero velocity. The magnitude of the external force must exceed the peak breakaway force ($\mu_0 F_n$) for sliding to occur. If μ_d is smaller than μ_s, the coefficient of friction function has a negative slope. The negative slope acts as a nonlinear negative damper and may cause the system to enter a stick-slip phase.

Examination of the forces acting on the disks representing the surfaces (J) suggests that, in order to have sliding contact, either the relative velocity (ω_r) must be other than zero or the internal force (F_i) generated by spring and damper must exceed the friction force ($\mu_0 F_n$). Conversely, in order to have a stick contact, the relative velocity must be zero and the internal force must be less than the friction force. In general, it is very difficult to capture the instant at which relative velocity equals zero. For simulation purposes, the velocity condition may be written in terms of a tolerance, ε. This can be summarized as follows:

$$\text{friction state} = \begin{cases} \text{slip}, & \text{if } F_i > \mu_0 F_n \text{ or } |\omega_r| > \varepsilon, \\ \text{stick}, & \text{if } F_i \le \mu_0 F_n, |\omega_r| \le \varepsilon, \end{cases} \tag{9}$$

where

$$F_i = \frac{1}{r}\left|c_t\dot{\phi}_1 + k_t\phi_1\right| \quad \text{stator surface,}$$
$$F_i = \frac{1}{r}\left|c_t\left(\dot{\theta} - \dot{\phi}_1\right) + k_t\left(\theta - \phi_1\right)\right| \quad \text{rotor surface.} \tag{10}$$

4. Results

4.1. Simulation Summary. In the simulations presented, the rotor is brought to a stop by the application of an axial actuating force. The axial force is assumed to increase linearly to a maximum value and to be kept constant thereafter.

$$F(t) = \begin{cases} F_0 \dfrac{t}{t_0}, & \text{if } t \le t_0, \\ F_0, & \text{if } t > t_0, \end{cases} \tag{11}$$

where F_0 and t_0 are the maximum axial actuating force and the application time, respectively. The rotor is assumed to rotate with an initial angular velocity ω_0 at $t = 0$, when the actuating force is applied. The object is to bring the rotor to a complete stop.

The parameters used in the simulations are as follows: equivalent axial stiffness, $k_e = 55.38 \cdot 10^6$ N/m ($3.161 \cdot 10^5$ lb/in); axial damping, $c_a = 8{,}760$ N s/m (50 lb-sec/in); disk mass, $m = 4.54$ kg (10 lbm); actuating force, $F_0 = 2{,}225$ N (500 lb); force application time, $t_0 = 1$ second; torsional stiffness, $k_t = 7.71 \cdot 10^6$ Nm/rad ($6.822 \cdot 10^7$ lb-in/rad); torsional damping, $c_t = 22.606$ Nm s/rad (200 lb-in-s/rad); effective radius of tangential force, $r = 0.052388$ m (2.0625 in); rotor inertia, $I = 2.262$ kg m² (20 slug-in-ft); equivalent surface inertia, $J = 0.0113$ kg m² (0.1 slug-in-ft); breakaway friction coefficient, $\mu_0 = 0.6$; friction coefficient for low velocities, $\mu_s = 0.6$; friction coefficient for high velocities, $\mu_d = 0.3$; relative angular velocity tolerance, $\varepsilon = 0.00001$. Runge-Kutta 5 with step size from 10^{-8} to 10^{-4} and integration tolerance 10^{-11} is used. Before generating the final results, simulations were run at various accuracies by setting the step size. It was determined that the most appropriate settings for step size in the range involving smaller numbers (e.g., from 10^{-9} to 10^{-5}) gave identical results to the ranges mentioned. Initial conditions at $t = 0$ are $z = \dot{z} = 0$, $\dot{\theta} = \dot{\phi}_2 = 100$ rad/s, and $\theta = \phi_1 = \phi_2 = \dot{\phi}_1 = 0$.

4.2. Time Response. Figure 5 illustrates the axial response of the system, resulting in rotor displacement of about 0.04 mm. Figures 6 and 7 depict the rotor angular velocity. The close-up view in Figure 7 shows rotor velocity during a very small window of time of (<0.04 seconds) that includes a small portion of slip state, all the stick-slip state, and a small portion of stick state. This is a decaying oscillation that eventually gets attenuated during the stick state of the system. The angular velocities of the rotor and stator surfaces (inertias J) are illustrated in Figures 8–11. It is observed that during braking the system attains three distinct dynamic states. From the start of braking at $t = 0$ to $t = 6.975$, the rotor-stator system is in slip state. This is followed by stick-slip from $t = 6.975$ seconds to $t = 6.99$ seconds. The third is the stick state during which the vibration amplitude decays to zero. The relative surface angular velocity is plotted in Figure 12.

4.3. Stick-Slip Response. As enumerated, the entire braking can be viewed as consisting of three dynamic states: slip, stick-slip transition, and stick. Figures 13 and 14 illustrate

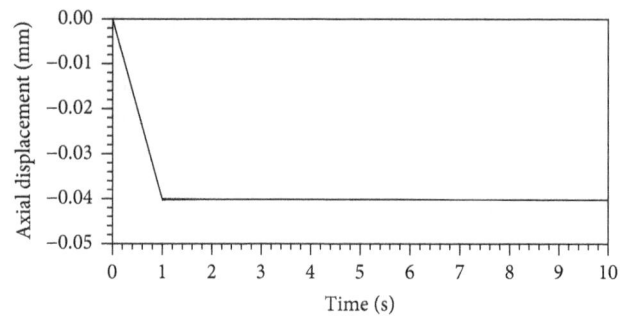

FIGURE 5: Axial motion.

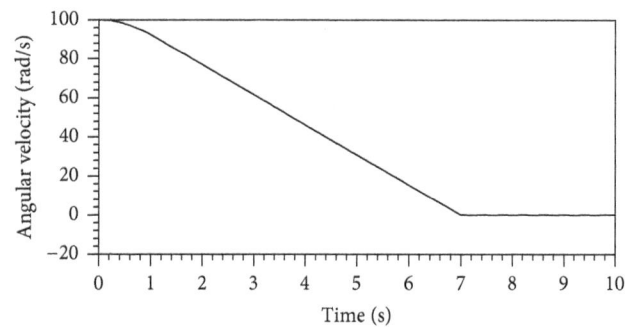

FIGURE 6: Angular velocity of the rotor.

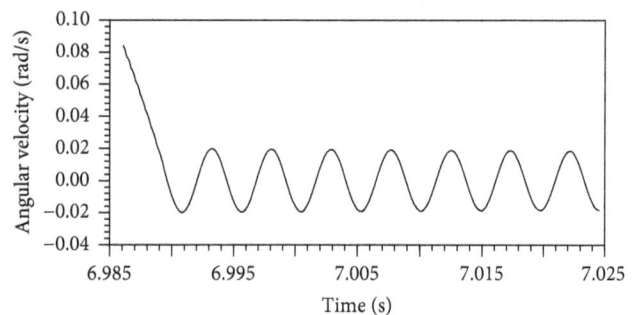

FIGURE 7: Angular velocity of the rotor; close-up view.

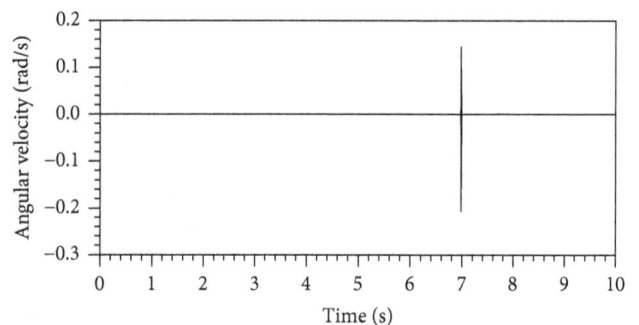

FIGURE 8: Angular velocity of the stator surface.

these three dynamic states. The close-up view of the stick-slip transition region in Figure 14 clearly shows that the duration of stick increases within the stick-slip region until a permanent state of sticking is achieved. The stick-slip behavior is presented in greater detail in Figure 15, where the

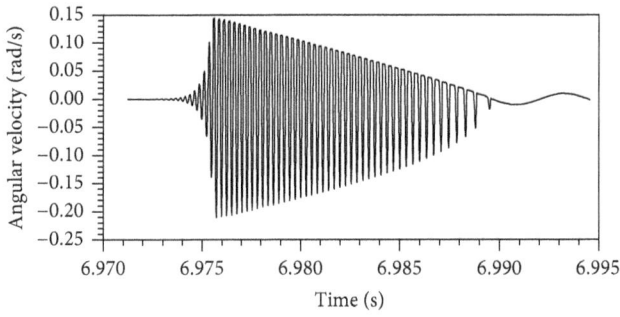

FIGURE 9: Angular velocity of the stator surface; close-up view.

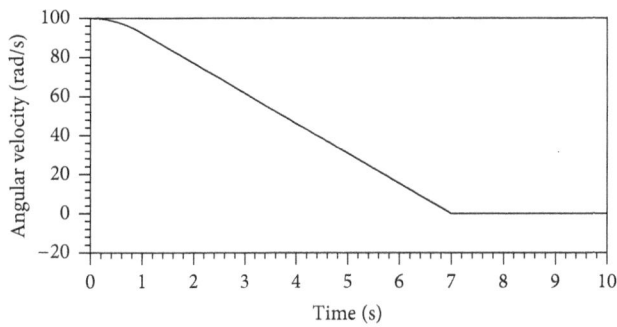

FIGURE 10: Angular velocity of the rotor surface.

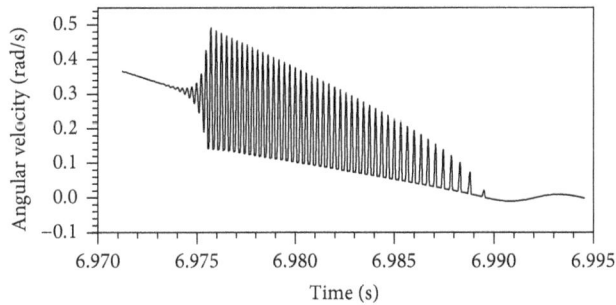

FIGURE 11: Angular velocity of the rotor surface; close-up view.

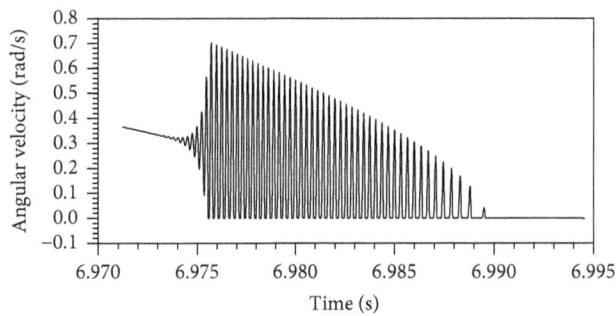

FIGURE 12: Relative angular velocity at the friction interface (rotor surface and stator surface); close-up view.

duration of slip and stick is shown as a function of the number of transitions from slip to stick (index).

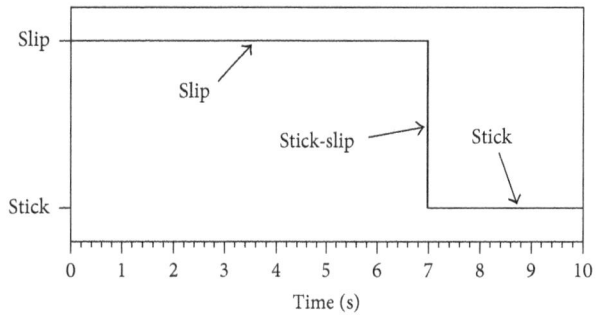

FIGURE 13: Stick-slip transition.

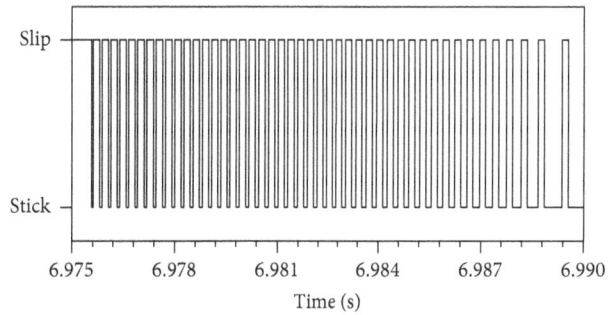

FIGURE 14: Stick-slip transitions; close-up view.

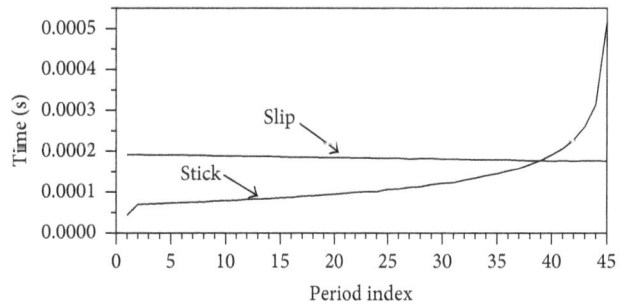

FIGURE 15: Duration of stick and slip period for the transition region.

Examination of the stick-slip transition region has led to the realization that duration of slip is relatively invariant and that stick increases exponentially with time until achieving a steady state of stick (Figure 15).

In the stick-slip transition region a period may be defined as a summation of slip and subsequent stick durations. As mentioned, slip duration is relatively constant at 0.00019 seconds. Stick duration starts at 0.00005 seconds and increases to 0.00053 seconds after which the transition region terminates. Considering the periods of stick-slip, the initial period at the onset of the stick-slip response is

$$T_i = 0.00005 + 0.00019 = 0.00024 \text{ s} \qquad (12)$$

and the final stick-slip period is given by

$$T_f = 0.00053 + 0.00019 = 0.00072 \text{ s}. \qquad (13)$$

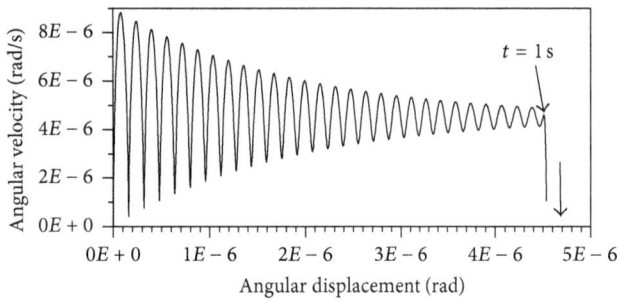

FIGURE 16: Phase plot $\dot{\theta} - \dot{\phi}_2$ versus $\theta - \phi_2$; 0 to 1.008 s.

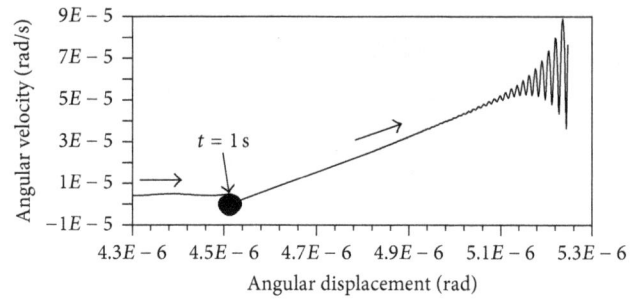

FIGURE 18: Phase plot $\dot{\theta} - \dot{\phi}_2$ versus $\theta - \phi_2$; 0.9483 to 6.9714 s.

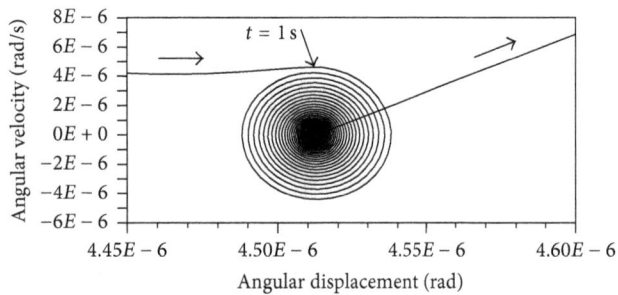

FIGURE 17: Phase plot $\dot{\theta} - \dot{\phi}_2$ versus $\theta - \phi_2$; 0.9483 to 6.9467 s.

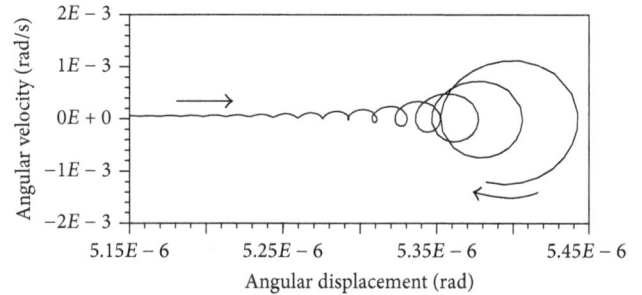

FIGURE 19: Phase plot $\dot{\theta} - \dot{\phi}_2$ versus $\theta - \phi_2$; 6.9697 to 6.9735 s.

Considering the highest component natural frequency of the system,

$$\omega_n = \sqrt{\frac{k_t}{J}}. \tag{14}$$

With $k_t = 7.71 \cdot 10^6$ Nm/rad ($6.822 \cdot 10^7$ lb-in/rad) and $J = 0.0113$ kg m^2 (0.1 slug-in-ft), we obtain the period

$$T_n = \frac{2\pi}{\omega_n} = 0.00024 \text{ s}. \tag{15}$$

We note that the period of stick-slip at the onset is nearly equal to the period for the highest component natural frequency of the system, $T_i \approx T_n$, and the final period is nearly equal to three times the period corresponding to the highest component natural frequency of the system, $T_f \approx 3T_n$.

4.4. Phase Plot Response. Figures 16–22 depict the phase plots for the lumped parameters representing the rotor and its contact surface. At the start of braking, material elastic deformation (at contact) occurs with decaying oscillatory behavior in rate of deformation (Figure 16). At about $t = 1$ second, monotonic increase in material deformation terminates and a stable behavior is observed in the form of a convergent spiral motion, as shown in Figure 17. This stable behavior continues until about $t = 6.97$ seconds, where a sudden increase in both material deformation and deformation rate is observed (Figure 18). This turns into an unstable vibration behavior in the form of a divergent spiral (Figure 19) until the onset of the stick-slip behavior (Figure 20). Figure 21 illustrates the stick-slip state in which the oscillation amplitude decreases. Finally, at $t = 6.9896$ s, the stick state prevails and a convergent spiral

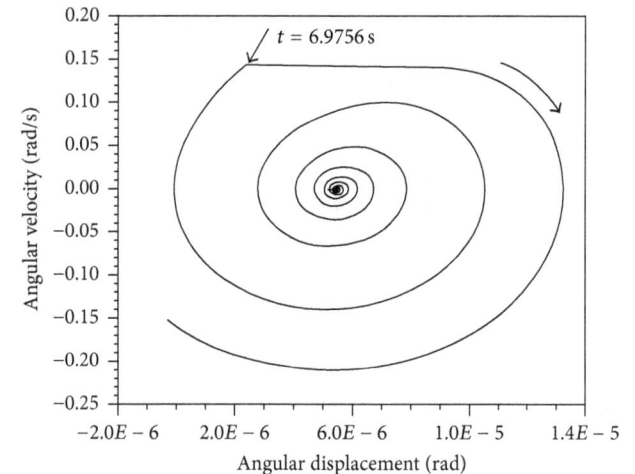

FIGURE 20: Phase plot $\dot{\theta} - \dot{\phi}_2$ versus $\theta - \phi_2$; 6.9697 to 6.9757 s.

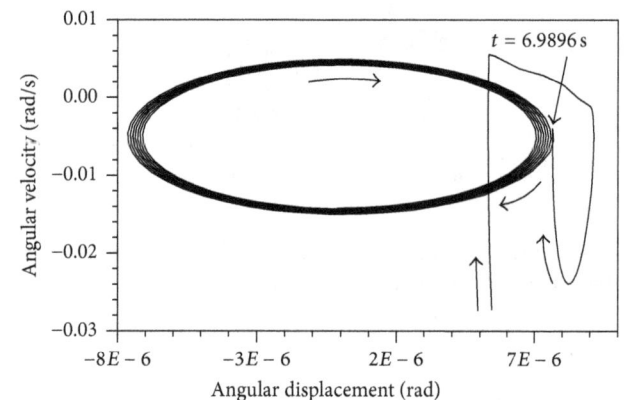

FIGURE 21: Phase plot $\dot{\theta} - \dot{\phi}_2$ versus $\theta - \phi_2$; 6.9755 to 6.9916 s.

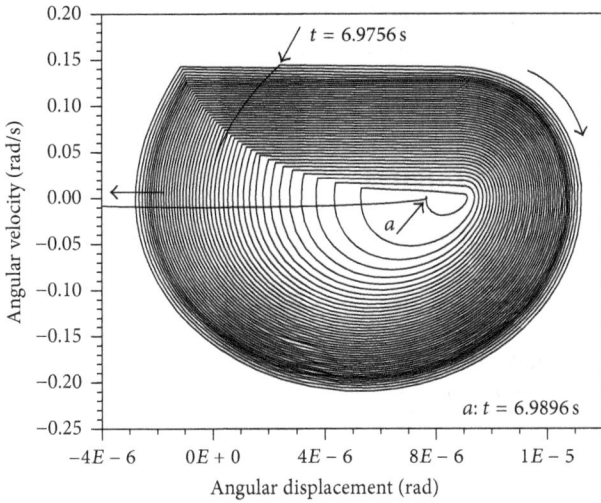

FIGURE 22: Phase plot $\dot{\theta} - \dot{\phi}_2$ versus $\theta - \phi_2$; 6.9888 to 7.0246 s.

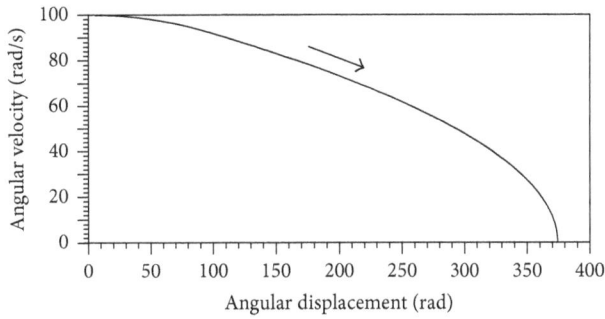

FIGURE 23: Phase plot $\dot{\phi}_2 - \dot{\phi}_1$ versus $\phi_2 - \phi_1$; 0 to 10 s.

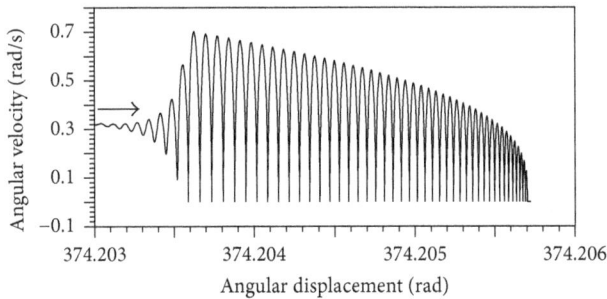

FIGURE 24: Phase plot $\dot{\phi}_2 - \dot{\phi}_1$ versus $\phi_2 - \phi_1$; 6.9728 to 10 s.

behavior is observed, as depicted in Figure 22. Figures 23 and 24 present the phase plots for the relative surface velocity.

4.5. Critical Parameters for Stick-Slip. The investigation of stick-slip behavior of the two-disk friction system results in establishing pairs of material damping and friction coefficient decay rate values for which the occurrence of stick-slip can be predicted.

Here a parametric study for different α and c_t values is presented. The number of stick-slip transitions is counted and is used to determine the critical pairs of α and c_t.

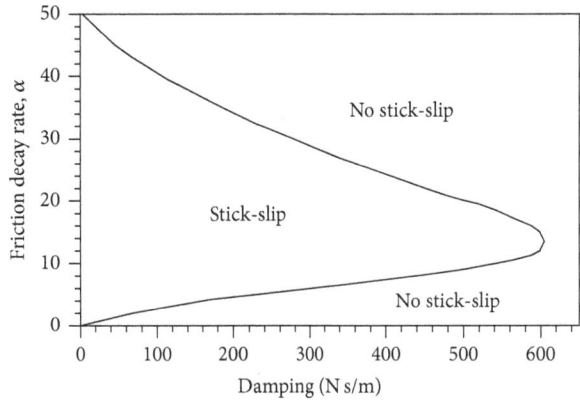

FIGURE 25: Critical decay rate, α, and damping values, c_t, for the two-disk brake system.

The critical values are defined to be those for which stick-slip transition region starts to disappear. Figure 25 depicts the resulting curve corresponding to critical α-c_t pairs. As illustrated, the domain is divided into two regions. The space enclosed within the curve and the α-axis provide the α-c_t pairs for which stick-slip behavior will take place. For the region outside the critical α-c_t curve stick-slip behavior does not exist.

The main utility of the result in Figure 25 lies in illustrating the connection between material and interface properties and the dynamic behavior of the two-disk system. If such properties can be traced back to the parameters used in fabrication of the friction material, then a figure such as Figure 25 can serve as a valuable design aid in deciding material composition and fabrication methods to be used so as to achieve a desired dynamic behavior of the resulting system.

5. Conclusion

A lumped parameter model for the prediction of vibration response of two-disk brake system has been developed. The model is capable of predicting brake response, including the stick-slip transition phase. Using this capability an investigation of friction-induced vibration and stick-slip phenomenon has been performed.

Examination of stick-slip transition region revealed that duration of slip is relatively invariant and that duration of stick increases nonlinearly with time until achieving a steady state of stick.

Critical values of the material damping and friction coefficient decay rate are found which divide the domain of the two parameters into two regions: one in which stick-slip behavior occurs and the other in which stick-slip will not occur.

The result of the investigation may be summarized as follows:

(1) Duration of slip in the stick-slip transition region is relatively invariant.

(2) Duration of stick increases almost exponentially with time until a final stick state is achieved.

(3) The period of stick-slip at the beginning of the stick-slip region is approximately equal to the period corresponding to the highest component natural frequency of the system.

(4) The final period is approximately equal to three times the period corresponding to the highest component natural frequency of the system.

(5) A critical curve (Figure 25), defined by set of material damping and friction coefficient decay rate values, is found that defines the region where stick-slip occurs. The resulting curve shows that, for structural damping in the range 0–600 N s/m, there are two limit values for friction-velocity decay rate within which stick-slip would occur. The range of friction-velocity decay rate corresponding to stick-slip response increases for lower structural damping within the brake.

Nomenclature

t: Time
t_0: Application time for the actuating axial force
F_0: Actuating normal force
$F(t)$: Axial actuating force as a function of time
F_n: Normal surface interface force
m: Mass of the rotor
k_e: Equivalent axial stiffness of a disk pair
c_a: Axial damping of a disk pair
z: Axial displacement of the rotor
I: Inertia of the rotor
J: Equivalent inertia lumped at the friction surface contributed by a disk
k_t: Equivalent torsional contact stiffness contributed by a disk
c_t: Equivalent torsional contact damping coefficient contributed by a disk
D: Outer diameter of each friction disk
d: Inner diameter of each friction disk
r: Effective radius of tangential force
θ: Angular rotation of the rotor
ϕ_1: Angular rotation of the friction surface on the stator
ϕ_2: Angular rotation of the friction surface on the rotor
ω_r: Relative angular velocity
ω_n: Highest component natural frequency for the rotational system
F_i: Internal force generated by spring and damper
T_i: Initial stick-slip period
T_f: Final stick-slip period
T_n: Period corresponding to ω_n
μ_H: Friction coefficient function
μ_0: Breakaway friction coefficient
μ_s: Friction coefficient for low relative velocities
μ_d: Friction coefficient for high relative velocities

α: Friction coefficient decay rate with increase in relative velocity
ε: Relative angular velocity tolerance
δ: Dirac delta function.

Competing Interests

The authors declare that they have no competing interests.

Acknowledgments

The authors would like to acknowledge the financial support for this project provided by the Center for Advanced Friction Studies at Southern Illinois University Carbondale.

References

[1] R. A. Ibrahim, "Friction-induced vibration, chatter, squeal, and chaos: part I—mechanics of friction," in *Friction-Induced Vibration, Chatter, Squeal, and Chaos*, vol. 49, pp. 107–121, Transactions of the ASME, 1992.

[2] R. A. Ibrahim, "Friction-induced vibration, chatter, squeal, and chaos—part II: dynamics and modeling," *Applied Mechanics Reviews*, vol. 47, no. 7, pp. 227–253, 1994.

[3] B. Armstrong-Hélouvry, "Friction: experimental determination, modeling and compensation," in *Proceedings of the International Conference on Robotics and Automation (ICRA '88)*, pp. 1422–1427, IEEE, Philadelphia, Pa, USA, 1988.

[4] B. Armstrong-Hélouvry, "Stick-slip arising from Stribeck friction," in *Proceedings of the IEEE International Conference on Robotics and Automation*, pp. 1377–1382, IEEE, Cincinnati, Ohio, USA, May 1990.

[5] B. Armstrong-Hélouvry, "A perturbation analysis of stick-slip," in *Friction-Induced Vibration, Chatter, Squeal, and Chaos*, vol. 49, pp. 41–48, Transactions of the ASME, 1992.

[6] B. Armstrong-Hélouvry, P. Dupont, and C. C. De Wit, "A survey of models, analysis tools and compensation methods for the control of machines with friction," *Automatica*, vol. 30, no. 7, pp. 1083–1138, 1994.

[7] D. A. Crolla and A. M. Lang, "Brake noise and vibration—the state of the art," *Vehicle Tribology*, vol. 18, pp. 165–174, 1990.

[8] D. Tabor, "Friction—the present state of our understanding," *Journal of Lubrication Technology*, vol. 103, no. 2, pp. 169–179, 1981.

[9] J. T. Oden and J. A. C. Martins, "Models and computational methods for dynamic friction phenomena," *Computer Methods in Applied Mechanics and Engineering*, vol. 52, no. 1–3, pp. 527–634, 1985.

[10] J. A. Martins, J. T. Oden, and F. M. Simões, "A study of static and kinetic friction," *International Journal of Engineering Science*, vol. 28, no. 1, pp. 29–92, 1990.

[11] U. Galvanetto and S. R. Bishop, "Stick-slip vibrations of a two degree-of-freedom geophysical fault model," *International Journal of Mechanical Sciences*, vol. 36, no. 8, pp. 683–698, 1994.

[12] K. Popp, "Some model problems showing stick-slip motion and chaos," in *Friction-Induced Vibration, Chatter, Squeal, and Chaos*, vol. 49, pp. 1–12, Transactions of the ASME, 1992.

[13] C. Gao, "Stick-slip motion in boundary lubrication," *Tribology Transactions*, vol. 38, no. 2, pp. 473–477, 1995.

[14] W. W. Tworzydlo, E. B. Becker, and J. T. Oden, "Numerical modeling of friction-induced vibrations and dynamic instabilities," in *Friction-Induced Vibration, Chatter, Squeal, and Chaos*, vol. 49, pp. 13–32, Transactions of the ASME, 1992.

[15] A. Soom and C. Kim, "Rougness induced dynamic loading at dry and boundary-lubricated sliding contacts," *ASME Journal of Lubrication Technology*, vol. 105, no. 4, pp. 514–517, 1983.

[16] A. Soom and C. Kim, "Interactions between dynamic normal and frictional forces during unlubricated sliding," *ASME Journal of Lubrication Technology*, vol. 105, no. 2, pp. 221–229, 1983.

[17] A. H. Dweib and A. F. D'Souza, "Self-excited vibrations induced by dry friction, part 1: experimental study," *Journal of Sound and Vibration*, vol. 137, no. 2, pp. 163–175, 1990.

[18] P. E. Dupont and D. Bapna, "Perturbation stability of frictional sliding with varying normal force," *Journal of Vibration and Acoustics*, vol. 118, no. 3, pp. 491–497, 1996.

[19] J. A. C. Martins, L. O. Faria, and J. Guimarães, "Dynamic surface solutions in linear elasticity with frictional boundary conditions," in *Friction-Induced Vibration, Chatter, Squeal, and Chaos*, vol. 49, pp. 33–39, American Society of Mechanical Engineers, 1992.

[20] C. Gao, D. Kuhlmann-Wilsdorf, and D. D. Makel, "The dynamic analysis of stick-slip motion," *Wear*, vol. 173, no. 1-2, pp. 1–12, 1994.

[21] H.-I. You and J.-H. Hsia, "Influence of friction-speed relation on the occurrence of stick-slip motion," *Journal of Tribology*, vol. 117, no. 3, pp. 450–455, 1995.

[22] Y. Yuan, "A study of the effects of negative friction-speed slope on brake squeal," in *Proceedings of the ASME Design Engineering Technical Conferences*, vol. 84, pp. 1153–1162, 1995.

[23] R. J. Black, "Self-excited multi-model vibrations of aircraft brakes with nonlinear negative damping," in *Proceedings of the ASME Design Engineering Technical Conferences*, vol. 84, no. 1, pp. 1241–1245, 1995.

[24] M. T. Bengisu and A. Akay, "Stability of friction-induced vibrations in multi-degree-of-freedom systems," in *Friction-Induced Vibration, Chatter, Squeal, and Chaos*, vol. 49, pp. 57–64, American Society of Mechanical Engineers, 1992.

[25] R. Avilés, G. Hennequet, E. Amezua, and J. Vallejo, "Low frequency vibrations in disc brakes at high car speed. Part II: mathematical model and simulation," *International Journal of Vehicle Design*, vol. 16, no. 6, pp. 556–569, 1995.

[26] L. C. Bo and D. Pavelescu, "The friction-speed relation and its influence on the critical velocity of stick-slip motion," *Wear*, vol. 82, no. 3, pp. 277–289, 1982.

[27] A. F. D'Souza and A. H. Dweib, "Self-excited vibrations induced by dry friction—part 2: stability and limit-cycle analysis," *Journal of Sound and Vibration*, vol. 137, no. 2, pp. 177–190, 1990.

[28] J. L. Swayze and A. Akay, "Effects of system dynamics on friction-induced oscillations," *Journal of Sound and Vibration*, vol. 173, no. 5, pp. 599–609, 1994.

[29] H. A. Sherif and S. S. Kossa, "Relationship between normal and tangential contact stiffness of nominally flat surfaces," *Wear*, vol. 151, no. 1, pp. 49–62, 1991.

[30] H. A. Sherif, "Effect of contact stiffness on the establishment of self-excited vibrations," *Wear*, vol. 141, no. 2, pp. 227–234, 1991.

[31] C. Gao, D. Kuhlmann-Wilsdorf, and D. D. Makel, "Fundamentals of stick-slip," *Wear*, vol. 162–164, pp. 1139–1149, 1993.

[32] S. Jang and J. Tichy, "Rheological models for stick-slip behavior," Tech. Rep. 96-TRIB-52, 1996.

[33] A. Polycarpou and A. Soom, "Transitions between sticking and slipping at lubricated line contacts," in *Friction-Induced Vibration, Chatter, Squeal, and Chaos*, vol. 49 of *Transactions of the ASME*, pp. 139–148, American Society of Mechanical Engineers, 1992.

[34] G. G. Adams, "Self-excited oscillations in sliding with a constant friction coefficient," in *Proceedings of the ASME Design Engineering Technical Conferences*, vol. 84, no. 1, pp. 1171–1177, 1995.

[35] D. A. Haessig Jr. and B. Friedland, "On the modeling and simulation of friction," *ASME Journal of Dynamic Systems, Measurement and Control*, vol. 113, no. 3, pp. 354–362, 1991.

[36] J. E. Shigley and C. R. Mischke, *Mechanical Engineering Design*, McGraw-Hill, New York, NY, USA, 5th edition, 1989.

13

Vibration Sideband Modulations and Harmonics Separation of a Planetary Helicopter Gearbox with Two Different Configurations

Nader Sawalhi

Mechanical Engineering Department, Prince Mohammad Bin Fahd University, AL-Khobar, Saudi Arabia

Correspondence should be addressed to Nader Sawalhi; nadersaw@hotmail.com

Academic Editor: Kim M. Liew

This paper examines the spectrum and cepstrum content of vibration signals taken from a helicopter gearbox with two different configurations (3 and 4 planets). It presents a signal processing algorithm to separate synchronous and nonsynchronous components for complete shafts' harmonic extraction and removal. The spectrum and cepstrum of the vibration signal for two configurations are firstly analyzed and discussed. The effect of changing the number of planets on the fundamental gear mesh frequency (epicyclic mesh frequency) and its sidebands is discussed. The paper explains the differences between the two configurations and discusses, in particular, the asymmetry of the modulation sidebands about the epicyclic mesh frequency in the 4 planets arrangement. Finally a separation algorithm, which is based on resampling the order-tracked signal to have an integer number of samples per revolution for a specific shaft, is proposed for a complete removal of the shafts harmonics. The results obtained from the presented separation algorithms are compared to other separation schemes such as discrete random separation (DRS) and time synchronous averaging (TSA) with clear improvements and better results.

1. Introduction

Vibration signals originating from a helicopter transmission gearbox represent a rich source for monitoring its health. Many failures that occur in rotating components such as gears and bearings often show their signature in the vibration signal and can be well detected at early stages. Monitoring these vibrations often requires an extensive interpretation by a trained diagnostician, due to the complexity of such systems [1]. A major part of this involves the correct understanding and identification of the frequency content of the vibration signal. Understanding the frequency content of the signal and the different families of harmonics and sidebands would enable a correct analysis of the health of the machine.

Signals are mixtures of different sources. For successful handling and interpretation of signals, analysts often need to separate these different sources and process them separately. One of the most successful ways of interpreting signals is the use of Fast Fourier Transformation (FFT), which transforms the signal from the time domain into the frequency domain by using sines and cosines as base functions for the signal decomposition. FFT requires the transformed signal to be stationary; that is, it has some statistical parameters which do not change with time. For nonstationary signals (have time dependent statistics), the use of time-frequency presentation such as the spectrogram (short time-frequency analysis), the wavelets, the winger vile transform, and so forth is commonly used. Stationary signals are mainly composed of deterministic (discrete) components and random components. Random components contain all nonstationary signals in addition to any nondeterministic part. Deterministic components are those which can be expressed as a series of discrete sinusoidal signals (thus they are predictable and periodic). Deterministic component can be interchangeably referred to as discrete signals. They generally fall into two main categories:

(i) Periodic (cyclic): they are composed of sinusoids whose frequencies are all integer multiples of some fundamental frequency like the shaft speed in rotating

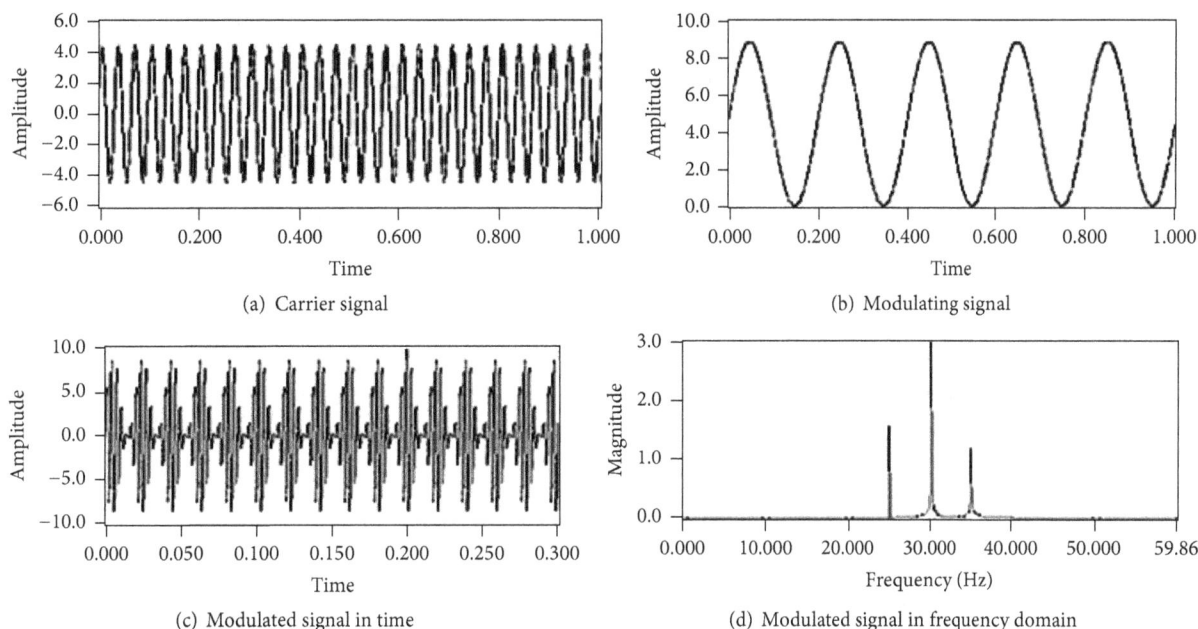

FIGURE 1: An example of amplitude modulation and sidebands.

machinery; The multiples of the fundamental frequencies are known as harmonics, with the fundamental being the first harmonic; periodic components can also manifest themselves as sidebands around a carrier frequency in the case of a modulated signal (e.g., a gearbox signal where the shaft speed (low frequency) modulates the gear mesh frequency (high frequency))

(ii) Quasi-periodic: they have at least two frequency components that are not rationally related and thus never repeated themselves exactly

Signal modulation or distortion occurs when the amplitude, frequency, or phase of a waveform is altered by the introduction of another physically related periodic signal or disturbance. The high frequency signal is known as the "carrier." The spectrum of the combined signal exhibits a discrete and dominant frequency (carrier) bounded by "sidebands" or peaks (modulator) spaced on either side of the carrier at the modulation frequency. Figure 1 illustrates the amplitude modulation. Figure 1(a) is the carrier signal, a pure sine wave with 30 Hz frequency in the time domain. Figure 1(b) displays the modulating signal, a pure sine wave with 5 Hz frequency. Figure 1(c) shows the modulated signal in the time domain. Figure 1(d) is the modulated signal in the frequency domain. The two sidebands can be identified easily.

For example, gear mesh frequency is a fixed multiple of rotating speed and is displayed as a distinct peak in the spectrum. If there are defects associated with the gearing (physically related) such as gear eccentricity or excessive wear that generate a synchronous force variation, a disturbance will accompany the gear mesh frequency. The high frequency gear mesh is the carrier. The low frequency gear defects will appear as "sidebands" on either side of the gear mesh peak.

The spectral distance from the peak to the side band peak is equal to frequency modulation or the rotating speed of the defective gear. Multiple gear defects are reflected as an extra set of sidebands.

Helicopter transmission gearboxes have been investigated in detail to understand the patterns of baseline frequencies and failure indicators that can be detected by monitoring vibrations [2, 3]. Those efforts resulted in a better understanding of the way in which different signals manifest themselves and mix with others and led to the development of a variety of diagnostic and prognostic techniques [4–6].

In general, interpreting the vibration signal transmitted through those gearboxes often requires more than the traditional inspection of the time signal and/or its frequency content. This is due mainly to the existence of a large number of rotating components, all of which contribute and mix in different ways [7], making it very hard to track changes in a certain component. In a planetary gearbox, it is quite important to investigate the different setups and determine the expected frequencies in the spectrum. Two examples are presented here and discussed for the Bell 206 helicopter gearbox.

Another important aspect which is presented in this paper is the separation of the gear mesh frequencies and shafts harmonics form the spectrum, using the information from only one tachometer, to enable further processing of the signal to detect gears and bearings faults. The results obtained from this analysis are compared to time synchronous averaging (TSA) processing and discrete random separation (DRS) [8].

When using TSA, the signal is resampled into the angular domain (rather than the temporal domain) to have the same number of samples for each shaft revolution. This removes any speed fluctuation from the signal. The shaft harmonics, now called orders, become locked to the shaft rotation and appear as discrete components in the frequency domain (with no order tracking, the higher harmonics usually smear and become broad). The ensemble average for all the rotations is calculated to give the so-called "synchronous average," which represents one shaft rotation and captures the deterministic part of the signal. If the synchronous average is subtracted from the signal, the result will be a residual, which contains noise, nonstationary signals, and any nondeterministic signal. TSA requires the presence of a tachometer signal for the order tracking stage.

DRS works well in the cases of slight (small) speed variation but requires a number of parameters to select and leaves notches in the signal at the locations of the discrete components. A linear transfer function (similar to the H1 transfer function estimate in modal analysis) is generated between the signal and a delayed version of it (using FFT). This gives a value of 1 for the discrete frequency bins and 0 elsewhere. This filter (amplitude of the transfer function) is then used to filter the signal and separate out the discrete components. The filtration is all based on the efficient FFT methods and thus the processing is computationally fast. The two main parameters required are the filter length and the amount of delay to capture the deterministic signals. There are a number of recommendations and a visual examination is usually required to set up these values to give the required separation. This method blindly removes all the discrete components from the whole frequency bandwidth and leaves notches in the spectrum at the location of these frequencies.

This paper is organized as follows. In Section 2, Bell 206 transmission setup and expected frequencies are presented. Section 3 compares the fundamental epicyclic mesh frequency and sidebands between the three- and four-planet arrangements. Section 4 presents the algorithm for the removal of synchronous shaft signals through resampling and signal truncation. The paper is concluded in Section 5.

2. Bell 206 Transmission Setup and Expected Frequencies

Although it is not necessary to be an expert in gear system design, it is essential to fully understand the power flow through the gears, the rate of rotation of each component as a function of input shaft speed, the number of teeth on each gear, and the placement/identification of bearings. Due to the variety of components, gear system frequencies typically populate a wide portion of the spectrum from less than shaft speed (tooth repeat frequencies) to multiples of gear mesh frequency.

Gear mesh frequency is defined as the number of teeth on the gear multiplied by the rotational speed of the shaft. Gear mesh is the key parameter to monitor, as any anomaly in the transfer of power through the gears will be reflected at this frequency.

Figure 2 illustrates a schematic presentation of the Bell 206 planetary gearbox with three-planet arrangement. Three accelerometers are fitted on the ring gear (front, rear, and left hand side). The different parameters of the gear system as well

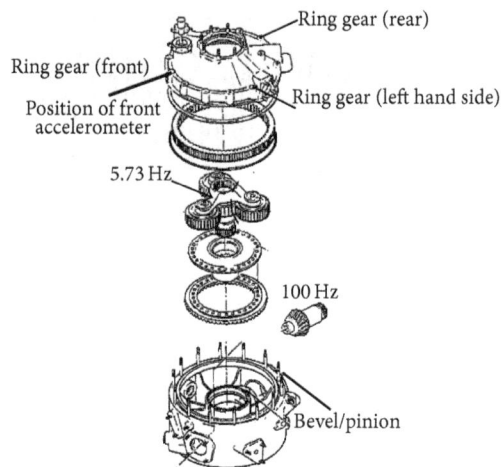

FIGURE 2: Bell 206 transmission with 3 planets.

TABLE 1: Bell 206 gears, speed, and data acquisition parameters.

Parameter	Value	Reduction ratios
Shaft input speed (Hz)	100	
Number of stages	2	
Number of teeth of the bevel pinion	19	First-stage reduction ratio:
Number of teeth of the bevel gear	71	3.73 : 1 (71/19)
Number of teeth of the sun gear	27	Second-stage reduction ratio:
Number of teeth of the planet gear	35	4.67 : 1 (99/27 +1)
Number of teeth of the ring gear	99	
		Overall reduction ratio: (1st stage times 2nd stage) 17.45 : 1
Carrier (arm) output speed (Hz)	5.73	
Sampling frequency (Hz)	51200	
Length of records	30 seconds	

as the data acquisition parameters are presented in Table 1. The data from the front accelerometer was used in examining the vibration content and observes variation between the three- and four-planet arrangements.

Shaft speeds and gear mesh frequencies and their calculations for the 3- and 4-planet arrangements are presented in Table 2. Both have the same frequencies except for the planet pass frequency (number of planets times the carrier speed). References [6, 7] can be consulted for detailed calculations of the kinematics of planetary gearboxes (both gears and bearings).

TABLE 2: Shaft and gear mesh frequencies.

	Frequency of interest (Hz)	Three planets	Four planets	Relationships and calculations
Stage 1	Input shaft frequency	100.00	100.00	
	Pinion/bevel mesh frequency	1900.00	1900.00	Number of teeth of the bevel pinion times input shaft frequency
	Bevel gear shaft speed	26.76	26.76	Input shaft frequency/1st-stage reduction ratio
Stage 2	Carrier frequency (output arm speed)	5.73	5.73	Input shaft frequency/overall reduction ratio
	Epicyclic mesh frequency	567.71	567.71	Number of teeth of the ring gear times carrier frequency
	Planet pass frequency	17.20	22.94	Number of planets times the carrier frequency

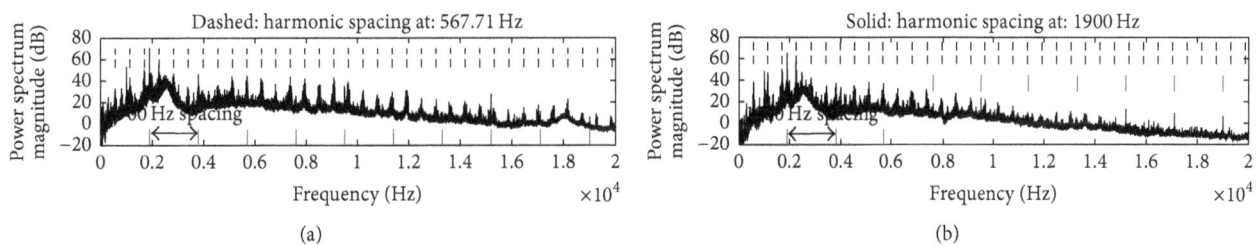

FIGURE 3: (a) Power spectrum density (PSD) of 3-planet arrangement. (b) Power spectrum density of 4-planet arrangement.

3. Fundamental Epicyclic Mesh Frequency and Sidebands Comparisons between the Three- and Four-Planet Arrangements

High quality gears are designed to transfer power by a combined sliding and rolling motion from one gear to the next as smoothly, quietly, and efficiently as possible. This statement means that, starting from the proper metal, the gear teeth are cut to precise dimensions about the geometric center. The tooth surfaces are ground smooth and any imperfections are removed. Each gear is properly fitted to a straight shaft to eliminate eccentricity. The shafts are spaced to optimize the tooth engagement. Any errors in gear manufacture or assembly and/or deterioration will result in a disturbance at gear mesh frequency with adjacent sidebands reflecting once per revolution modulation caused by pitch abnormalities in one or both gears. Since perfection is an asymptotic endeavor, gear systems always display some gear mesh activity. Thus, the presence of the frequency is acceptable and should be apparent. The amplitude increases as the load on the gearing increases. Hence, unless the load is steady, monitoring the amplitude can be a bit tricky. However, an unexplained and significant increase in amplitude at gear mesh frequency is cause of concern.

A large increase in sideband amplitude suggests that something is changing in the geometry and is also cause of concern. The relative amplitude of sideband to gear mesh peak is a good parameter to watch. Scrutinize the waveform for periodic impacts that relate to rotational speed of the gears. Since bearing failure will permit unexpected shaft displacement thereby upsetting gear engagement, always be on the alert for bad bearings. Often, bearing failure precedes

gear damage. Increased impacting from deterioration in the transfer of power may excite gear natural frequencies.

A comparison between the three planets and four planets frequency contents from the front accelerometer is shown in Figure 3. The pinion/bevel mesh frequency harmonics are denoted by the dotted cursor, while epicyclic mesh frequency harmonics are indicated using solid line (these can be seen at the x-axis). The general overall spectra for both configurations are the same, but this is investigated in further zoom-in analysis around the epicyclic mesh frequency as shown in Figure 4.

The number of planets in the system affects the planet pass frequency (number of planets times the carrier speed) and may also cause an asymmetry of the modulation sidebands about the epicyclic mesh frequency (carrier speed times the number of teeth on the ring gear). In some cases, this may also cause complete suppression of the component at the epicyclic mesh frequency [5]. Thus, although 567.71 Hz (99 times 5.73 [99x]) is the listed epicyclic mesh frequency in Table 2, it is not necessary that this will be the dominant harmonic. For *evenly spaced planets*, the epicyclic mesh frequency will be the dominant only if the number of teeth on the ring gear is an integer multiple of the number of planets (ratio of number of teeth on the ring/number of planets is integer) as discussed and detailed in reference [5]. This implies that the energy of the signal will be concentrated at integer multiples of the planets and this what will determine the suppression/nonsuppression of certain sidebands. For Bell-206-B3, this ratio is integer (99/3), and the epicyclic mesh frequency and its harmonics are not expected to be suppressed; thus, the epicyclic mesh frequency will have the highest amplitude with sidebands around it. For the Bell-206-B4,

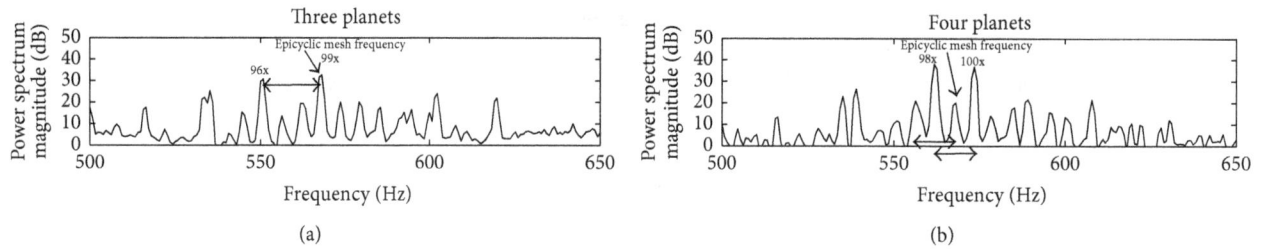

FIGURE 4: Zoom in PSD showing the epicyclic mesh frequency and its side bands: (a) 3 planets and (b) 4 planets (epicyclic mesh suppressed).

TABLE 3: Bell-206-B4 gearbox modulation sidebands (assuming planets are evenly spaced).

Sidebands	$m = 1$	$(N_{ring} + \text{Sideband})/N_{planet}$
−5	0	$(99 − 5)/4$ (NI)
−4	0	$(99 − 4)/4$ (NI)
−3 (550.52 Hz)	**1**	**$(99 − 3)/4$ (I)**
−2	0	$(99 − 2)/4$ (NI)
−1 (561.97 Hz)	0	$(99 − 1)/4$ (NI)
0 (567.71 Hz)	0	$(99 − 0)/4$ (NI)
1 (573.74 Hz)	**1**	**$(99 + 1)/4$ (I)**
2	0	$(99 + 2)/4$ (I)
3	0	$(99 + 3)/4$ (NI)
4	0	$(99 + 4)/4$ (NI)
5 (596.36 Hz)	**1**	**$(99 + 5)/4$ (I)**

0: suppressed. 1: dominance: nonzero vibration. N_{ring}: number of teeth on the ring gear (99 teeth). N_{planet}: number of planetary gears (4 planets). I: integer. NI: noninteger.

this ratio (99/4) is not integer and it is expected then that most of the harmonics of the epicyclic mesh frequency will be suppressed (other sidebands will appear strongly). A prediction based on McFadden and Smith's paper [2], for the Bell-206-B4 gearbox, assuming that the planets are evenly spaced is given in Table 3 for the first harmonic of the epicyclic mesh frequency ($m = 1$). Table 3 suggests that the epicyclic mesh frequency will be suppressed and that the sidebands (550.52, 573.74, and 596.36 Hz) will appear strongly in the spectrum. Note that only the vibration at the first upper sideband and every forth sideband around is nonzero.

To see if this agrees with the actual frequency content of the signal, a zoom-in around the fundamental epicyclic for the 3- and 4-planet arrangements is plotted and presented in Figure 4. It is noticed that, in the 3-planet arrangement, there is no suppression for the epicyclic mesh frequency as predicted earlier (ring number of teeth is exact multiple of the number of teeth on the planet).

For the four-planet arrangements, there is suppression of the epicyclic mesh frequency. The suppression of the epicyclic mesh frequency is quite obvious; however, the sidebands with the highest dB values (lower and upper sidebands, with a maximum dB at the 1st lower sideband at 562 Hz) do not agree with the predictions of Table 3 (third sideband to the left and 1st upper sideband). This could be explained by the fact that the derivation of Table 3 is based on the assumption

that the planets are evenly spaced, which could not be the case for the Bell-206-B4. In reference [6], a system completely similar to Bell-206-B4, referred to as OH-58 C, is discussed. It is stated that the system contains four *nonuniformly spaced* planet gears and that the planet gears come in *pairs spaced 180 degrees apart* with the angle between the pairs other than 90 degrees. This explains the observation of Figure 4, as it is the vibration of the pairs (two planet gear sets) rather than the individual planets (4 planets) that determine the nonzero vibration. In this case the highest amplitude is expected to be for integer multiples of 2 rather than 4 and this explains the lower sideband (98 times the carrier speed (a multiple integer of 2 not 4)) and the upper sideband (100 times the carrier speed (multiple integer of both 2 and 4)).

A further useful summary of these discussed observations can be seen clearly through inspecting the real cepstrum of the two signals as shown in Figure 5. The cepstrum [9] is the inverse Fourier transform of the log spectrum; thus, it has the units of time (Figure 5 shows the time in samples). Cepstrum is a very useful tool to characterize families of sidebands and modulation "in a clear-easy" to interpret format as these become concentrated in harmonics. The dominance of the planet pass frequency in the 3 planets configuration is obvious from Figure 5(a), while a half planet pass presence is present in the 4 planets, which clearly agrees with observation and discussions deduced from the spectrum analysis.

4. Extraction and Removal of Synchronous Shaft Signals through Resampling and Signal Truncation

In order to extract and then remove the harmonics related to each shaft (input, intermediate, and output shafts), three tachometers are required if the shafts are independent, for example, aero engines. In such cases, the angular resampling process should be repeated to allow the order tracking of the speeds of the three shafts, and their harmonics could be removed by subtracting the synchronous average after each resampling step. Order tracking (angular resampling) involves resampling the signal at equal intervals of shaft rotation rather than equal time intervals. This removes any speed fluctuation so that the harmonics of the shaft are genuinely discrete frequencies. This enables their removal by performing synchronous averaging of the order-tracked signal and subtracting it from the latter. Order tracking can be performed by phase demodulating a tacho or shaft encoder

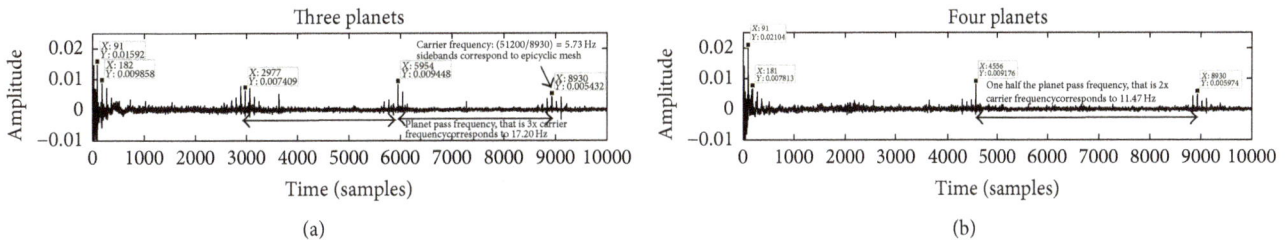

FIGURE 5: Real cepstrum for the three- and four-planet arrangements: (a) 3 planets and (b) 4 planets.

signal and using the mapping between the shaft angle and time to make interpolations in either direction. The process includes the use of cubic spline interpolation of the vibration signal to calculate the values at the required sample points, based on the tachometer signal.

In gearboxes, only one speed reference signal is provided, which is usually sufficient for removing the harmonics of the three related shafts.

A synchronous averaging separation algorithm [10], which is based on signal resampling, is utilized here for the extraction of the shafts' synchronous signals and separating other nonsynchronous signals. In this sense, there is no difference between the 3- and 4-planet setups as the processing algorithm will remove families of harmonics regardless of what the level of the sidebands is.

This algorithm works by resampling the order-tracked signal to obtain an integer number of samples per revolution for a specific shaft. This enables the removal of the shaft harmonics without much disruption of the vibration signal. As there are only three main shafts in the gearbox arrangement (input, intermediate, and output shafts), this will provide a quick yet efficient way of removing all the harmonics of the three shafts (complete removal of shaft/gear related components). The removal of the harmonics of a specific shaft can be achieved by one of two methods. The first is by finding the synchronous average and subtracting it (repeated periodically) from the signal. The second is by truncating the signal to an integer number of revolutions (preferably a power of 2) as described next and setting the lines corresponding to the harmonics of that shaft (after FFT analysis) to the mean value of the adjacent frequencies. To avoid treating the negative frequency components, it is recommended to set them to zero after the FFT step and double the positive frequency components and then take the real part of the resulting analytic signal in the time domain. Both methods give the same results but with less processing time when using the synchronous averaging method.

The steps included in the algorithm are listed below and presented schematically in Figure 6:

(1) Order track the raw signal based on the input shaft tacho. Ensure that the number of samples per revolution of the input shaft (SPR_i) is an integer number (next power of 2 of the nominal numbers of samples between two shaft rotations).

(2) Find an integer number of periods (p) to truncate the signal to.

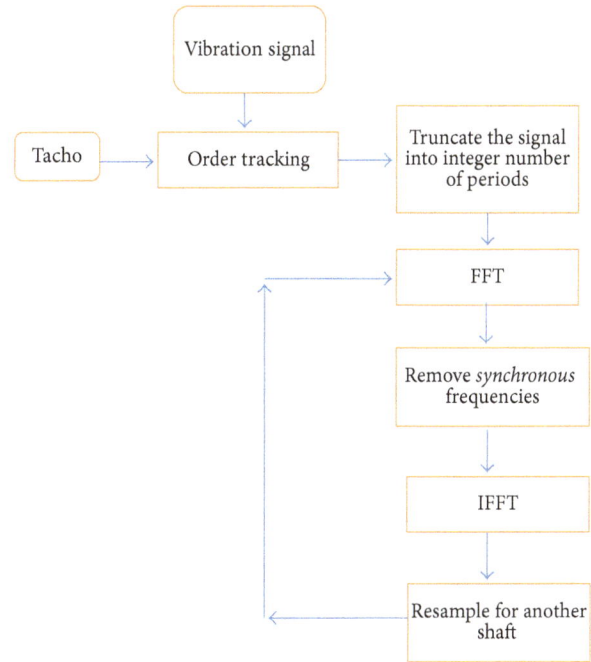

FIGURE 6: Signal separation algorithm.

(3) Truncate the signal to (Nfft) samples (preferably a power of 2), which is equivalent to p periods of the input shaft (i.e., Nfft = $p \times$ SPR_i).

(4) Take the Fast Fourier Transform (FFT) of the truncated signal.

(5) Remove synchronous frequencies related to the input shaft by setting the frequency lines of ($np + 1$) of the fundamental frequency and its harmonics (up to SPR_i − 1) to the mean value of its adjacent frequencies. The result to this stage is illustrated in Figure 7 (further illustrated through Figure 8). Figures 7 and 8 show the removal of the harmonics of the input shaft (i.e., the 100 Hz harmonic). In particular, the pinion/bevel mesh frequency at 1900 Hz (19x 100), which is shown in dark color in both figures, is completely removed. To avoid treating the positive and negative frequency, it is advised to set the negative frequencies to zeros.

(6) Perform an Inverse Fast Fourier Transform (IFFT) on the resulting frequency content obtained in step (5)

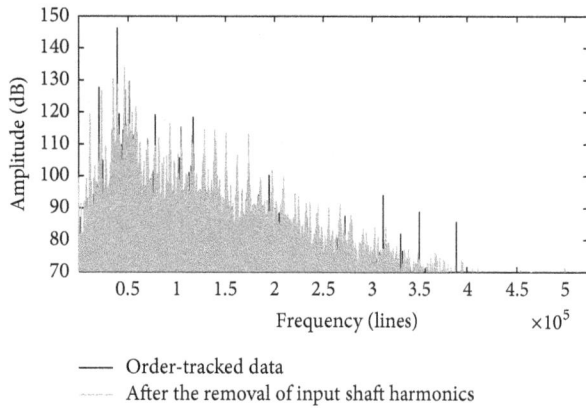

— Order-tracked data
---- After the removal of input shaft harmonics

FIGURE 7: Removing the harmonics of the input shaft speed: dark: before the removal; light: after the removal.

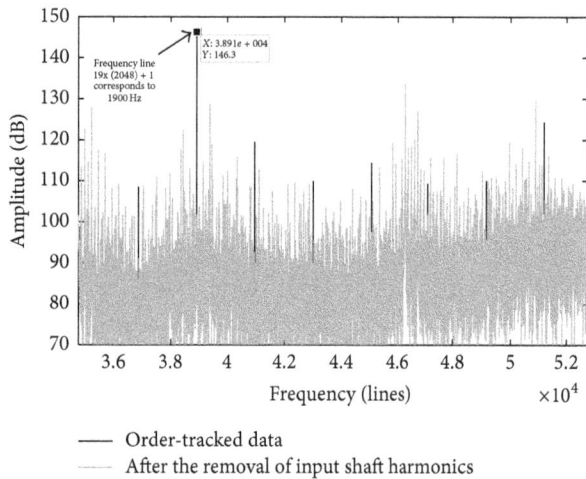

— Order-tracked data
---- After the removal of input shaft harmonics

FIGURE 8: Zooming-in around the pinion mesh frequency: dark: before the removal; light: after the removal.

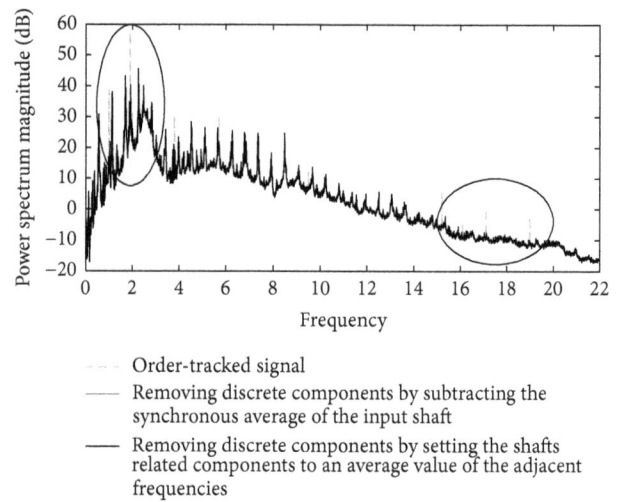

---- Order-tracked signal
— Removing discrete components by subtracting the synchronous average of the input shaft
— Removing discrete components by setting the shafts related components to an average value of the adjacent frequencies

FIGURE 9: Power spectrum showing the effectiveness of the discrete removal algorithm.

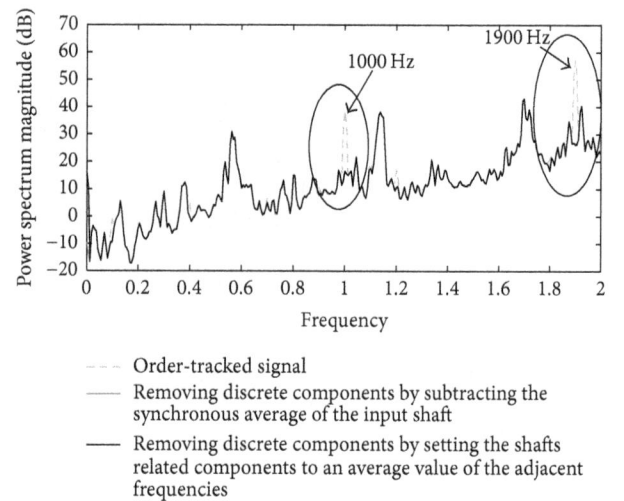

---- Order-tracked signal
— Removing discrete components by subtracting the synchronous average of the input shaft
— Removing discrete components by setting the shafts related components to an average value of the adjacent frequencies

FIGURE 10: Zoom-in (0–2000 Hz): light: signal before processing; dark: signals obtained through different approaches (synchronous averaging subtraction and harmonic removals in the frequency domain).

back to the time domain. If the negative frequencies were set to zero, the real part of the signal should be obtained.

(7) Resample so that there is an integer number of samples (power of 2) for each revolution of the intermediate shaft (this can be achieved by working out the gear ratio and using that to resample the signal to give an integer number of samples per revolution of the intermediate shaft).

(8) Repeat steps (4) to (6) to remove the harmonics related to the intermediate shaft.

(9) Resample the obtained signal so that there are an integer number of samples (power of 2) for each revolution of the output shaft (use gear ratio to resample the signal so that an integer number of samples per revolution of the output shaft are achieved).

(10) Repeat steps (4) to (6) to remove the harmonics related to the intermediate shaft.

The output until step number (6) is equivalent to removing the synchronous average after order tracking the signal with respect to the input shaft (i.e., only one set of harmonics, related to the input shaft speed (100 Hz), is removed at this stage). This is shown (by inspecting the power spectrum) in Figures 8 (whole frequency range) and 9 (0–2000 Hz). Note how removing the input shaft speed harmonics by subtracting the synchronous average of the order-tracked signal gives a similar result to the one obtained when processing the signal through steps (1)–(7).

Note the circled parts of Figures 9 and 10, which correspond to the harmonics of the input shaft. Note that the epicyclic mesh frequency, its sidebands, and harmonics are still present at this stage and further processing is needed to remove them.

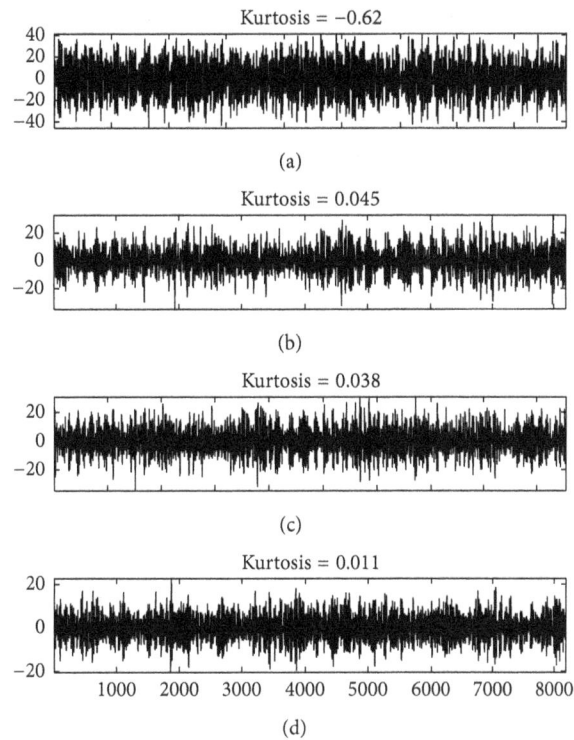

FIGURE 11: (a) Raw signal [order tracked]. (b) Harmonics of input shaft removed. (c) Harmonics of intermediate shaft removed. (d) Harmonics of output shaft removed.

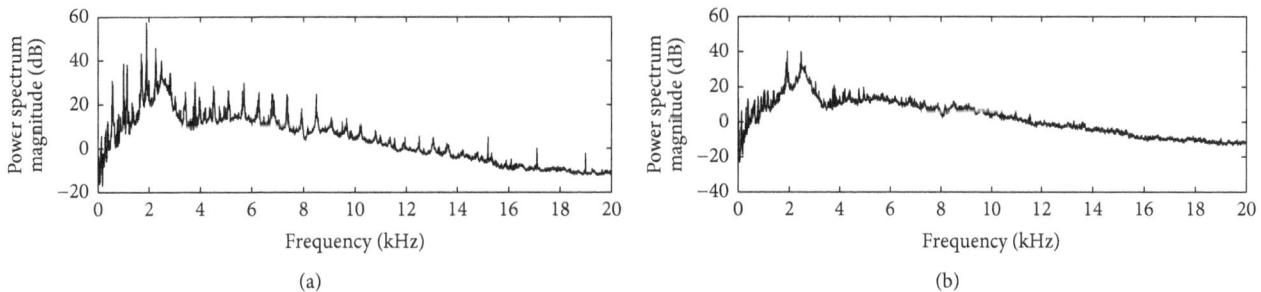

FIGURE 12: Power spectrum of (a) order-tracked signal and (b) signal processed using the new algorithm.

The end result of steps (1)–(10) is illustrated in Figure 11, which shows the time domain signals (one complete rotation of the output shaft) for each step of removing the shafts' harmonics, that is, the input, intermediate, and output shafts.

Note the disappearance of the discrete components from the spectrum in the processed signal as illustrated in Figure 12.

Figure 13 shows how the new processing algorithm compares between the synchronous averaging approach and the DRS algorithm. Note that the synchronous averaging algorithm on one hand only removed harmonics related to the input shaft. DRS on the other hand gives better results in terms of the removal of all discrete frequencies but disrupts the power spectrum by introducing holes instead of the discrete components.

5. Summary and Conclusions

This paper has compared the effect of changing the number of planets of Bell 206 helicopter planetary gearbox on the modulation sidebands around the epicyclic mesh frequency. The planet pass frequency strongly modulates the epicyclic mesh in the 3-planet arrangement. In the case of the 4 planets, the epicyclic mesh frequency and its harmonics are suppressed (other sidebands appear strongly). McFadden and Smith's model [2] was used to predict the highest sidebands/harmonics, which was predicted as 96 and 100. However, the spectrum and cepstrum showed that modulation in fact happens at two times the carrier frequency (2x) and thus the highest harmonics were 98x and 100x. This has been attributed to the idea that planet gears come in pairs of spaced 180 degrees apart with the angle between the

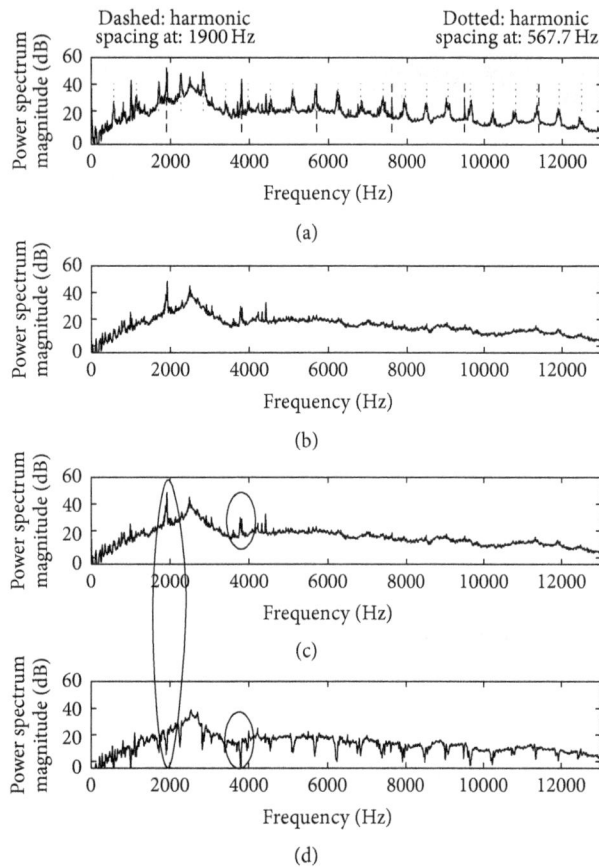

FIGURE 13: Power spectrum of (a) order-tracked signal, (b) residual signal obtained by setting the shaft' related harmonics to the mean of adjacent (noise) lines, (c) residual signal obtained by subtracting the synchronous averages after resampling, and (d) DRS residual (removing discrete components using DRS).

pairs other than 90 degrees. For the three-planet arrangement, the epicyclic mesh frequency and its harmonics are not suppressed; thus the epicyclic mesh frequency has the highest amplitude with sidebands around it. For the four-planet arrangement, the epicyclic mesh frequency and its harmonics are suppressed (other sidebands appear strongly). A synchronous/nonsynchronous separation algorithm has been presented to enable the extraction and then removal of the harmonics of the different shafts in the gearbox using only one input tachometer. The algorithm which relies on successive resampling and signal truncation according to the number of shafts gives superior result to those obtained from the discrete random separation algorithm.

Competing Interests

The author declares that there is no conflict of interests regarding the publication of this paper.

Acknowledgments

Data used in this paper was provided by Australian Defence Science and Technology Group (DSTG).

References

[1] J. Pouradier and M. Trouvé, "An assessment of eurocopter experience in HUMS development and support," in *Proceedings of the 57th AHS International Annual Forum*, American Helicopter Society, Washington, DC, USA, May 2001.

[2] P. D. McFadden and J. D. Smith, "An explanation for the asymmetry of the modulation sidebands about the tooth meshing frequency in epicyclic gear vibration," *Proceedings of the Institution of Mechanical Engineers*, vol. 199, no. 1, pp. 65–70, 1985.

[3] D. Lewicki and J. Coy, "Vibration characteristics of OH58a helicopter main rotor transmission," NASA Technical Paper NASA TP-2705/AVSCOM TR 86-C-42, 1987.

[4] P. D. McFadden and I. M. Howard, "The detection of seeded faults in an epicyclic gearbox by signal averaging of the vibration," Propulsion Report 183, Department of Defence, Aeronautical Research Laboratory, 1999.

[5] E. Huff and I. Tumer, "Using triaxial accelerometer data for vibration monitoring of helicopter gearboxes," *Journal of Vibration and Acoustics*, vol. 128, no. 4, pp. 120–128, 2003.

[6] P. Dempsey, D. Lewicki, and H. Decker, "Transmission bearing damage detection using decision fusion analysis," NASA Technical Paper NASA/TM-2004-213382, 2004.

[7] G. Lang, "S&V geometry 101," in *Sound and Vibration*, pp. 1–12, 1999.

[8] J. Antoni and R. B. Randall, "Unsupervised noise cancellation for vibration signals—part II: a novel frequency-domain algorithm," *Mechanical Systems and Signal Processing*, vol. 18, no. 1, pp. 103–117, 2004.

[9] D. G. Childers, D. P. Skinner, and R. C. Kemerait, "The cepstrum: a guide to processing," *Proceedings of the IEEE*, vol. 65, no. 10, pp. 1428–1443, 1977.

[10] C. L. Groover, M. W. Trethewey, K. P. Maynard, and M. S. Lebold, "Removal of order domain content in rotating equipment signals by double resampling," *Mechanical Systems and Signal Processing*, vol. 19, no. 3, pp. 483–500, 2005.

14

Simulating the Effects of Surface Roughness on Reinforced Concrete T Beam Bridge under Single and Multiple Vehicles

Rahul Kalyankar and Nasim Uddin

Department of Civil Construction and Environmental Engineering, University of Alabama at Birmingham, Birmingham, AL, USA

Correspondence should be addressed to Rahul Kalyankar; krahul2807@gmail.com

Academic Editor: Lucio Nobile

This research focuses on the application of the spatial system of finite element modeling for the vehicle-bridge interaction on reinforced concrete US Girder Bridge in order to obtain the effect of surface roughness. Single vehicle and multiple vehicles on reinforced concrete T beam bridge were studied with variable surface roughness profiles. The effects of six different surface roughness profiles (very good, good, measured, average, poor, and very poor) were investigated for vehicle-bridge interaction. The values of the Dynamic Amplification Factor (DAF) were obtained for single and multiple vehicles on T Beam Bridge for different surface roughness profiles, along with the distances between the axles of heavy vehicle. It was observed that when the bridge has very good, good, measured, and average surface roughness, the DAF values for the single vehicle over the bridge were observed to be within acceptable limits specified by AASHTO. However, for the bridge with multiple vehicles only very good and measured surface roughness profiles showed a DAF and vehicle axle distances within the acceptable limits. From the current studies, it was observed that the spatial system showed reliable responses for predicting the behavior of the bridge under variable road surface roughness conditions and was reliable in vehicle axle detection, and therefore, it has a potential to be use for realistic simulations.

1. Introduction

The vehicle's interaction with the bridge is very complicated issue. The vehicles' rotating tires exert load on the bridge while vehicle in motion. Since the vehicle is a complex assembly of various components, the vehicle induced motion creates dynamic forces in the suspension and tire assembly which gets transferred to the bridge. The interaction between vehicle wheels and bridge is also dominated by bridge surface irregularities. The interaction between the vehicle's tire and the rough bridge surface causes the vehicle to move in different directions, thus causing vibration in the bridge. The schematic of surface irregularities under a vehicle during vehicle-bridge interaction is as shown in Figure 1.

As shown in Figure 1, the vehicle passing over the irregular surface swings the vehicle axles, thus creating significant amount of shaking in the bridges [1–4]. It was observed that, as a result of surface roughness present during vehicle-bridge

interaction, the dynamic forces could increase up to 90% [5]. The surface roughness profile used for obtaining dynamic forces is created using vehicle-structure mathematical model under full probabilistic formulation in frequency domain that considers dynamics of vehicle-bridge interaction. The vehicle velocities, combined with surface irregularities, significantly increase the magnitude of the vibration. Along with the velocities, the number of vehicles present over the bridge also severely affects the dynamic forces [6].

As per AASHTO bridge specification, the static responses of bridges are magnified by 33% to approximate dynamic responses of the bridge for vehicle induced dynamic forces [7]. This magnification factor is generally known as Dynamic Amplification Factor (DAF) as per Euro code or Dynamic Impact Factor (DIF) as per AASHTO and is given by (1) and (2):

$$\text{DAF} = \frac{S_d}{S_s} = 1 + \text{DIF}, \tag{1}$$

TABLE 1: Road surface roughness coefficients [10–12].

Road surface roughness classification	Road surface roughness coefficient
Very good surface roughness (VGSR)	2×10^{-6}–8×10^{-6}
Good surface roughness (GSR)	8×10^{-6}–32×10^{-6}
Average surface roughness (ASR)	32×10^{-6}–128×10^{-6}
Poor surface roughness (PSR)	128×10^{-6}–512×10^{-6}
Very poor surface roughness (VPSR)	512×10^{-6}–2048×10^{-6}

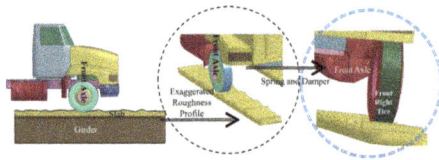

FIGURE 1: Vehicle on rough bridge surface during vehicle-bridge interaction.

where S_d: dynamic strain at the center or sensor location, S_s: static strain at the center or sensor location:

$$I = \frac{50}{L + 125},\qquad (2)$$

where I: impact factor and L: length of bridge (m).

Using (1) and (2), the impact factor can be obtained for any bridge. However, the vehicle bouncing as a result of surface irregularities is neglected in the DAF and a uniform factor is applied over the bridge regardless of bridge type, nature of vehicular traffic, and amount of surface irregularity over it [7–9].

There were numerous efforts taken by researchers in the past to determine the surface roughness on the bridge by classifying them into different profiles [10–12]. Based on previous studies, the classification of road roughness was done into five categories using a road surface roughness coefficient as given in Table 1 [10–12].

The coefficients as listed in Table 1 were obtained using a Power Spectral Density (PSD) function. In addition to this, efforts were also taken by past researchers to measure the road surface roughness on actual bridge [13]. The road surface roughness profile measured on actual site (MSR), along with five road surface profiles (VGSR, GSR, ASR, PSR, and VPSR) obtained using PSD, is as shown in Figures 2(a)–2(f).

From Figures 2(a)–2(f), it can be observed that the profiles obtained from measured road surfaces on actual bridges closely resemble the average road surface profile in terms of magnitude of trough or crest. Numerous studies were conducted by the past researchers using these road surface roughness profiles to obtain bridge responses [9, 13]. Numerous simulations were carried out on the bridges for vehicle-bridge interaction with variable roughness surface profiles using finite element modeling (FEM). AASHTO HS20-44 along with a simply supported T beam bridge modeled using orthotropic plate theory was used for the analysis [9]. However, while attempting to obtain the DAF,

these studies did not address multiple vehicle scenarios, and the vehicle path was assumed to be in the center of the bridge. In addition to this, the variable roughness surface was not addressed in the study [9]. To this end, another study on the application of roughness with multiple vehicles with the change in roughness profile from very good to very poor showed an increase in the DAF [13]. However, the grillage method was used for the bridge FEM which possesses some drawbacks in case of complicated bridge structures [14]. During previous studies for obtaining the DAF, the focus was only on the bridge strain responses, and the effect of surface roughness on axle detection was not addressed in the previous studies.

Therefore, due to these limitations of previous analysis, an alternate approach of Simplified 3 Dimensional Finite Element Model (3D SFEM) using spatial system was proposed in the previous work for obtaining the effect of surface roughness with single and multiple vehicle scenarios [15]. For the application of the spatial method, the most common type of reinforced concrete T beam bridge on US 78 was selected, along with the ALDOT 5 Axles Truck as heavy vehicle, as described in Section 2.

2. Spatial System for US Girder Bridges and Heavy Vehicle

The simplified 3D FEM (3D SFEM) of the US 78 T Beam Bridge (designated as US 78 for the rest of the article) was developed using spatial method. For the US 78 bridge, the geometric information is provided by Alabama Department of Transportation. The geometry of US 78 used for the development of 3D SFEM is as shown in Figure 3.

The 3D SFEM of US 78 was developed using information provided by ALDOT as shown in Figure 3. Along with the geometric information, the material properties were supplied by ALDOT and are as listed in Table 2.

Using the properties and geometric information provided in Figure 3 and Table 2, the 3D SFEM of US 78 was developed. The shell elements used for slab and for the girder bottom shell are assigned with six degrees of freedom per nodes. Similarly, the beam elements used for reinforcements and for the web are also assigned with six degrees of freedom per nodes. For the support, five and six degrees of freedom were assigned at each support, thus allowing the movement in longitudinal direction at one end. The 3D SFEM of US 78 was verified for deflection and natural frequencies in order to use it for vehicle-bridge interaction [15]. The 3D SFEM of US 78 is as shown in Figure 4.

As shown in Figure 4, the discretization of various components of US 78 was carried out using different types of elements. For the sake of brevity, the 3D SFEM of only one girder is shown in Figure 4. Along with US 78, the 3D SFEM of ALDOT 5 Axles Truck, representing the common heavy vehicle on US highways, was modeled using spatial method. The schematic of ALDOT 5 Axles Truck used for finite element modeling is as shown in Figure 5.

The 3D SFEM of ALDOT 5 Axles Truck developed using properties as shown in Figure 5 was verified for mass

TABLE 2: Properties of Components of US 78.

S. Number	Component	Actual size (mm)	3D SFEM size (mm)	Element type	ρ (kg/m^3) Actual	FEM	E (GPa) Actual	FEM	ν
1	Deck slab (t)	150	225.5	Shell	2400	2020	25	24	0.22
2	T-Beam (t)	940	947	Beam	7850	2140	25	31.3	0.22
3	Beam MR (Φ)	36	51	Beam	7850	7850	210	210	0.3
4	Beam Tr R (Φ)	36	51	Beam	7850	7850	210	210	0.3
5	Girder BS (t)	—	10	Shell	—	—	—	25	0.22

MR: main reinforcement; Tr R: top reinforcement; Girder BS: girder bottom shell; t: thickness.

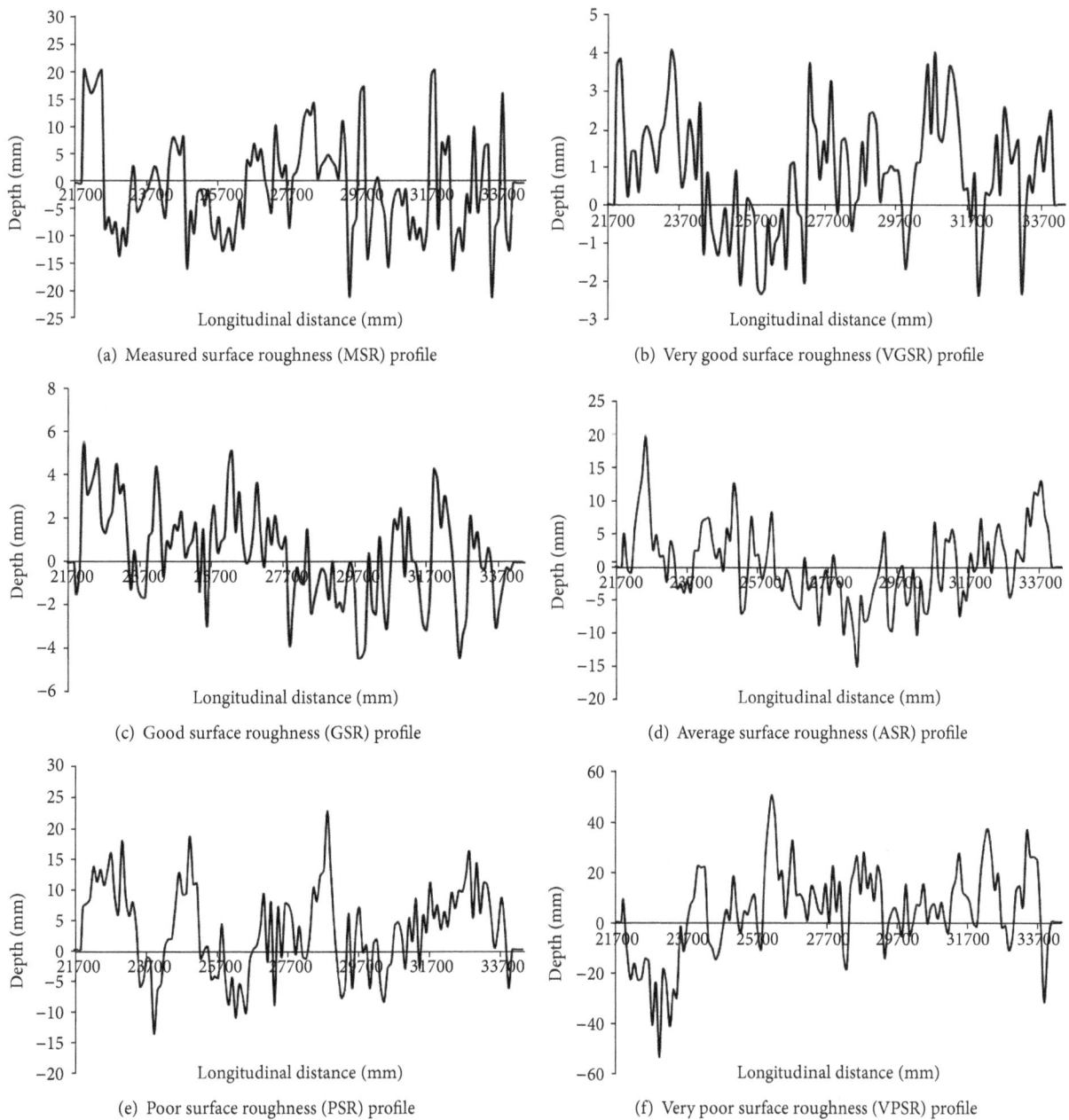

(a) Measured surface roughness (MSR) profile
(b) Very good surface roughness (VGSR) profile
(c) Good surface roughness (GSR) profile
(d) Average surface roughness (ASR) profile
(e) Poor surface roughness (PSR) profile
(f) Very poor surface roughness (VPSR) profile

FIGURE 2: Road surface roughness profiles for vehicle-bridge interaction.

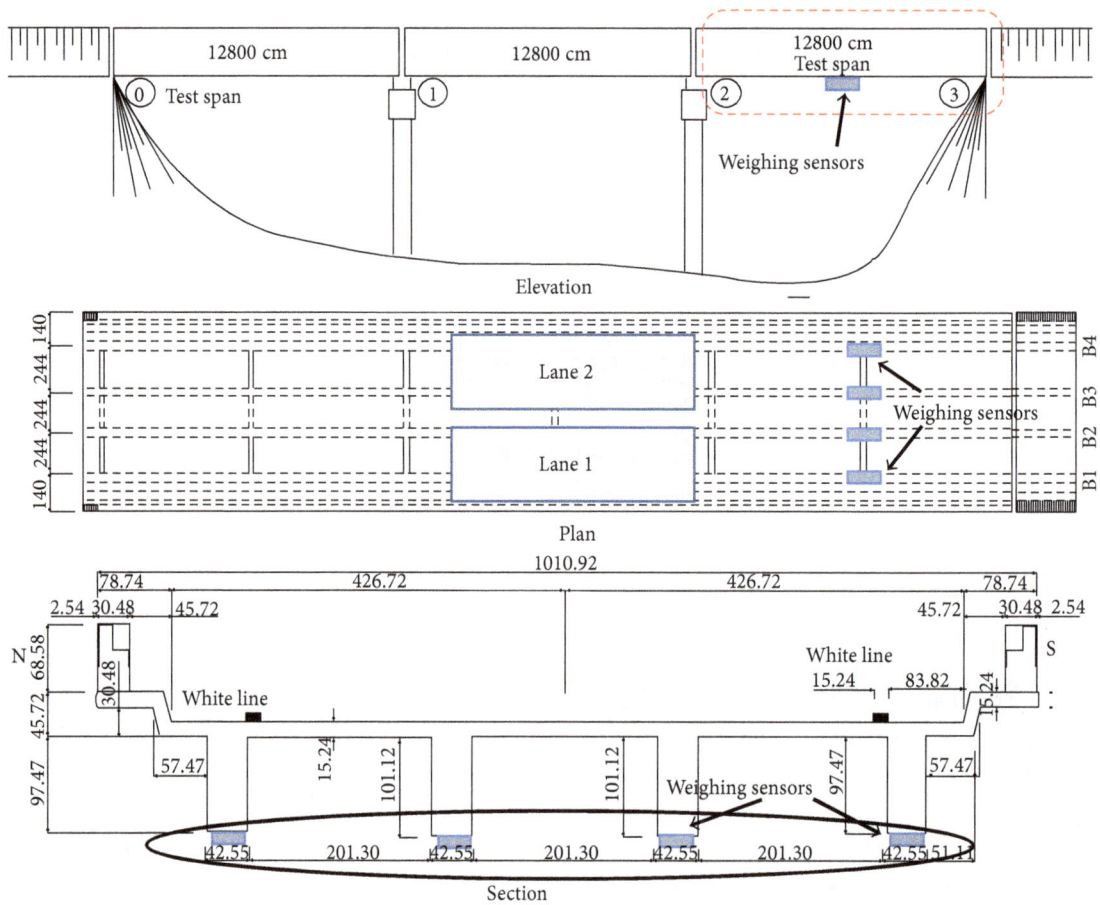

FIGURE 3: Detailed schematic of US 78.

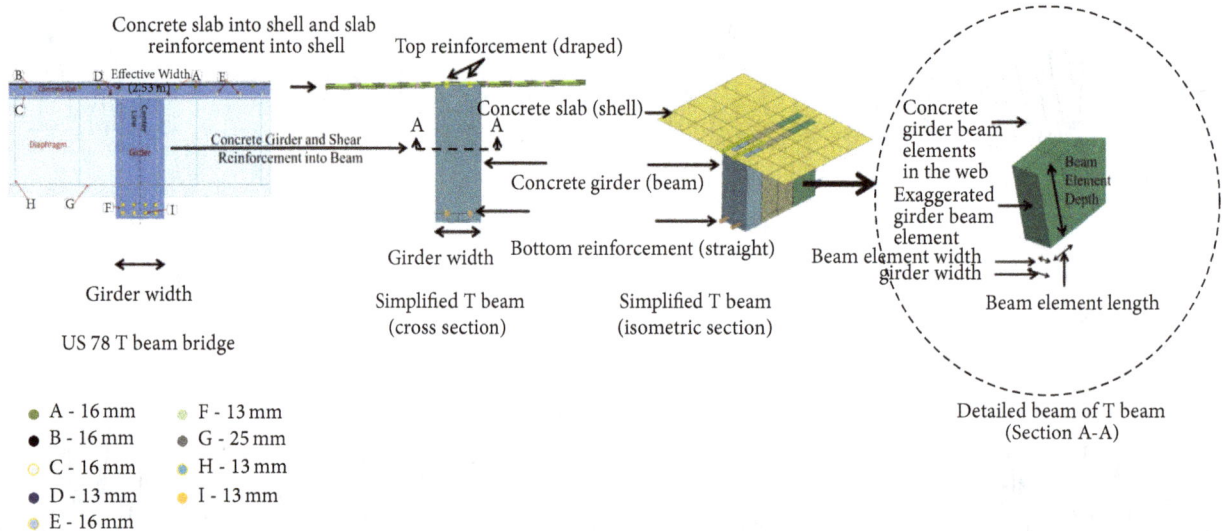

FIGURE 4: Detailed views of discretized US 78 used for 3D SFEM.

distribution. For the sake of conciseness, the properties of various components of the vehicle are not included in this work and can be obtained from previous research [15]. The 3D SFEM of heavy vehicle used for vehicle-bridge interaction is as shown in Figures 6(a) and 6(b).

As shown in Figures 6(a) and 6(b), the vehicle was discretized using mass, beam, and discrete elements. The detailed view of the front axle of a heavy vehicle consisting mass, beam, and discrete elements is as shown in Figure 6(b). The vehicle-bridge interaction was carried out using US 78

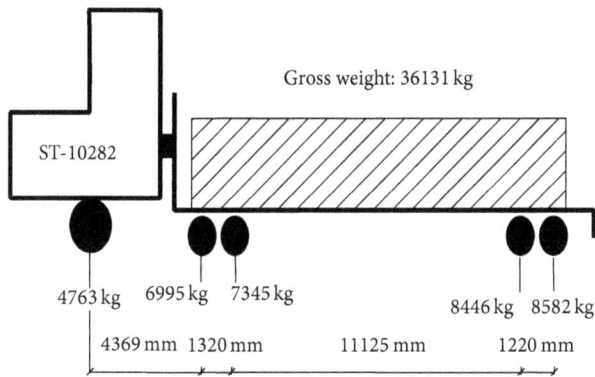

FIGURE 5: Schematic of ALDOT 5 axles fully loaded truck used as heavy vehicle.

as shown in Figure 4 and heavy vehicle as shown in Figures 6(a) and 6(b). For the vehicle's interaction with the bridge, RAILTRAIN contacts were used. The RAILTRAIN contact is a node-to-beam type contact which describes the interaction of the node with the beam elements [16]. To facilitate vehicle's passage over US 78, a set of beam elements was created on the bridge surface underneath M1. This set of beam elements serves as a path for the vehicle, which transfers the dynamic vehicular axle loads to the bridge. The RAILTRAIN contacts were defined between beam elements under the vehicle wheels with the node located at M1. The RAILTRAIN track provides a mode to assign different types of surface roughness profiles to the bridges that are used for obtaining roughness effects on bridge responses. To this end, using different road surface roughness profiles (as shown in Figure 2), and the 3D SFEM of bridge and vehicle (as shown in Figures 4 and 6), the vehicle-bridge interaction was initiated and verified using experimental results as described in Section 3.

3. Dynamic Vehicle-Bridge Interaction of US 78 with Heavy Vehicle Using 3D SFEM

Dynamic vehicle-bridge interaction was carried out using already verified 3D SFEMs of US 78 and heavy vehicles in order to obtain dynamic responses of vehicle-bridge interaction in terms of time strain histories [15]. To obtain girder strain responses, four strain gauges known as weighing sensors were installed under the girder at the mid-span. Similarly, to obtain vehicular parameters such as number of axles, axle spacing, and vehicle velocity, four strain gauges known as Free of Axle Detector (FAD) sensors were placed under the slab as shown in Figure 7.

As shown in Figure 7, W represents a weighing sensor at the mid-span. The FAD sensors were placed under the slab so that there are two FAD sensors falling in the same lane and line, 3.7 m away from each other. From Figure 7, it can be observed that, for the US 78, FAD sensors were placed in L1 and L2 (L = lane; G = girder). As shown in Figure 7, the position of the heavy vehicle in L1 of US 78 was used for obtaining weighing and FAD sensor responses. The vehicle position in Figure 7 accurately reflects the vehicle position

used for the experimental test. The times strain histories were obtained for US 78 with the vehicle velocity being 88 km/h (55 mph). In case of the US 78, the vehicle is over G1 and G2 and therefore, using the 3D SFEM, the strains were obtained from the weighing sensors for these girders and compared with experimental strain as shown in Figures 8(a) and 8(b).

From Figures 8(a) and 8(b), it can be observed that the 3D SFEM showed strain patterns similar to the experimental strain for US 78. The G1 of US 78 showed 6% less response, compared to experimental strain, and the G2 showed 18% higher strain responses than the experimental strain. Since the values of first peak responses were consistent in all the cases, only these values were considered for the comparison [15]. Overall, the responses of the first peak values obtained from 3D SFEM for US 78 were within acceptable limits.

In addition to these responses, the responses of FAD sensors were obtained for identification of vehicle parameters. The goal of the FAD sensor is to identify the peak responses in order to represent the individual axles. Therefore, the magnitudes of the peak responses were neglected and only the peak response pattern was used for vehicle and velocity identification. For the FAD sensor responses, it was observed from previous analysis, using 3D Finite Element Model (3D FEM) of US 78 and heavy vehicle that when the vehicle path is near or above the girder, the responses are poor [15]. For this scenario, the sensor optimization was carried out to determine the ideal sensor location near the girder for detecting vehicle parameters [17]. Using already optimized sensor locations on the slab from previous studies, further analysis was carried out on a scenario where the vehicle paths for US 78 were over or near the girder, as shown in Figure 7. The sensor locations selected for FAD response on US 78 bridge are as shown in Figure 9.

From Figure 9, it can be observed that, for US 78, the FAD sensors were located at 0.25L from each end (L, length of test span) and at 0.25WS (WS, Width of test span or distance between the girders). The responses for the vehicle position (as shown in Figure 7) and sensor position (as shown in Figure 9) were obtained using FAD sensors, as shown in Figure 10.

As shown in Figure 10, the responses obtained for US 78 are the bridge responses before applying any filters on the strain responses. From Figure 10 it can be observed that the unfiltered sensor data showed a significant amount of vibration and the axles were unrecognized (designated as UR). Both sensors showed significant vibration thus making distinction of rear axles impossible. In order to remove the excessive vibration and noise due to vehicle movement on US 78, the Butterworth Filter, also known as the maximum flat magnitude filter, was selected for filtering the vibration responses of the FAD sensors. The Butterworth Filter is designed to provide flat frequency responses and the maximum flatness can be achieved for the filtered data with the use of Butterworth Filter [18–20]. The Butterworth Filter provides smooth responses for the frequencies and maximum flatness to the curve which was helpful in minimizing the vibration responses near axle locations [21–23]. Using Butterworth Filter, the excessive vibration was filtered out from the strain responses as shown in Figure 11.

(a)

(b)

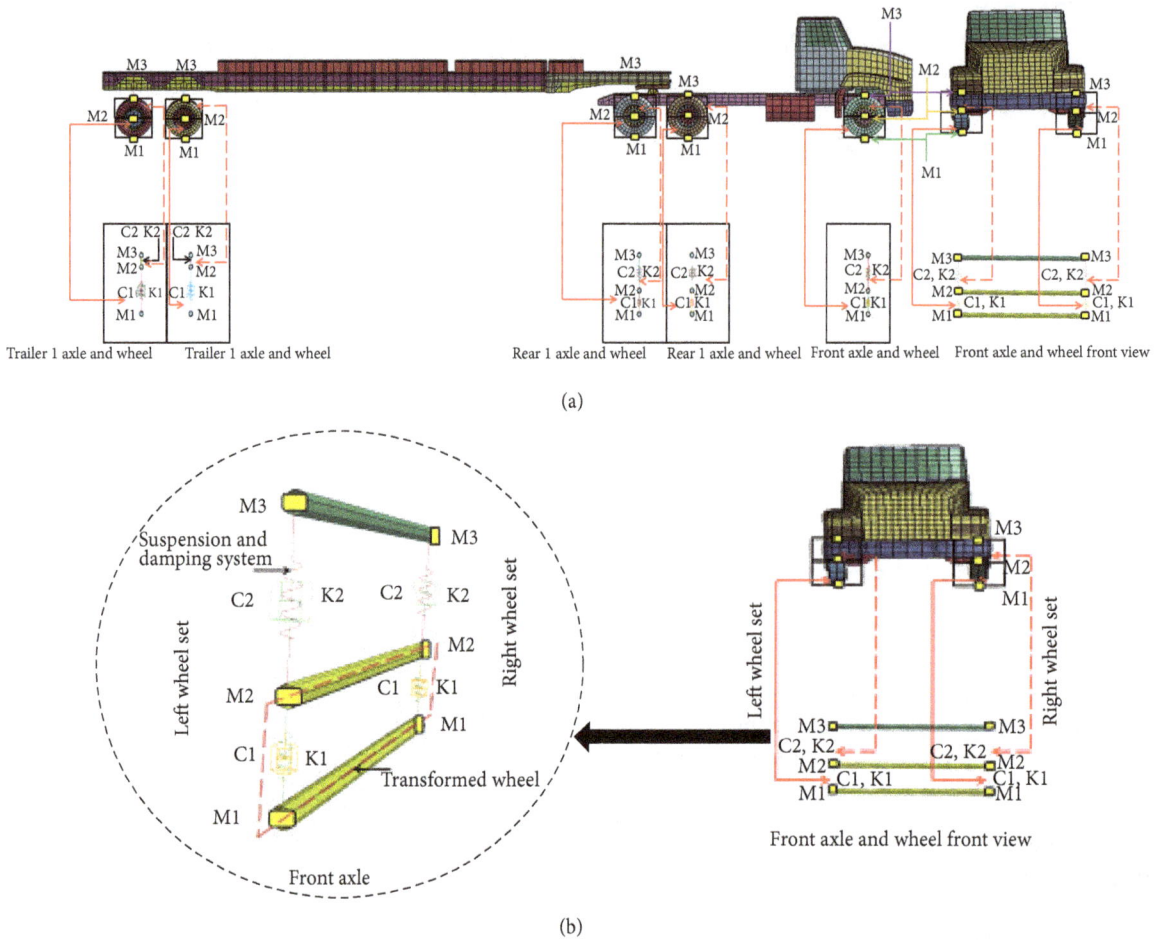

FIGURE 6: (a) 3D SFEM spatial model of ALDOT 5 axles truck created from 3D FEM. (b) Detailed schematic of the 3D SFEM of the front axle (spatial system).

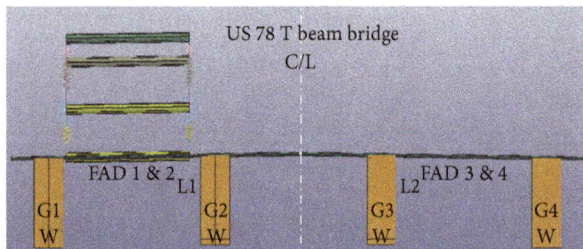

FIGURE 7: Positions of heavy vehicle on US 78 for vehicle-bridge interaction.

The filtered responses as shown in Figure 11 for US 78 showed clear peak values for the axles of heavy vehicles. All five axles were detected with the velocity within acceptable limits, Moreover, the axle distances were measured and verified with the measured distances as shown in Table 3.

As observed from Table 3, the FAD sensor responses were in acceptable limit for MSR profile, with a maximum of 3% difference observed between measured responses and responses obtained from 3D SFEM for A3-A4. All the responses are in the acceptable limit, however, the bridge

road surface is always subjected to continuous tire friction which increases the road surface roughness. Therefore, it is very important to address the issues of effect of variable road surface roughness on bridge performance. To this end, the investigation of effect of roughness on the bridge was carried out using road surface roughness profiles as shown in Figure 2. The responses of weighing sensors and FAD sensors were obtained under variable road surface roughness profiles in order to identify the axles over the bridge while the vehicle is in motion as described in Section 4.

4. Effect of Surface Roughness on the Responses of the Bridge

In order to obtain the effect of the road surface roughness on the bridge's response, the 3D SFEM was assigned the roughness profiles as shown in Figure 1. The weighing sensor responses for G1 were obtained for the scenario of a single vehicle over US 78. The weighing sensor responses on G1 of the US 78 bridge are as shown in Figures 12(a)–12(f).

From Figures 12(a)–12(f), it can be observed that, with the increase in the surface roughness from very good to

TABLE 3: Comparison of FAD sensor response.

Type	D (m)	V (m/s)	A1-A2	A2-A3	A3-A4	A4-A5
Measured	6.25	—	4.27	1.32	11.54	1.22
3D SFEM (MSR)	6.125	24.5	4.43	1.35	11.15	1.23

A1-A2, A2-A3, A3-A4, and A4-A5 are the distances between Axle 1-Axle 2, Axle 2-Axle 3, Axle 3-Axle 4, and Axle 4-Axle 5 in meters.

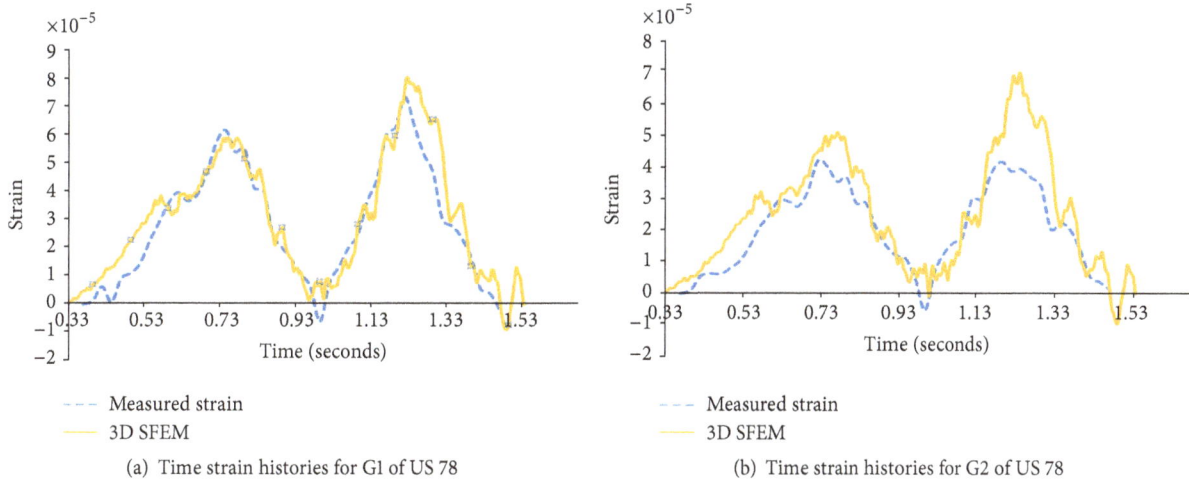

(a) Time strain histories for G1 of US 78

(b) Time strain histories for G2 of US 78

FIGURE 8: Time strain histories for US 78.

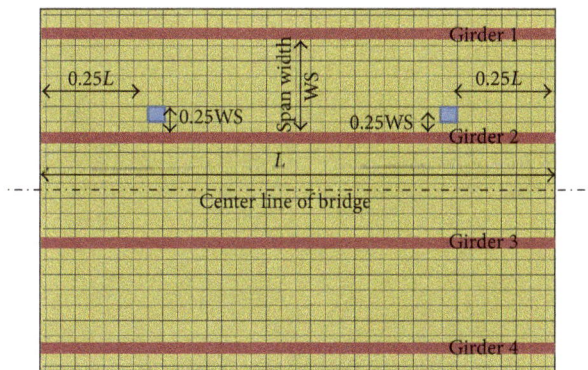

FIGURE 9: FAD sensor locations on US 78.

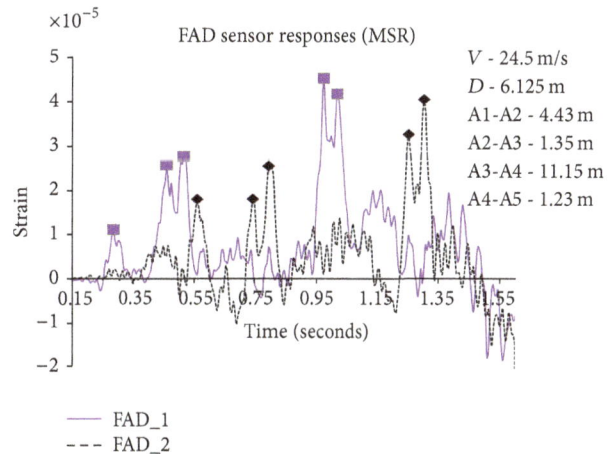

FIGURE 11: Filtered FAD sensor responses for US 78.

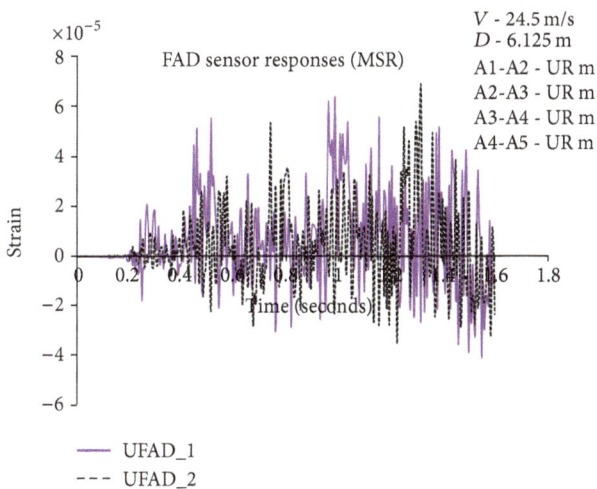

FIGURE 10: FAD sensor responses for US 78.

very poor, the vibration in the strain responses increased significantly. The strain obtained from the examples of good surface roughness (GSR), very good surface roughness (VGSR), measured surface roughness (MSR), and average surface roughness (ASR) showed less vibration in the curve.

The strain obtained from the examples of good surface roughness (GSR); very good surface roughness (VGSR); and measured surface roughness (MSR); and average surface roughness (ASR) showed less vibration in US 78 compared to poor surface roughness (PSR) and very poor surface roughness (VPSR). The DAF was obtained based on weighing sensor responses as shown in Figure 12. For the DAF, the static strain responses were obtained by positioning the vehicle in such a way that the maximum load accumulates at the

(a) Weighing sensor response with MSR profile

(b) Weighing sensor response with VGSR profile

(c) Weighing sensor response with GSR profile

(d) Weighing sensor response with ASR profile

(e) Weighing sensor response with PSR profile

(f) Weighing sensor response with VPSR profile

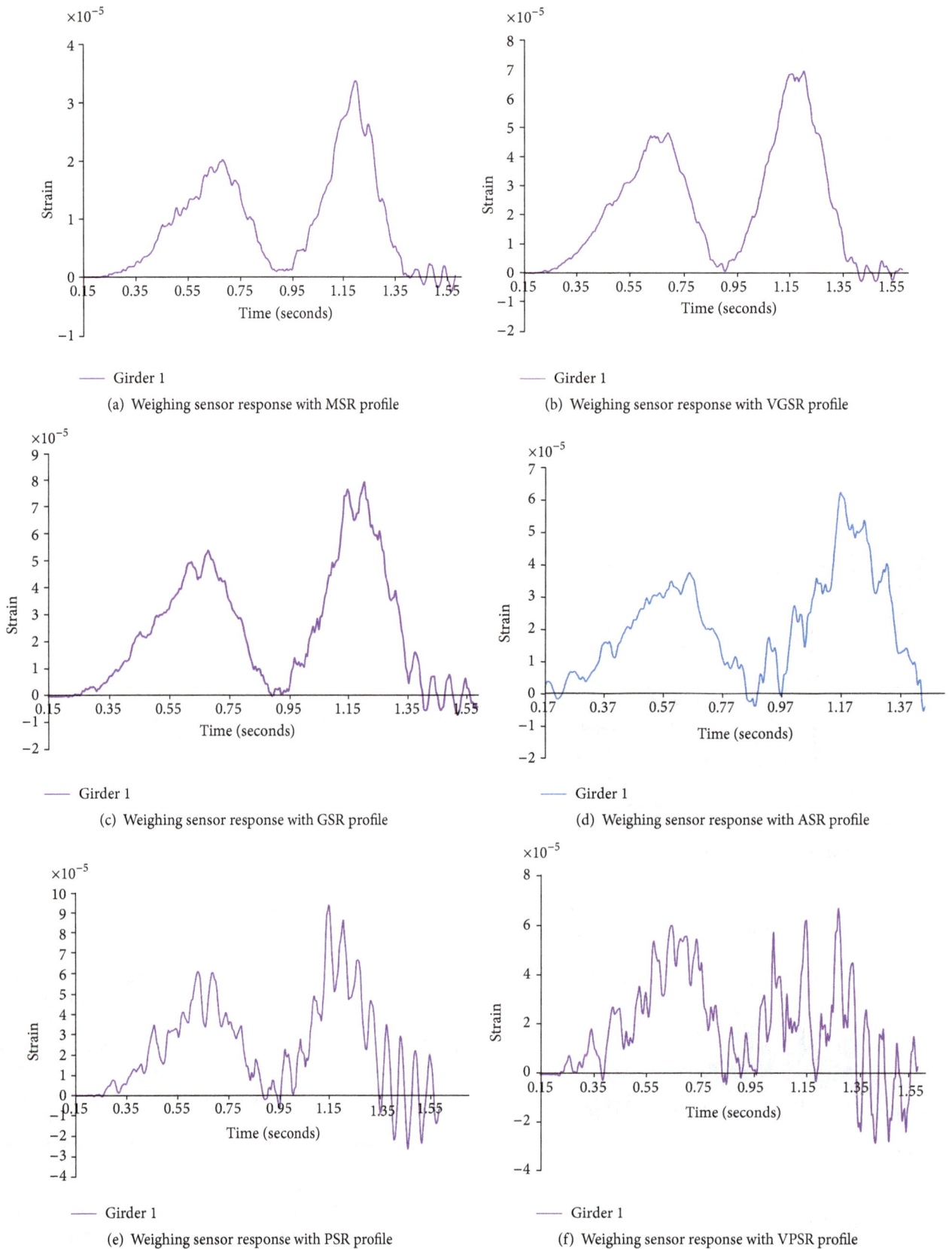

FIGURE 12: Weighing sensor responses for different surface roughness profiles on US 78.

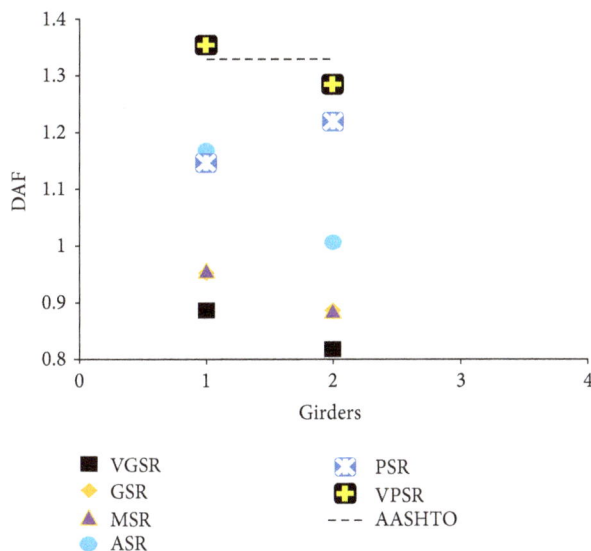

FIGURE 13: Effect of surface roughness on DAF of US 78.

center [21]. Using this approach to static strain, the DAF was obtained for each case of surface roughness as shown in Figure 13.

As seen from Figure 13 for US 78, the value of the DAF was increased significantly with the increase in surface roughness. The increase in the surface roughness was found to be proportional to the vibration of the bridge, where the DAF is the ratio of dynamic and static strain. From Figure 13 the VPSR produced the largest increase in DAF (45% for G1 of US 78). To amend this study, in order to investigate the effect of surface roughness on axle detection, further analysis was carried out on US 78 bridge and strain responses were obtained for the FAD sensors. Figures 14(a)–14(f) show the FAD sensor responses for FAD_1 and FAD_2 using variable roughness profiles.

As shown in Figures 14(a)–14(f), for the VGSR, GSR, MSR, and ASR profiles, at least one FAD sensor showed clear peak responses and satisfactory axial distances. However, in the case of PSR and VPSR, the axial distances were unrecognized due to excessive vibrations even after applying the filter. From Figures 14(a)–14(f), it can be observed that for the US 78 the difference between Rear Axle 1 and Rear Axle 2 is 32% when comparing ASR to MSR. In the case of PSR and VPSR, the axle is undetectable. Although the axle detection for a single vehicle with VGSR, GSR, MSR, or ASR is valid, a bridge is always subjected to multiple vehicles which has made axle detection a great challenge. Therefore, to account for and resolve this difficulty, further analysis was carried out on multiple vehicles on US 78 using four surface roughness profiles (MSR, VGSR, GSR, and ASR). The arrangement of multiple vehicles over the bridges is as shown in Figure 15.

From Figure 15, it can be observed that the heavy vehicle path in US 78 is above the girders. In the case of the US 78, the strain responses were obtained for G1 to G4. For the sake of brevity, only the weighing sensor responses located on G2 and G3 of US 78 were obtained and interpreted for

further analysis. The strain responses for US 78 are as shown in Figures 16(a)–16(d).

From Figures 16(a)–16(d), it can be observed that, for US 78 with multiple vehicles over it, bridge responses showed significant vibration compared to the single vehicle scenario. Based on the weighing sensor responses for US 78 with multiple vehicles, the DAF was obtained for the variable road surface roughness profiles as shown in Figure 17.

From Figure 17, it can be observed that with the surface roughness changed from very good to average, the DAF values for each girder increased. In all the cases, the DAF values were found to be less than specified by the AASHTO design code, except in the case of ASR, where DAF values obtained were on the higher side. Based on this information, the following conclusions can be made as described in Section 5.

5. Conclusion

Based on the studies carried out on US 78 using the spatial system, the following conclusions were drawn:

(i) The spatial system of finite element modeling was successfully implemented in the analysis of the reinforced concrete US 78 T beam bridge, using vehicle-bridge interaction with variable road surface roughness profiles.

(ii) The strain responses observed from the 3D SFEM showed close resemblance with the experimental strain responses from the weighing sensors.

(iii) In the case of single heavy vehicle over the bridge the peak strain responses were obtained within satisfactory limits from the FAD sensors.

(iv) The application of variable surface roughness was carried out and the strain responses were obtained for a single vehicle over the US 78 using weighing sensors and FAD sensors. Values for DAF were obtained for the scenario of a single heavy vehicle over US 78 where 45% increase in the DAF for the G1 was observed when surface roughness changed to VPSR from MSR.

(v) The vehicle axle detection for the single vehicle over the bridge was observed to be within acceptable limits for the road surface profiles VGSR, GSR, MSR, and ASR. The PSR and VPSR failed to allow for recognition of the distance between axles.

(vi) Since the road surface roughness profiles VGSR, GSR, MSR, and ASR successfully detected the axles and distances between them, these surface roughness profiles were analyzed for a multiple vehicle scenario, and weighing and FAD sensor responses were observed.

(vii) From weighing sensor responses with multiple vehicles over the bridge the values of the DAF were obtained for US 78 bridge. In case of US 78, an extreme DAF value was obtained for G3 using ASR, where the DAF was 3.75% lower than specified by AASHTO. Based on the simulations and observations

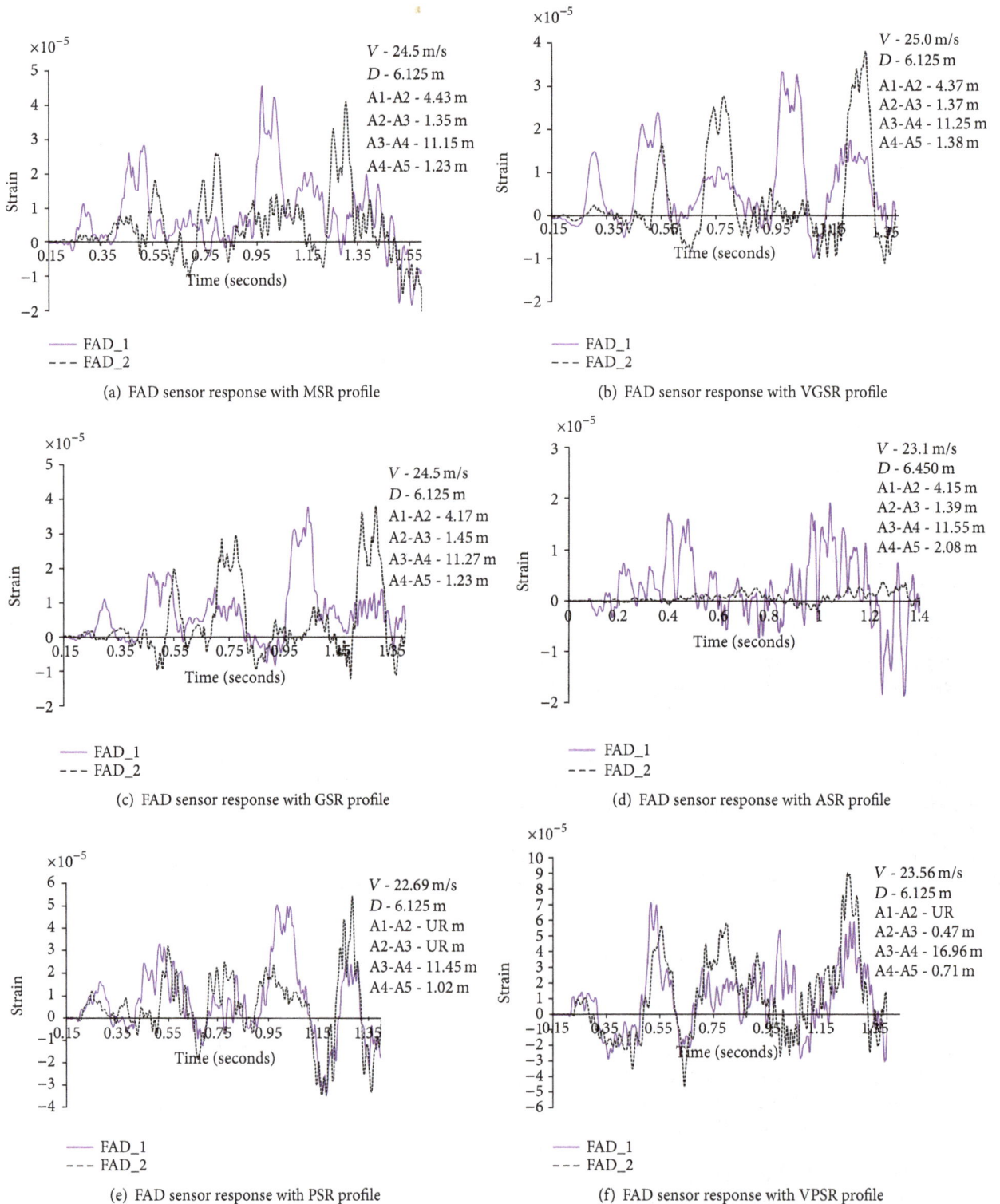

(a) FAD sensor response with MSR profile

(b) FAD sensor response with VGSR profile

(c) FAD sensor response with GSR profile

(d) FAD sensor response with ASR profile

(e) FAD sensor response with PSR profile

(f) FAD sensor response with VPSR profile

FIGURE 14: FAD sensor responses of US 78 with poor road surface roughness.

conducted, it can be concluded that the surface roughness effects on US 78 T Beam Bridge produced reliable responses for the DAF and for axle detection for the VGSR, GSR, and MSR profiles, and for the ASR profile the reliable responses were observed only

when subjected to one vehicle. In case of multiple vehicle scenarios, the ASR showed significant vibration in the strain responses, and the values for the DAF were observed to increase radically, and the axle detection was observed to be unreliable.

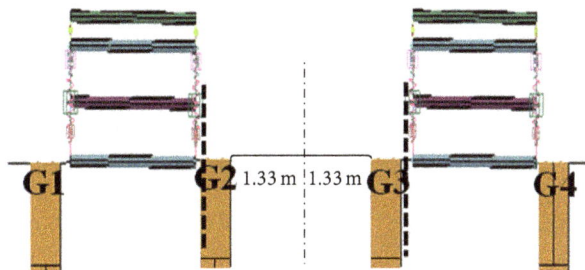

FIGURE 15: Multiple vehicles over US 78 bridge.

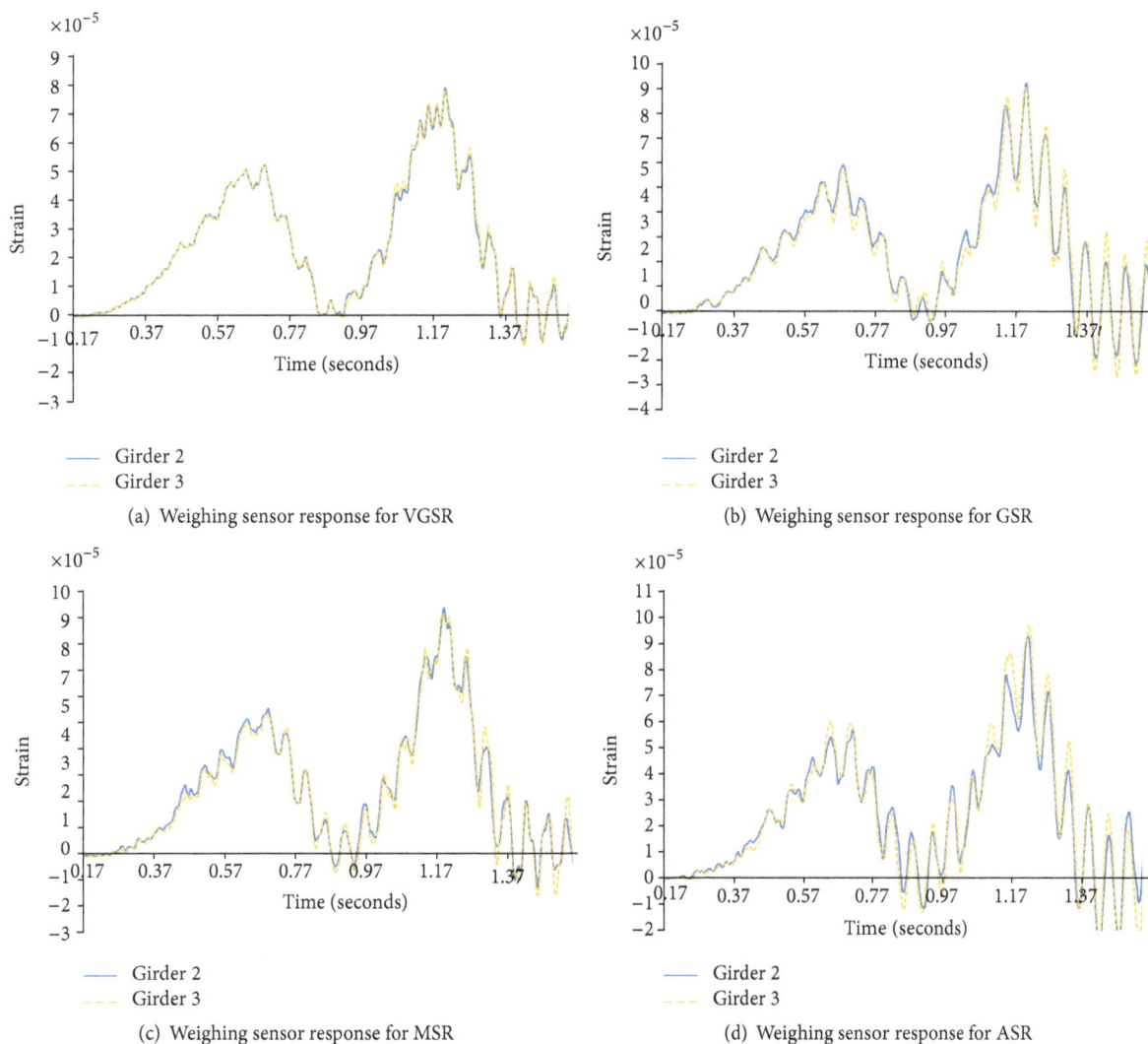

(a) Weighing sensor response for VGSR

(b) Weighing sensor response for GSR

(c) Weighing sensor response for MSR

(d) Weighing sensor response for ASR

FIGURE 16: Weighing sensor responses on US 78 with multiple vehicle scenario.

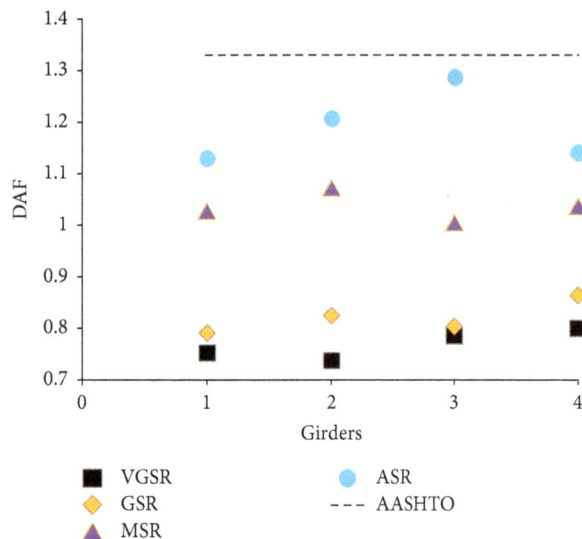

FIGURE 17: Effect of surface roughness on the DAF of US 78 with multiple vehicles.

Using the data presented in this research article, the responses of the bridges can be used for obtaining the effects of surface roughness on any type of bridges in terms of DAF and axle detectability. With these studies, the implementation of different surface roughness profiles can be possible for accurately predicting bridge responses. Moreover, with the development of 3D SFEM vehicle-bridge interaction and the implementation of surface roughness in the 3D SFEM, the resulting responses can be used in selection of the bridge for experimental B-WIM implementation.

Competing Interests

The authors declare that they have no competing interests.

Acknowledgments

The authors would like to express their gratitude for the financial support received from National Science Foundation (NSF-CMMI-1100742) towards this investigation.

References

[1] M. Yener and K. Chompooming, "Numerical studies of vehicle-bridge systems," Structural Engineering and Mechanics Division Report CEE-SEMD-91-13, Utah State University, 1991.

[2] J. Fleming and P. Romualdi, "Dynamic response of highway bridges," Journal of Structural Engineering (ASCE), vol. 87, pp. 31–61, 1961.

[3] E. C. Ting and M. Yener, "Vehicle—structure interactions in bridge dynamics," Shock and Vibration Digest, vol. 15, no. 12, pp. 3–9, 1983.

[4] M. Yener and K. Chompooming, "An overview on vehiclebridge interaction problems and optimum design of bridge systems," Tech. Rep. CEE-SEMD-91-09, Utah State University, 1991.

[5] J. G. S. da Silva, "Dynamical performance of highway bridge decks with irregular pavement surface," Computers & Structures, vol. 82, no. 11-12, pp. 871–881, 2004.

[6] X. Q. Zhu and S. S. Law, "Dynamic load on continuous multi-lane bridge deck from moving vehicles," Journal of Sound and Vibration, vol. 251, no. 4, pp. 697–716, 2002.

[7] American Association of State Highway and Transportation Officials, AASHTO LRFD Bridge Design Specifications, American Association of State Highway and Transportation Officials, Washington, DC, USA, 4th edition, 2010.

[8] R. K. Gupta, "Dynamic loading of highway bridges," American Society of Civil Engineers, Journal of Engineering Mechanics, vol. 106, pp. 377–393, 1980.

[9] S. Jerath and S. B. Gurav, "Impact factors for highway bridges using road surface roughness and vehicle dynamics," Modern Traffic and Transportation Engineering Research, vol. 2, no. 2, pp. 113–120, 2013.

[10] International Standard Organization, "Mechanical vibration: road surface profiles-reporting of measured data," Tech. Rep. ISO 8068 (E), International Standard Organization, Geneva, Switzerland, 1995.

[11] S. Abdalmaged, Dynammic analysis and fatigue assessment of bridge decks subjected to traffic and corronsion effects [M.S. thesis], Department of Civil and Environmental Engineering, Colorado State University Fort Collins, Fort Collins, Colo, USA, 2013.

[12] Cebon, Handbook of Vehicle-Road Interaction, Swets and Zeitlinger, 1999.

[13] W. Han, J. Wu, C. S. Cai, and S. Chen, "Characteristics and dynamic impact of overloaded extra heavy trucks on typical highway bridges," Journal of Bridge Engineering, vol. 20, no. 2, 2015.

[14] I. Vayas, A. Iliopoulos, and T. Adamakos, "Spatial systems for modelling steel-concrete composite bridges—comparison of grillage systems and FE models," Steel Construction, vol. 3, no. 2, 2010.

[15] R. Kalyankar, Simulation of bridge responses to heavy vehicles [Ph.D. thesis], Department of Civil Construction and Environmental Engineering, the University of Alabama at Birmingham, Birmingham, Ala, USA, 2015.

[16] LS-DYNA Keyword User's Manual Volume 1, Version 971, Livermore Software Technology Corporation, Livermore, Calif, USA, 2007.

[17] R. Kalyankar and N. Uddin, "Optimization of free of axle detector sensor location using 3D finite element simulations for bridge weigh in motion system on US girder bridges," Journal of Structural Control and Health Monitoring, In press.

[18] S. Butterworth, "On the theory of filter amplifiers," Experimental Wireless and the Wireless Engineer, vol. 7, pp. 536–541, 1930.

[19] K. Lacenette, A Basic Introduction to Filters—Active, Passive, and Switched Capacitors, National Semiconductor Application Note 779, 1991.

[20] G. Bianchi and R. Sorrentino, Electronic Filter Simulation and Design, McGraw-Hill, 2007.

[21] Web article, https://zone.ni.com/reference/en-XX/help/371361J-01/lvanlsconcepts/lvac_butterworth_filters/.

[22] http://www.etc.tuiasi.ro/cin/Downloads/Filters/Filters.htm.

[23] http://www.radio-electronics.com/info/rf-technology-design/rf-filters/butterworth-rf-filter-basics.php.

15

Comparison of the Time Domain Windows Specified in the ISO 18431 Standards Used to Estimate Modal Parameters in Steel Plates

Jhonatan Camacho-Navarro,[1] **R. Guzmán-López,**[2] **Sergio Gómez,**[2] **and Marco Flórez**[1]

[1]*Escuela de Ingeniería Electrónica, Universitaria de Investigación (UDI), Grupo de Investigación en Procesamiento de Señales (GPS), Bucaramanga, Colombia*
[2]*Universidad Pontificia Bolivariana, Bucaramanga, Colombia*

Correspondence should be addressed to Jhonatan Camacho-Navarro; camacho.navarro.jhonatan@gmail.com

Academic Editor: Massimo Viscardi

The procedures used to estimate structural modal parameters as natural frequency, damping ratios, and mode shapes are generally based on frequency methods. However, methods of time-frequency analysis are highly sensible to the parameters used to calculate the discrete Fourier transform: windowing, resolution, and preprocessing. Thus, the uncertainty of the modal parameters is increased if a proper parameter selection is not considered. In this work, the influence of three different time domain windows functions (Hanning, flat-top, and rectangular) used to estimate modal parameters are discussed in the framework of ISO 18431 standard. Experimental results are conducted over an AISI 1020 steel plate, which is excited by means of a hammer element. Vibration response is acquired by using acceleration records according to the ISO 7626-5 reference guides. The results are compared with a theoretical method and it is obtained that the flat-top window is the best function for experimental modal analysis.

1. Introduction

Physical behavior of complex engineering systems can be studied through prediction and simulation analysis by means of specialized software [1, 2]. Thus, it is possible to check abnormal performance with the help of monitoring methods based on simulation tools. In particular, the analysis of structural dynamics can be addressed by determining modal parameters defined by mode shapes, natural frequencies, and damping ratios. In this sense, by means of Operational Modal Analysis it is possible to conduct nondestructive testing, fatigue analysis, and issues concerned with field in structural analysis [3]. These tasks also involve updated finite element models used to predict the dynamic behavior of the structure reliability [4]. For instance, an application of modal analysis is the assessment of highway-bridge by dynamic testing and finite-element model updating [5]. Also, damage detection has been carried out in reinforced concrete beams by using modal flexibility residuals [6].

Because of the importance of modal analysis in the field of structural analysis, well established procedures to obtain proper estimations of modal parameters are required. Although there exists a huge documentation about methods used for modal analysis [7, 8], their practical implementation is still difficult because there are many parameters involved in the procedure, which must be decided by engineers of different areas and knowledge. In this sense, the 7626 standard ISO specifies a guideline of methodological steps to conduct experiments in order to obtain the frequency response measurement and the 18431 ISO describes the procedures for time-frequency analysis of vibrational records. Thus, by considering the recommendations of the ISO standards, the further estimation of modal model parameters by means of well-known modal analysis methods (as, e.g., peak-picking or least squares among others) is facilitated.

In this paper, a practical implementation of the above-mentioned ISO standards with a special emphasis on computing the natural frequency values is demonstrated. Thus, the influence of using three domain window functions (*Hanning, flat-top, and rectangular*) to estimate natural frequencies is discussed. Experimental results are conducted over an AISI 1020 steel plate, which is excited by means of a hammer element.

2. Theoretical Framework

The procedure used in this paper to estimate the natural frequencies of a modal model is based on the analysis of measurements from Frequency response functions (FRFs). In this sense, the extraction of relevant frequency information is performed by applying spectrum estimation techniques to structural vibrational records. Thus, FRFs are approximated by the cross-power spectral density (PSD) between the vibrational responses.

In this section, the conceptual issues involved in the estimation of structural natural frequencies by means of FRFs are detailed. Also, fundamentals of methods used to estimate the frequency decomposition are presented, focusing on given details about the parameters with great influence for its implementation.

2.1. Frequency Response Functions (FRFs). Dynamical model of structures is constructed on physical knowledge and fundamental laws of motion according to [9]

$$M\ddot{\mathbf{X}} + C\dot{\mathbf{X}} + K\mathbf{X} = F, \tag{1}$$

where K, C, and M are the stiffness, damping, and mass matrices, respectively. Likewise, $F(t)$ and $X(t)$ denote the forcing vector and displacement. Assuming that $F(t)$ is a delta-correlated exciting force and using properties demonstrated in the Natural Excitation Technique (NExT), it is possible to write the law motion of (1) in the form of [9]

$$M\ddot{\mathbf{R}}_{\mathbf{X}\dot{\mathbf{X}}_i}(\tau) + C\dot{\mathbf{R}}_{\mathbf{X}\dot{\mathbf{X}}_i}(\tau) + K\mathbf{R}_{\mathbf{X}\dot{\mathbf{X}}_i}(\tau) = 0, \tag{2}$$

where $\mathbf{R}_{\mathbf{X}\dot{\mathbf{X}}_i}$ is a vector of correlation functions for all positive lags $\tau > 0$, between the displacement vector and a reference signal, which must be uncorrelated with respect to excitation signal. Thus, (2) establishes that correlation function between acceleration records and a reference signal can be treated as free vibration data, which allows determining modal characteristics of the structures.

Moreover to law motion and NExT equations used in methodologies for modal parameter identification, the FRF relationship in structures must be specified. The FRF for one *n-degree* of freedom system represented by (1) or (2) can be written in the form of partial fractions as is expressed in (3) (classical pole/residue) [10, 11].

$$\left[G_{yy}(\omega)\right]$$

$$= \sum_{k=1}^{m} \frac{[A_k]}{j\omega - \lambda_k} + \frac{[A_k]^*}{j\omega - \lambda_k^*} + \frac{[B_k]}{-j\omega - \lambda_k} + \frac{[B_k]^*}{-j\omega - \lambda_k^*}, \tag{3}$$

where $G_{yy}(\omega)$ is the output PSD matrix, m is the total number of modes of interest, $\lambda_k = -\sigma_k + j\omega_{dk}$ is the pole of

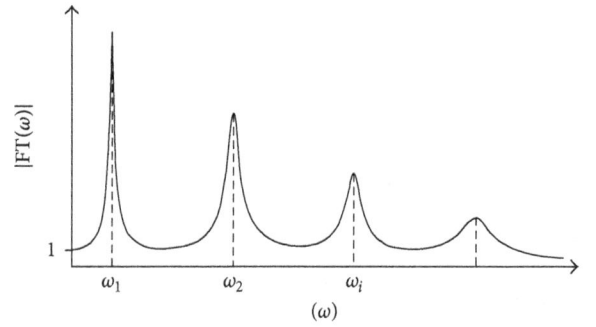

FIGURE 1: Typical FRF of an n-degree of freedom system.

the kth mode, σ_k is the modal damping (constant decay), ω_{dk} is the damped natural frequency, and the $*$ superscript denotes complex conjugate. The typical PSD curve for the FRF determined by (3) is depicted in Figure 1.

According to Figure 1, the modes (ω_k) located at the peak can be identified from spectral density through common signal processing as discrete Fourier transform.

2.2. Frequency Domain Decomposition (FDD). Classical approach used to estimate modal parameters is often referred to as Peak-Peaking (PP), which is a nonparametric method essentially developed in frequency domain. The main advantages of this method are its user-friendliness, simple use, and fast results obtaining. In this method, average normalized power spectral densities and frequency response functions between all the measurement points of the structure are evaluated [12]. As a result, first the natural frequencies are simply taken from the observation of the peaks on the graphs of the magnitude of the response (Figure 1). Next, damping ratios (σ_k) are calculated from the sharpness of the peaks obtained by half power band method. Then, the mode shapes are calculated from the ratios of the peak amplitudes at various points in the structure [13]. Finally, modal participation factors are computed to scaled mode shapes by using measures of force exciting. In order to apply the PP method, its procedure can be summarized as follows:

  (i) Determine the natural frequencies by means of the PSD computed from acceleration records by identifying all frequencies present at peaks (ω_r).

 (ii) Estimate damping ratios (ξ) by means of the loss factor (n_r) which depends of the half power band frequencies (ω_a, ω_b) as is illustrated in Figure 2.

According to Figure 2, the half power band comprises the frequencies where the PSD amplitude decays by 3 dB with respect to its maximum value. At once the half power band is determined; the loss and damping factors are computed by using

$$n_r = \frac{\omega_b - \omega_a}{\omega_r};$$

$$\xi = \frac{\omega_b - \omega_a}{2\omega_r} = \frac{n_r}{2}. \tag{4}$$

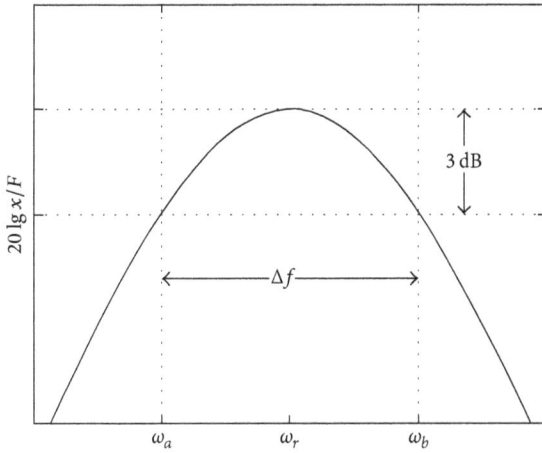

FIGURE 2: Half power band frequencies.

In this paper an enhanced method known as frequency domain decomposition (FDD) is used instead, which is an extension of PP method. It consists of three main steps [14]:

(i) Estimate the power spectral density matrix at discrete frequencies.

(ii) Do a singular value decomposition of the power spectral density.

(iii) For an *n-degree* of freedom system, then pick the *n* dominating peaks in the power spectral density. These peaks correspond to the mode shapes.

FDD technique removes disadvantages associated with classical PP method as, for example, the difficulty to detect close modes and bias estimation. Furthermore FDD reduces the uncertainty or hardness (even impossible) of damping estimation. Also, it keeps the user-friendliness giving a feeling of the data it is dealing with.

2.3. Spectrum Estimation. Frequency response functions between input and output are approximated as cross-power spectral densities between responses while the impulse response functions are approximated by cross-correlations between responses. Cross-PSDs are obtained using Welch method (FFT based method) [15]. Welch's periodogram averaging with overlapping is a method which introduces improved properties based on basic periodogram methods:

(i) *Simple Periodogram.* Quotient of the squared magnitude of the Fourier transform of the signal and length of the signal.

(ii) *Modified Periodogram.* Certain window other than rectangular window that is applied to the signal before taking the Fourier transform. Windows solve the leakage problem.

(iii) *Bartlett Periodogram Averaging.* Averaging of the different blocks of the signal, which decrease variance of the signal at the expense of resolution.

Thus, the cross-power spectra between $X(n)$ and some $Y(n)$ discrete signal, computed through Welch's method, are

expressed by (5). Therefore, the cross-PSD S_{XY} can be defined as the Fourier Transform of the cross-correlation function $R_{XY}(\tau)$.

$$S_{XY} = \frac{1}{K}\sum_{k=1}^{K} S_{XY}(k),$$

$$\text{where } S_{XY}(k) = \frac{1}{N}\sum_{n=0}^{L-1}\left|w(n)R_{XY}(\tau)^{(k)}e^{-j\omega n}\right|^2 \tag{5}$$

According to (5), the Welch method divides the signal into k blocks and then increases the averaging by taking overlaps of the blocks. Thus, the L samples of R_{XY} are divided into K overlapping sections with N samples in each block, each of which is then windowed by the WINDOW $w(n)$. Finally, it averages the periodograms of the overlapping sections to form S_{XY}, the power spectral density estimation of R_{XY}.

2.4. Windowing Effect. In estimating power spectral density (PSD) of a signal, there are two tradeoffs. One is frequency resolution and the other is noise in the signal. To obtain a good estimation of PSD, we should have large length of the signal but during measurements we have finite length of signal. If we take small block size, bad frequency resolution could introduce leakage in the spectrum. To reduce leakage, signals are windowed, that is, multiplied with a window in time domain. Many windows are available, each one having specific application in signal processing [16, 17]. Figure 3 illustrates the spectral characteristics of time windows recommended by the ISO standard: *Hanning, flat-top, and rectangular.*

According to Figure 3, the frequency estimation for modal parameter identification is limited by the frequency resolution of the spectral density. Resolution refers to the ability of distinguishing narrowband spectral components. In this sense, the spectrum of the ideal window with high frequency resolution and good noise suppression is narrower than the main lobe width and lesser than the side lobe level. Consequently, in order to select the suitable window function, a compromise between noise suppression and resolution is required. In this sense, the flat-top and rectangular windows produce the worst/best leakage factors (96.79%–9.14%), respectively, which corresponds to low and high relative side lobe attenuations. Also, it is noted that Hanning window offers balanced characteristics.

2.5. The ISO Standards. The 7626-5 and 18431 ISO standards give recommendations about recording protocol and selection of window function in order to estimate modal parameters [18]. Specifically, in the document entitled "Vibration and Shock Experimental Determination of Mechanical Mobility Part 5," it is found that the guidelines for measurements using impact excitation with an exciter which are not attached to the structure. Thus, the ISO 7626-5:1994 specifies procedures to obtain acceleration records using impact excitation. Also, signals analysis based on the discrete Fourier transform is covered where a recommendation about *Hanning* window

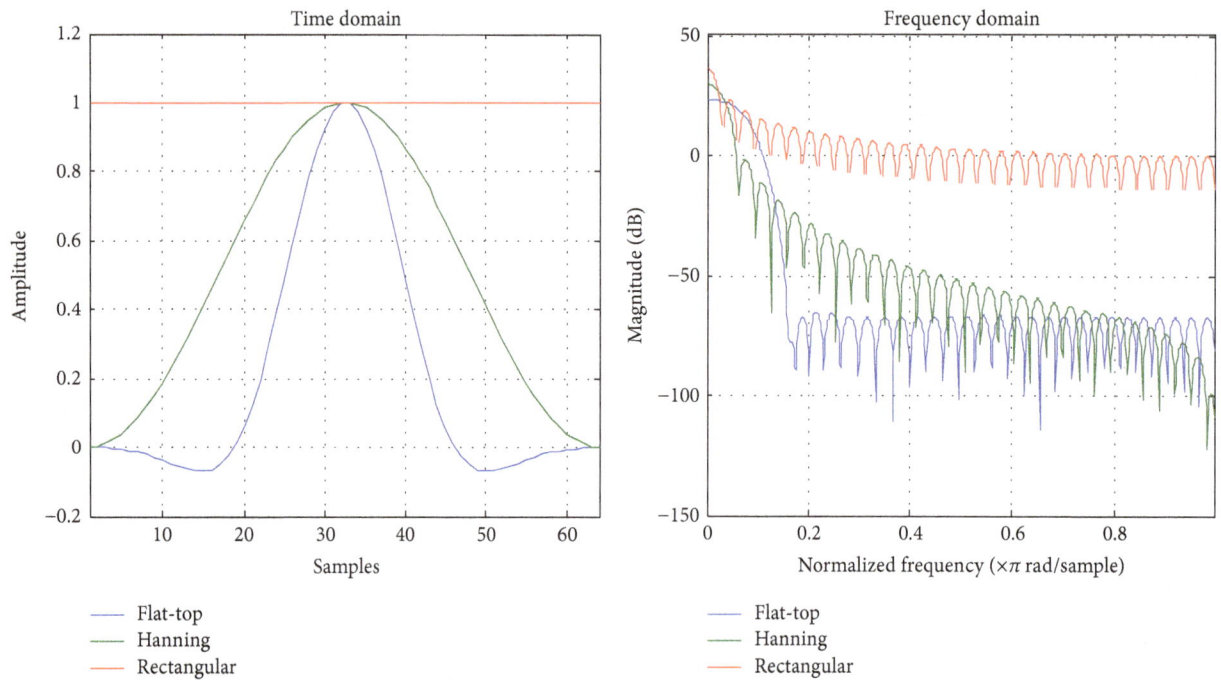

FIGURE 3: Spectral characteristics of recommended ISO-time windows.

can be justified. Similarly, the document "Mechanical Vibration and Shock Signal Processing Part 2: Time Domain Windows for Fourier Transform Analysis" which contains the ISO 18431-2:2004 standard, describes three time domain windows consisting of *Hanning*, flat-top, and rectangular that are suggested to be used for preprocessing samples vibration.

Considering the selected set of time windows specified in the ISO standards, in this paper the results of applying *Hanning*, flat-top, and rectangular windows for modal modes estimation are discussed.

3. Results and Analysis

The next sections show the results obtained by applying experimental and theoretical procedures. For experimental analysis the three time windows specified in the ISO standard are considered: Hanning, flat-top, and rectangular functions, while the theoretical approach is based on finite element method.

3.1. Experimental Setup. A steel plate was used in order to analyze the influence in the estimation of natural frequencies when different window functions are considered. In Figure 4, a scheme of the structure is shown.

Measurements from acceleration response were conducted over 4 points of the structure (P6, P7, P8, and P12). Vibration data were recorded under impulse hammer excitation type with sample time $T_s = 20\,\mu s$. Then, 25.000 samples for each accelerometer were processed which correspond to 0.5 seconds of data. Figure 5 shows the recorded signals.

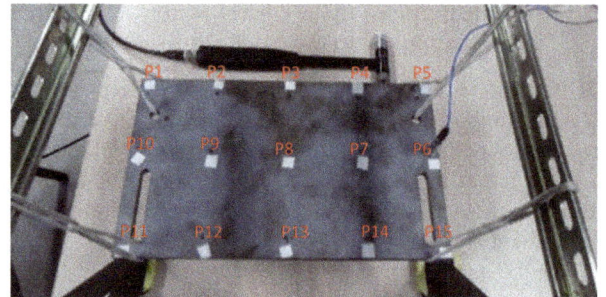

FIGURE 4: Bench structure.

3.2. Numerical Analysis. In order to calculate modal frequency of different deformation modes, finite element simulation through a numerical model was performed by using ANSYS software. The model includes the plate detailed geometry and it is characterized by density = 7733.75 Kg/m^3, $E = 200.000$ Mpa, and Poison = 0.29 parameters. The used finite elements mesh is constituted by total number of 40052 reduced integration hexahedral elements of eight integration nodes C3D8R (see Figure 6).

The modal shapes studied by means of the simulation software are depicted in Figure 7.

3.3. Experimental Results. In order to evaluate the influence of windowing effect, experimental measurements of acceleration records were processed considering windows described in previous sections. As a first result, the 2 Khz range of interest for cross-power spectral density of the acceleration

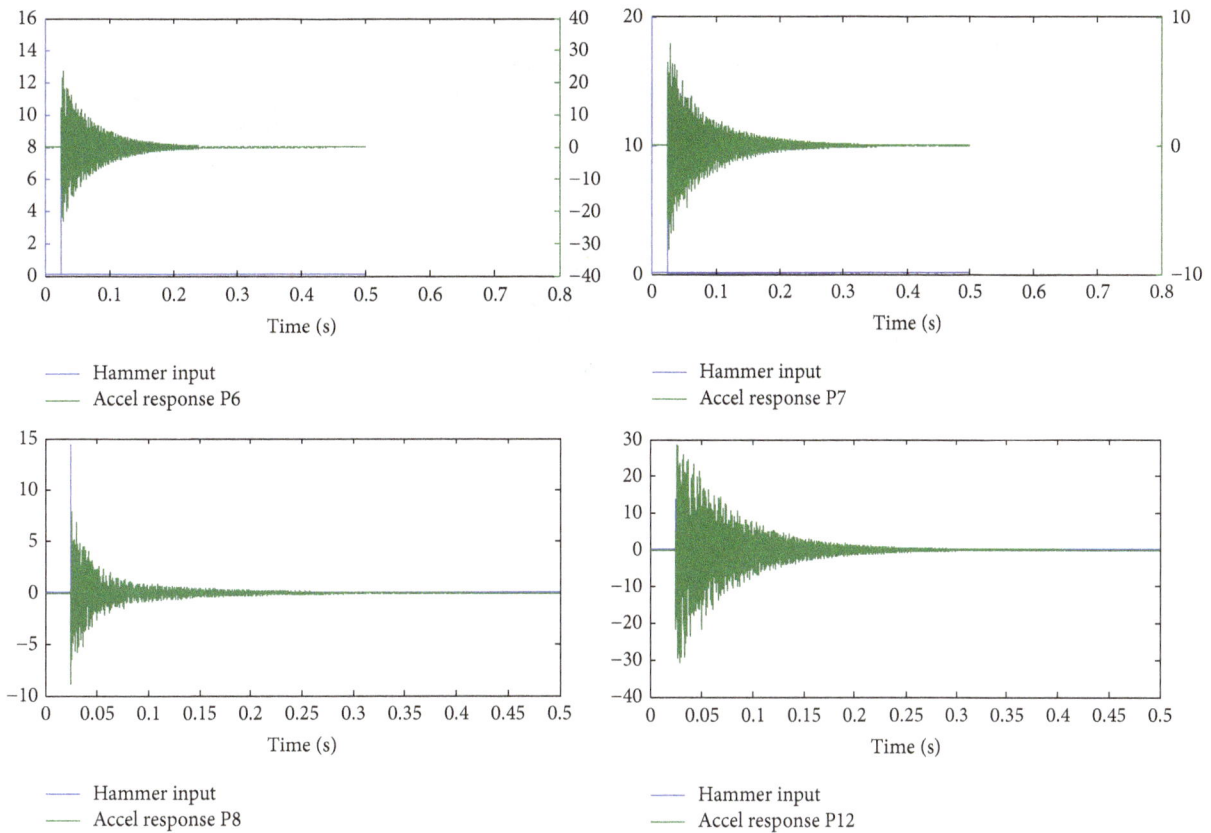

FIGURE 5: Recorded vibration signals.

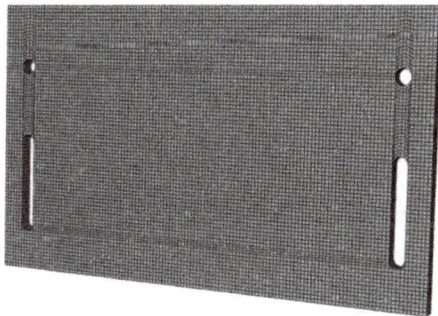

FIGURE 6: Structural finite element model.

TABLE 1: Natural frequencies [Hz] estimated by using three different windows.

Vibrational mode	Numerical model	Hanning	Flat TOP	Rectangular
First bending	636	640.9	634.8	636.3
First torsion	671	671.4	671.4	668.3
Second torsion	1472	1465	1469	1468
Second bending	1640	1697	1691	1694

TABLE 2: Percentage error (%) of mode estimation with respect to numerical model.

Vibrational mode	Hanning	Flat-top	Rectangular
First bending	0.7704	0.1887	0.0472
First torsion	0.0596	0.0596	0.4024
Second torsion	0.4755	0.2038	0.2717
Second bending	3.4756	3.1098	3.2927

measurements is depicted in Figure 8. The signals were divided into eight sections with 50% overlap, each section was windowed with a Hamming window, and eight modified *periodograms* were computed and averaged.

Also, the frequency response function computed by means of frequency domain decomposition after processing the CPSD of data in Figure 8 is depicted in Figure 9, where the amplitude in db corresponds to the first singular values of the PSD matrix estimated by using *Hanning* window function.

A comparison of the modes estimated according to ISO 7626-5 by using the three selected time window functions is presented in Table 1.

Finally, the percentage errors for each natural frequency with respect to theoretical numerical model are summarized in Table 2.

According to results in Table 2, the high error corresponds to the second bending mode, while the other modes maintain

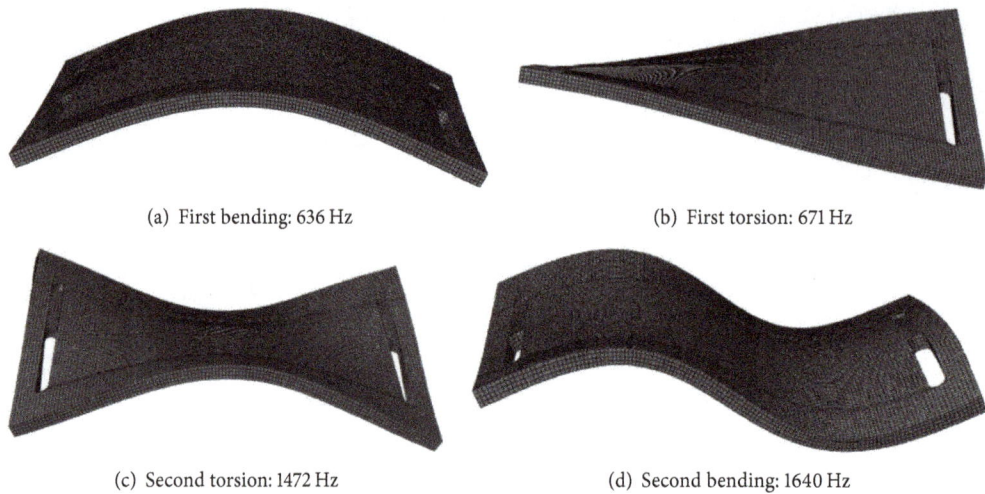

(a) First bending: 636 Hz

(b) First torsion: 671 Hz

(c) Second torsion: 1472 Hz

(d) Second bending: 1640 Hz

FIGURE 7: Modal frequency obtained by numerical analysis.

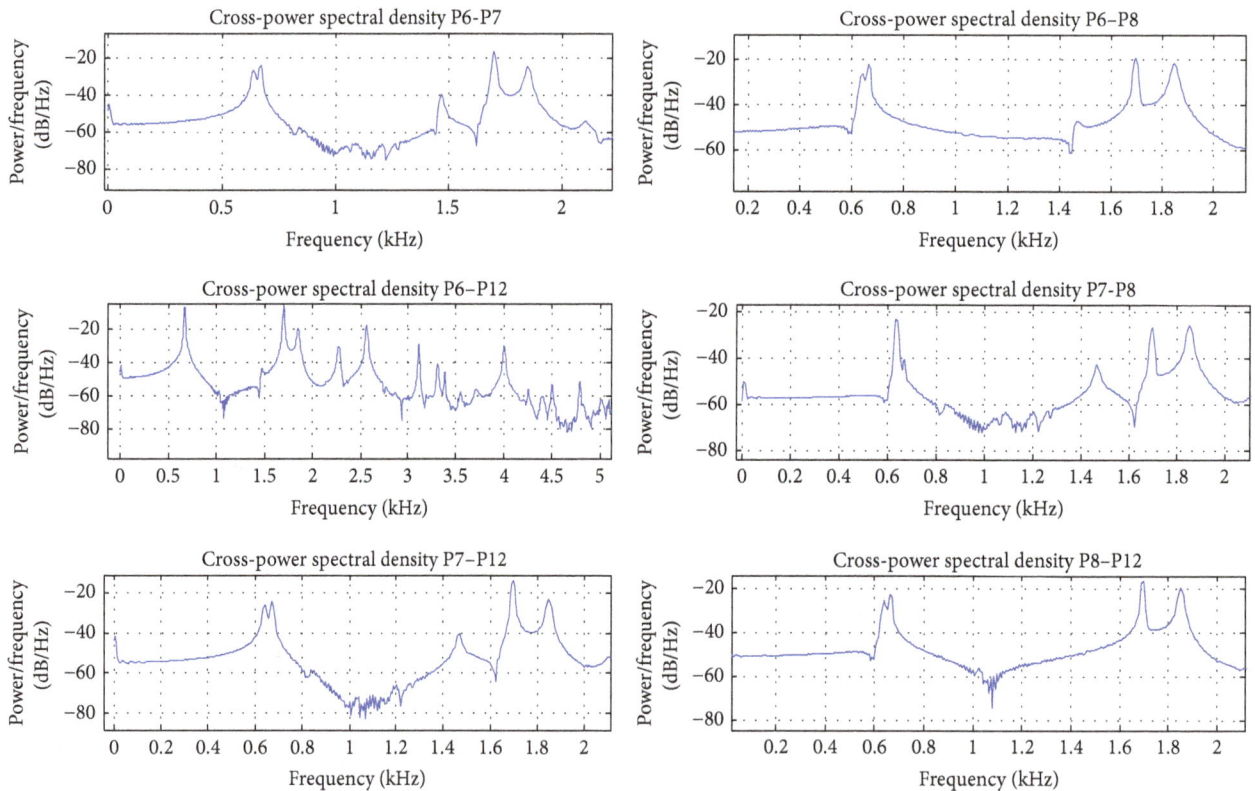

FIGURE 8: CPSD for acceleration records.

comparable values to the theoretical ones with errors lower than 1%.

4. Conclusion

Although no meaning differences were found when the three windows specified in the ISO standard were used to estimate natural frequencies, a slight better result is obtained for flat-top function. This implies that for modal parameter estimation purposes the selection of time windowing function has low influence, with major errors for the highest modes. However, the influence of the windowing preprocessing for the analysis of different modal parameters as shape mode and factor participation should be studied. Also, further analysis should be conducted with respect to additional parameters involved in the spectrum estimation such as overlap, FFT length, and segmentation. Moreover, it is recommended to include uncertainty analysis to evaluate the influence of using different time windows.

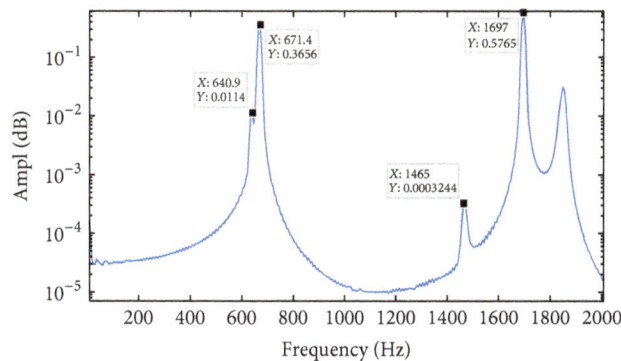

FIGURE 9: Frequency response function.

Competing Interests

The authors declare that they have no competing interests.

Acknowledgments

This work has been developed as part of a collaborative work between researches from Universitaria de Investigación y Desarrollo (UDI) and Universidad Pontificia Bolivariana (UPB), Bucaramanga, Colombia.

References

[1] E. Madenci and I. Guven, *The Finite Element Method and Applications in Engineering Using ANSYS®*, Springer, Berlin, Germany, 2015.

[2] H. H. Lee, *Finite Element Simulations with ANSYS Workbench 16*, SDC Publications, 2015.

[3] R. Brincker and C. Ventura, *Introduction to Operational Modal Analysis*, John Wiley & Sons, New York, NY, USA, 2015.

[4] M. Imregun, W. J. Visser, and D. J. Ewins, "Finite element model updating using frequency response function data: I. Theory and initial investigation," *Mechanical Systems and Signal Processing*, vol. 9, no. 2, pp. 187–202, 1995.

[5] J. M. W. Brownjohn, P. Moyo, P. Omenzetter, and Y. Lu, "Assessment of highway bridge upgrading by dynamic testing and finite-element model updating," *Journal of Bridge Engineering*, vol. 8, no. 3, pp. 162–172, 2003.

[6] B. Jaishi and W.-X. Ren, "Damage detection by finite element model updating using modal flexibility residual," *Journal of Sound and Vibration*, vol. 290, no. 1-2, pp. 369–387, 2006.

[7] D. J. Ewins, *Modal Testing: Theory and Practice*, vol. 6, Research Studies Press, Letchworth, UK, 1995.

[8] N. M. M. Maia and J. M. M. e Silva, Eds., *Theoretical and Experimental Modal Analysis*, Research Studies Press, Taunton, UK, 1997.

[9] J. M. Caicedo, *Two structural health monitoring strategies based on global acceleration responses: development, implementation, and verification [M.S. thesis]*, Washington University, 2001.

[10] S. Gade, N. B. Møller, H. Herlufsen, and H. Konstantin-Hansen, "Frequency domain techniques for operational modal analysis," in *Proceedings of the 1st International Operational Modal Analysis Conference*, pp. 261–271, St. Louis, Mo, USA, 2005.

[11] R. Brincker, L. Zhang, and P. Andersen, "Modal identification from ambient responses using frequency domain decomposition," in *Proceedings of the 18th International Modal Analysis Conference (IMAC '00)*, San Antonio, Tex, USA, February 2000.

[12] A. J. Felber, *Development of a hybrid bridge evaluation system [Ph.D. thesis]*, University of British Columbia, 1993.

[13] N. M. M. Maia, *Extraction of valid modal properties from measured data in structural vibrations [Ph.D. thesis]*, Imperial College London, University of London, 1988.

[14] R. Brincker, L. Zhang, and P. Andersen, "Modal identification of output-only systems using frequency domain decomposition," *Smart Materials and Structures*, vol. 10, no. 3, pp. 441–445, 2001.

[15] S. M. Kay, *Modern Spectral Estimation*, Prentice Hall, Upper Saddle River, NJ, USA, 1988.

[16] A. V. Oppenheim, R. W. Schafer, and J. R. Buck, *Discrete-Time Signal Processing*, vol. 2, Prentice-Hall, Englewood Cliffs, NJ, USA, 1989.

[17] C. Michael and A. F. Audrey, "The fundamentals of FFT-based signal analysis and measurement," Strategies for Choosing Windows, National Instruments Application Notes 041.

[18] S. B. Blaeser and P. D. Schomer, "Acoustical standards news," *The Journal of the Acoustical Society of America*, vol. 136, no. 1, pp. 439–448, 2014.

16

Optimization of Automotive Suspension System by Design of Experiments: A Nonderivative Method

Anirban C. Mitra, Tanushri Soni, and G. R. Kiranchand

Department of Mechanical Engineering, MES's College of Engineering, Wadia College Campus, 19 Late Prin,
V. K. Joag Path, Pune 411001, India

Correspondence should be addressed to Anirban C. Mitra; bile007m@yahoo.com

Academic Editor: Marek Pawelczyk

A lot of health issues like low back pain, digestive disorders, and musculoskeletal disorders are caused as a result of the whole body vibrations induced by automobiles. This paper is concerned with the enhancement and optimization of suspension performance by using factorial methods of Design of Experiments, a nonderivative method. It focuses on the optimization of ride comfort and determining the parameters which affect the suspension behavior significantly as per the guidelines stated in ISO 2631-1:1997 standards. A quarter car test rig integrated with a LabVIEW based data acquisition system was developed to understand the real time behavior of a vehicle. In the pilot experiment, only three primary suspension parameters, that is, spring-stiffness, damping, and sprung mass, were considered and the full factorial method was implemented for the purpose of optimization. But the regression analysis of the data obtained rendered a very low goodness of fit which indicated that other parameters are likely to influence the response. Subsequently, steering geometry angles, camber and toe and tire pressure, were included in the design. Fractional factorial method with six factors was implemented to optimize ride comfort. The resultant optimum combination was then verified on the test rig with high correlation.

1. Introduction

Automobiles travel at a high speed and as a consequence experience a broad spectrum of random noncyclic vibrations transmitted either by tactile, visual, or aural paths. The term "ride" is commonly used to represent tactile and visual vibrations, while the audibly perceptible vibrations are categorized as "noise." As per SAE J670e [1] terminology, the term ride is defined as the low frequency vibrations of the sprung mass up to 5 Hz. The lower frequency ride vibrations are manifestations of dynamic behavior. The motions and mechanical stresses resulting from the application of mechanical forces to the human body can have a variety of physiobiological effects. ISO 2631-1:1997 [2] standards imply that, for vertical vibrations, humans are most sensitive to those in the frequency range of 4 to 8 Hz, this being the resonant frequency range for human body. In its simplest form, a modern road vehicle suspension has been defined by Damian [3] as a linkage to allow the wheel to have relative motion with respect to the body and to support loads while allowing for that motion.

The RC has been shown to be affected by a variety of suspension parameters and a variety of analytical, numeric, computational, and experimental methods have been exercised for the optimization of suspension systems. Khajavi et al. [4] have developed a full car 8-DOF model by adopting ride comfort, handling, and suspension travel as the main criteria and using Multiobjective Programming Discipline to find the Pareto front. A 33-DOF multibody model of suspension system has been prepared by Zhang et al. [5] using ADAMS software. Even the stiffness coefficient of bushing has been considered. Step and sinusoidal excitations are provided to the model, and RC is optimized using GA in one of the research works by Farid et al. [6].

Kilian et al. [7] had worked on the optimization of torsion, bending, and swaying of suspension designs by using finite element methods like topology optimization and topography optimization in Altair Opti Struct software to maximize ride comfort during the design stage. In a similar work, a two-dimensional 8-DOF model was developed by Roy and Liu [8] to simulate and animate the response of a vehicle to different road, traction, braking, and wind conditions in a 3D VRML environment. A model validation was conducted by comparing a 0–100 kmph acceleration run against a Honda Accord car equipped with an accelerometer and an engine rpm recorder. Gonçalves and Ambrósio [9] have proposed a methodology for optimization of ride and stability of a vehicle based on the use of flexible multibody model. The ride optimization is achieved by finding the optimum of ride index by measuring acceleration in several key points of the vehicle. Time histories of acceleration, velocities, and displacement at the center of gravity have been considered by Naudé and Snyman [10, 11] along with the time histories of forces, deflections, and deflection rate of wheels and suspension components. The presence of noise discontinuities presents major challenges in optimization. Multiobjective Genetic Algorithm (MOGA) is used for Pareto optimization of a 4-DOF vehicle vibration model by Sharifi and Shahriari [12].

Bagheri et al. [13] has used GA for Pareto optimization of a 2-DOF vehicle vibration model considering sprung mass acceleration and relative displacement between sprung mass and tire. Evaluation of vibration transmitted from the road profile to the driver or a passenger in a moving vehicle has been done by Kuznetsov et al. [14]. The paper by Chi et al. [15] presents a comparative study of three optimization algorithms, namely, Genetic Algorithms (GAs), Pattern Search Algorithm (PSA), and Sequential Quadratic Program (SQP), for the design optimization of vehicle suspensions based on a quarter-vehicle model. Uys et al. [16] generated a Land Rover Defender model in MSC ADAMS and the spring damper settings were determined which ensures optimal ride comfort of an off-road vehicle at different speeds and over different road profiles.

A 7-DOF full car model has been developed and optimum ride comfort has been achieved by trying out different spring damper setting using DoE by Mostaani et al. [17]. Road surface has been simulated using power spectral density (PSD) and it was found that car spring stiffness is most sensitive. RMS acceleration and pitch angle for optimum setting at different speeds was generated and only ride comfort has been optimized. Marzbanrad et al. [18] performed optimization of passive suspension system on a 7-DOF model in MATLAB using DoE for speeds ranging from 60 kmph to 90 kmph.

In this work, a quarter car test rig has been developed so as to vary the influential parameters within the predetermined sampling range. The test rig has been integrated with NI LabVIEW DAQ system for evaluating and assimilating the raw data. Subsequently, the DoE methodology has been implemented to optimize the suspension system. This paper has been organised in the following manner: a brief description of recent development in the field of suspension performance, RC, and Optimization processes is given in Section 1. Section 2 describes the experimental setup and measurement technique used for the accumulation of raw data. Section 3 focuses on the optimization process

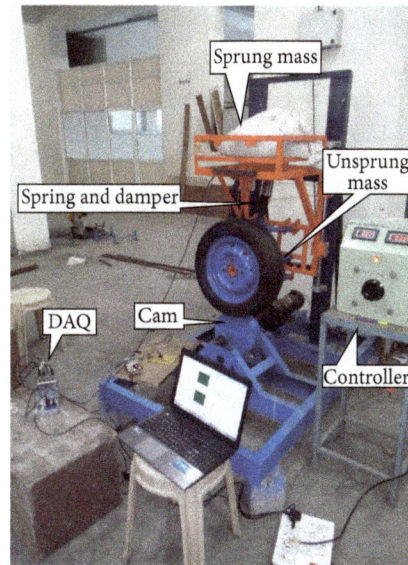

FIGURE 1: Quarter car test rig.

using DoE. Section 3.1 comprises the optimization using full factorial method; in which three fundamental suspension parameters, namely, the sprung mass (m), spring stiffness (ks), and damping coefficient (cs), are considered. The section also elaborates various aspects of the regression analysis used in optimization process. The regression analysis indicates the presence of additional influential factors which have been left uncontrolled. Hence, in Section 3.2, three more parameters were added and the fractional factorial methodology is used for optimizing RC, hence considering six parameters, that is, the sprung mass (m), spring stiffness (ks), damping coefficient (cs), camber angle (cma), and toe and tire pressure (typ). Then, Section 4 explains the method of response optimization and the model generated. The verification of results and the concluding remarks have been mentioned in Sections 5 and 6, respectively.

2. Quarter Car Suspension Test Rig

In order to study the suspension behavior in controlled environment and to evaluate the influence of various factors and their interactions over RC, a quarter car suspension test rig was designed and developed as shown in Figure 1. The excitation source is a sinusoidal bump profile and the wheel of quarter car model is considered as follower. The cam profile has been actuated with a motor so as to simulate the relative motion between the road and the wheel. A provision to vary a multitude of parameters within the predetermined sampling range has been incorporated.

Two highly sensitive accelerometers, of ICP (IEPE) make (model: 351B03), have been attached to sprung mass and the wheel assembly to measure the acceleration of the sprung mass and the unsprung mass, respectively, and, as per ISO 2631-1:1997 [2] standards, RC is expressed in terms of RMS acceleration of the sprung mass in m/s^2.

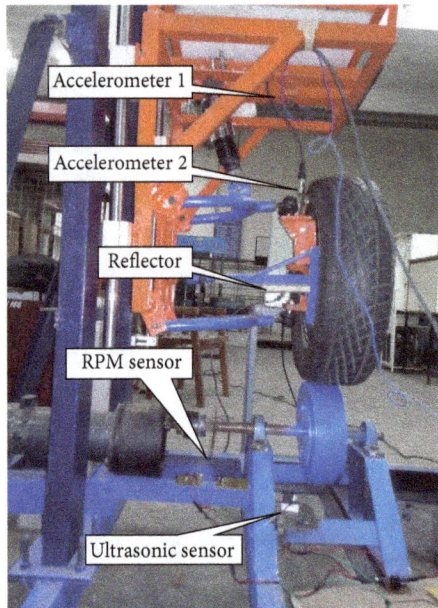

FIGURE 2: Sensors mounting.

TABLE 1: Full factorial design.

Design parameter	Value
Factors	3
Base design	3, 8
Runs	16
Replicates	2
Blocks	2

TABLE 2: Levels of influential parameters in full factorial design.

Factors	Low (−1)	High (+1)
Spring stiffness (ks), N/mm	18000	26000
Damping coefficient (cs), N-s/m	418	673
Mass (m), kg	41	81
Speed (N), RPM	150	250

The entire system has been integrated with NI LabVIEW data acquisition system, a development environment for a visual programming language from National Instruments. The mounting of accelerometer on control arms has been done as explained by Mohan et al. [19]. Arraigada and Partl [20] have discussed problems involving theoretical and analytical aspects of sensor calibration, data collection, and error identification and, as per authors' observations, the entire system has been integrated with NI LabVIEW data acquisition system and positions of sensors are shown in Figure 2.

3. Design of Experiments

DoE is a statistical technique used in this work to determine the various influential parameters affecting the response, that is, RC and the settings of those parameters in order to optimize the performance. The empirical relationship between the factors has also been derived in the form of equation by using the concept of regression analysis, which describes the relation between the factors and responses statistically. DoE quantifies the effects, variations, and uncertainty in the process and normalizes the data on a common scale in order to assist the analysis.

3.1. Full Factorial DoE Analysis with Three Parameters. In this section, the three fundamental variables sprung mass, spring stiffness, and the damping coefficient were considered to optimize the suspension system to enhance the RC of the vehicle. For the purpose of sampling range determination, an SAE BAJA buggy was taken as the reference. Accordingly, the softest and the hardest of springs and dampers as available in the Indian market were taken to be the extremities of the sampling range and the one-fourth of the total mass of a buggy was considered and the sprung mass was decided to be

varied between 40 kg and 80 kg in the test rig. The speed of the vehicle here determines the frequency with which excitations are encountered by the vehicle body. While designing this experiment, the speed of the vehicle was not considered as a controlled variable because it depends completely on the user's aspirations. But, due to its considerable impact on the response, that is, RC, it was necessary to arrange for the elimination of variability due to its effect. So, the rotational speed of wheel in rpm was incorporated within two blocks. Mohan et al. [19] have surmised that bump traversal for a car can occur within a speed range of 10 kmph to 20 kmph. Consequently, it was decided to vary the rotary speed of the cam from 150 rpm to 250 rpm. The damping of the bearings, tires, and the body are small enough not to be considered.

Since only three variables have been considered for optimization, the full factorial approach was implemented. In this approach, no aliasing or compounding effect occur between the parameters. For two levels of each factor, the design is denoted as 2^k full factorial design where k is the number of factors in the study. This method accounts for the effect of all the interactions and for reducing variability in the design; all the runs were replicated twice hence resulting in 16 runs. According to Montgomery [21], an obvious risk when conducting an experiment that has only one run at each test combination is that the model is fitted to noise and, with only one replicate in the design, the pure error and lack of fit cannot be estimated. The experiment was performed by considering two levels for each factor along with two blocks for speed. In a replicated blocked design, each replicate of the design is considered within a block and the design features have been depicted in Tables 1 and 2.

The experimentation was performed as per the orthogonal design matrix generated in MINITAB and the corresponding RC values were tabulated as shown in Table 3.

As evident from Table 3, the RMS values of sprung mass acceleration vary from 0.285 m/s^2 to 1.38 m/s^2. ISO 2631-1:1997 [2] clearly specifies that the sprung mass acceleration above 0.315 m/s^2 is a little uncomfortable whereas that above 0.5 m/s^2 is fairly uncomfortable for a passenger. Hence, the

TABLE 3: Observation table: full factorial design.

Run order	Speed	m	ks	cs	RC
1	150	41	26000	418	0.35
2	150	81	26000	418	0.285
3	150	81	26000	673	0.87
4	150	81	18000	418	0.34
5	150	41	18000	673	0.535
6	150	81	18000	673	1.38
7	150	41	26000	673	0.71
8	150	41	18000	418	0.41
9	250	81	26000	418	0.565
10	250	41	26000	673	0.99
11	250	81	18000	673	1.1
12	250	41	18000	418	0.66
13	250	41	26000	418	0.47
14	250	81	18000	418	0.665
15	250	81	26000	673	0.74
16	250	41	18000	673	0.72

TABLE 4: Effect coefficient table: full factorial design.

Term	Effect	Coef	SE Coef	t	P
Constant	0.67438	0.03624	18.61	0.000	
Block	−0.06437	0.03624	−1.78	0.114	
ks	−0.10375	−0.05188	0.03624	−1.43	0.190
cs	0.41250	0.20625	0.03624	5.69	0.000
m	0.13750	0.06875	0.03624	1.90	0.094
ks ∗ cs	−0.15250	−0.07625	0.03624	−2.10	0.069
cs ∗ m	0.14625	0.07313	0.03624	2.02	0.078
ks ∗ cs ∗ m	−0.17625	−0.08813	0.03624	−2.43	0.041

main aim of this work is to obtain the optimal combination of suspension parameters so as to restrict the magnitude of the sprung mass acceleration below 0.315 m/s^2.

3.1.1. Regression Analysis. Regression analysis is the science of fitting straight lines to patterns of data. In a linear regression model, the dependent variable (RC in this case) is predicted from "n" independent variables (here ks, cs, and m) using a linear equation and residuals are calculated in order to get the estimates of errors in model.

Before proceeding for the analysis, firstly the model is reduced and the unnecessary and insignificant parameters present in the model are removed. The presence of insignificant parameters inflates the biased error and decreases the model accuracy and adequacy. The estimated coefficient tables and the Pareto charts have been generated for this purpose.

The coefficient table gives the quantitative effect of each parameter on the response which is calculated by considering the average effects of the rest of the parameters. Table 4 gives the effect value of damping coefficient as 0.4125, which shows that it has the highest effect on RC. Here, stiffness and sprung mass are individually insignificant for RC as their respective P values are 0.190 and 0.094, which is more than 0.05. Also, a Pareto chart of the effects used to determine the magnitude and the importance of an effect visually is shown in Figure 3. There is a reference line on the chart which corresponds to the critical t value ($t = 2.306$) and any effect that exceeds this reference line is significant.

Figure 3 shows that the damping coefficient and the interaction of the three variables are significant; hence the parameters like mass and stiffness cannot be neglected even after being individually insignificant.

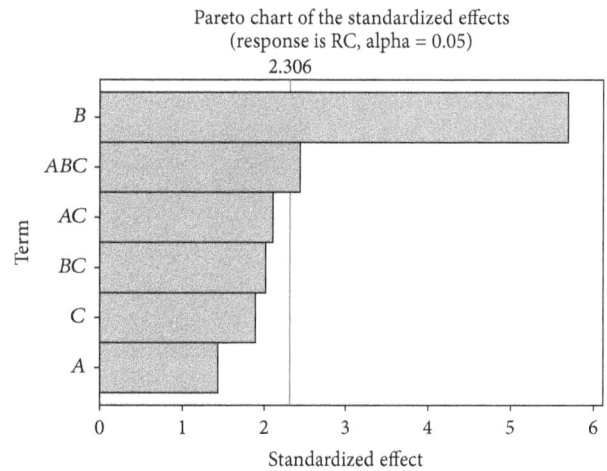

Pareto chart of the standardized effects
(response is RC, alpha = 0.05)

Factor	Name
A	ks
B	cs
C	m

FIGURE 3: Pareto chart: full factorial design.

3.1.2. Goodness of Fit. Now, the model of RC can be analyzed to check its adequacy and fitment. Three statistics are used in Ordinary Least Squares regression to evaluate model fit: R-Squared, standard error (S), and the overall F-test. Hence, the statistical terms such as standard error, coefficient of determination, and predictability values were calculated for the response as shown in Table 5.

S stands for standard error and is 0.145 for the model. It is an estimate of the standard deviation of the true noise; that is, the variations in the response that are not explained by the model. The smaller the value of S, the greater the closeness of experimental responses with the fitted line and the accuracy of the model. The R-Squared value (coefficient of determination) is 87.42%, which gives an estimate of explained variation in the model and determines how closely a certain function fits a particular set of experimental data. R-Squared adjusted value, 76.41%, is assessed. The perceptible difference between the values of R-Squared and R-Squared adjusted here shows that some influential parameters might not have been considered for this analysis.

Along with this, the PRESS value in the model is 0.672, which is comparatively larger than the ideal value. Similarly,

TABLE 5: Regression statistics: full factorial design.

Term		Value
S	Standard error	0.145
PRESS	Prediction sum of squares	0.672
R-Squared	Coeff of multiple determination	87.42%
R-Squared (pred)	Predicted coeff of determination	49.68%
R-Squared (adj)	Adjusted coeff of determination	76.41%

the R-Squared predicted value is 49.68% which depicts a very low predictability and that the model is unable to account for the variability in the model. Hence, it was surmised that, apart from the primary parameters, other factors are likely to influence the response and can be included in the design. Consequently, it was decided to take steering geometry parameters into consideration along with the tire pressure and obtain a setting of all the parameters in order to minimize the RC.

3.2. Fractional Factorial DoE Analysis with Six Parameters.
This section deals with identifying the other influential parameters and then optimizing the RC using fractional factorial method. The process comprises extensive literature survey and study of vehicle dynamics. Study of steering geometry reveals that the alignment and orientation of the wheels and the suspension linkages also play a nodal role in altering the area of contact at the tire-road interface. One such parameter is tire camber which determines the orientation along which the normal reaction generated at the tire-road interface is transferred to the vehicle body. Another aspect of the steering geometry, namely, the toe, is also said to have a profound impact on the directional stability and the self-straightening ability of the general passenger car. It has been mentioned in SAE J670e [1] that the toe determines the magnitude and orientation of lateral forces acting on the wheel assembly under dynamic conditions.

The generation of the forces necessary to initiate the turn, to constrain the vehicle at the correct sideslip angle, and to return it to the straight-running condition is the role of the tires. Moreover, as the tire is made of rubber, which is partially elastic, the tire is analogous to a spring. Pillai [22] has stated that the stiffness of the tire is likely to affect the nature of the vibrations transmitted to the vehicle and as the tire stiffness is said to be a direct function of tire pressure, it was decided to incorporate camber angle, toe, and tire inflation pressure among the parameters.

3.2.1. Fractional Factorial Design.
As the number of parameters increases to 6, in the subsequent section of this paper, fractional factorial methodology has been implemented to create and analyze the design by considering the 2 levels of six influential parameters. Fractional factorial design is an efficient alternative for full factorial design.

Here, the fractional factorial design is selected by defining the design generator such that it minimizes the aliasing effect

TABLE 6: Fractional factorial design.

Design parameter	Value
Factors	6
Base design	6, 16
Runs	32
Replicates	2
Blocks	2
Resolution	IV

TABLE 7: Levels of influential parameters in fractional factorial design.

Factors	Low (−1)	High (+1)
Speed (N), RPM	150	250
Tire pressure (typ), psi	35	40
Camber (cma), degree	1	3
Spring stiffness (ks), N/mm	18000	26000
Damping coefficient (cs), N-s/m	418	673
toe, mm	10	20
Mass (m), kg	41	81

in the design depending upon the resolution chosen. Here, to obtain a 2^{6-2} fractional factorial design, (1/4)th fraction of 2^6 design requires 2 design generators to be defined which are $E = ABC$ and $F = BCD$. Now, those runs are selected which give same sign for the generators and the identity column. Hence, the defining relation, which is the total collection of all the design generators, can be given as

$$I + ABCE + ADEF + BCDF, \quad (1)$$

where A, B, C, D, E, and F are parameters and I indicates the identity column.

Based on the defining relation, the alias structure is obtained which describes the confounded effects in a design. Aliasing or confounding in a design occurs when the estimate of an effect includes the influence of one or more other effects. The alias structures of this design show that all the main effects are confounded with third- and fifth-order effects but no main effect is aliased with any other main effect. The second-order effects are confounded with second- as well as fourth- and sixth-order effect and all the third order effects are confounded with each other. Consequently, the Resolution IV design was chosen as it does not confound the main effects with the two factor interactions.

Table 6 shows the overall summary of the fractional factorial design obtained as per the above discussion and the two levels of all the 6 factors have been shown in Table 7.

A Randomized Complete Block Design was formulated with two replicates and blocks, and the experimentation was performed accordingly as shown in Table 8.

3.2.2. Regression Analysis.
To evaluate the nature of influence of each parameter over the response, the effects coefficient table was generated as shown in Table 9.

TABLE 8: DoE matrix: fractional factorial design.

Run	N	typ	cma	ks	cs	toe	m	RC
1	155	35	3	18000	418	10	41	0.48
2	155	35	3	26000	673	20	81	0.67
3	155	35	1	26000	673	10	41	0.8
4	155	35	1	26000	673	10	41	0.72
5	155	35	1	18000	418	20	81	1.24
6	155	35	3	18000	418	10	41	0.45
7	155	35	3	26000	673	20	81	0.65
8	155	35	1	18000	418	20	81	1.3
9	250	35	3	18000	673	10	81	0.54
10	250	35	3	26000	418	20	41	1.2
11	250	35	3	26000	418	20	41	1.06
12	250	35	1	26000	418	10	81	1.7
13	250	35	1	26000	418	10	81	1.5
14	250	35	1	18000	673	20	41	1.15
15	250	35	1	18000	673	20	41	1.2
16	250	35	3	18000	673	10	81	0.62
17	155	40	3	26000	418	10	81	1.12
18	155	40	1	18000	673	10	81	0.55
19	155	40	1	26000	418	20	41	0.9
20	155	40	3	18000	673	20	41	0.54
21	155	40	1	26000	418	20	41	0.82
22	155	40	1	18000	673	10	81	0.57
23	155	40	3	26000	418	10	81	1.17
24	155	40	3	18000	673	20	41	0.44
25	250	40	1	18000	418	10	41	0.97
26	250	40	1	26000	673	20	81	1.6
27	250	40	3	26000	673	10	41	0.61
28	250	40	3	18000	418	20	81	1.75
29	250	40	1	18000	418	10	41	0.91
30	250	40	3	18000	418	20	81	1.75
31	250	40	1	26000	673	20	81	1.25
32	250	40	3	26000	673	10	41	0.69

TABLE 9: Effects and coefficients for RC: fractional factorial.

Term	Effect	Coef	SE Coef	t	P
Const		0.9663	0.01647	58.67	0
Blocks		−0.19	0.01647	−11.54	0
typ (psi)	0.0225	0.0113	0.01647	0.68	0.502
cma (deg)	−0.215	−0.1075	0.01647	−6.53	0.029
ks (N/m)	0.125	0.0625	0.01647	3.79	0
cs (N-s/m)	−0.3575	−0.1787	0.01647	−10.85	0
toe (mm)	0.2575	0.1288	0.01647	7.82	0
m	0.315	0.1575	0.01647	9.56	0
typ (psi) * cma (deg)	0.2775	0.1388	0.01647	8.42	0
typ (psi) * m	0.17	0.085	0.01647	5.16	0

It can be observed from Table 9 that the SE Coef value of 0.01647 is much lower than that obtained when three parameters were considered. This is a direct indication that the

TABLE 10: Regression statistics: fractional factorial design.

Term		Value
S	Standard error	0.093
PRESS	Prediction sum of squares	0.0403
R-Squared	Coeff of multiple determination	96.21%
R-Squared (pred)	Predicted coeff of determination	91.98%
R-Squared (adj)	Adjusted coeff of determination	94.66%

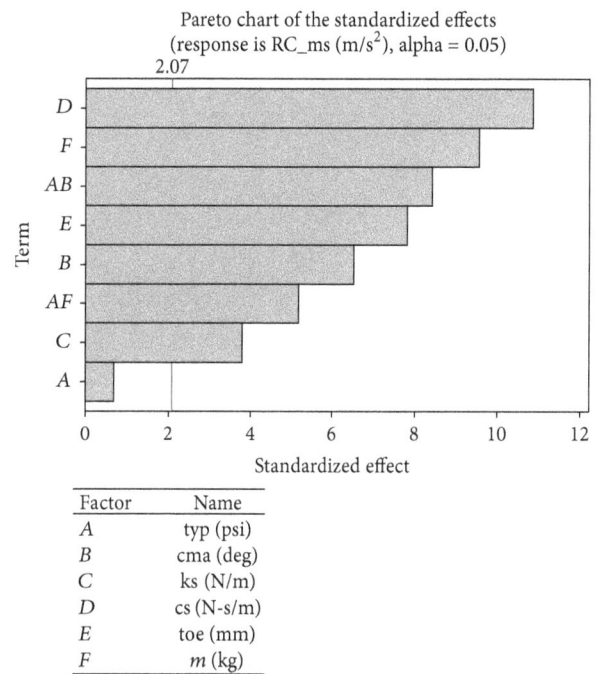

FIGURE 4: Pareto chart: fractional factorial design.

residual error in evaluating the system has been minimized after considering additional parameters. Moreover, all the individual parameters other than the tire pressure are found to be significant. In the previous instance, while taking three factors, the effect of the noise factors caused the individual influence of parameters like mass and spring stiffness to be clouded. This further emphasizes the importance of selection of influential variables in a DoE optimization procedure. A visual perception of the same can be derived from the Pareto charts which clearly demarcate the significant and insignificant influences, as shown in Figure 4.

The critical t value is 2.07. Tire pressure is insignificant individually but it cannot be neglected as its interactions are significant. The Pareto chart obtained also conforms to the Effect-Heredity principle, which states that, for an interaction to be significant, at least one of its parent factors should be significant. After reducing the model of RC to its significant variables, the model was checked for goodness of fit by evaluating the R sq statistics shown in Table 10.

The S value obtained here is 0.093 which is relatively smaller and accounts for greater accuracy of the resulting model. The high value of R-Squared, that is, 96.21%, and the closeness of R-Squared and adjusted R-Squared (94.66%)

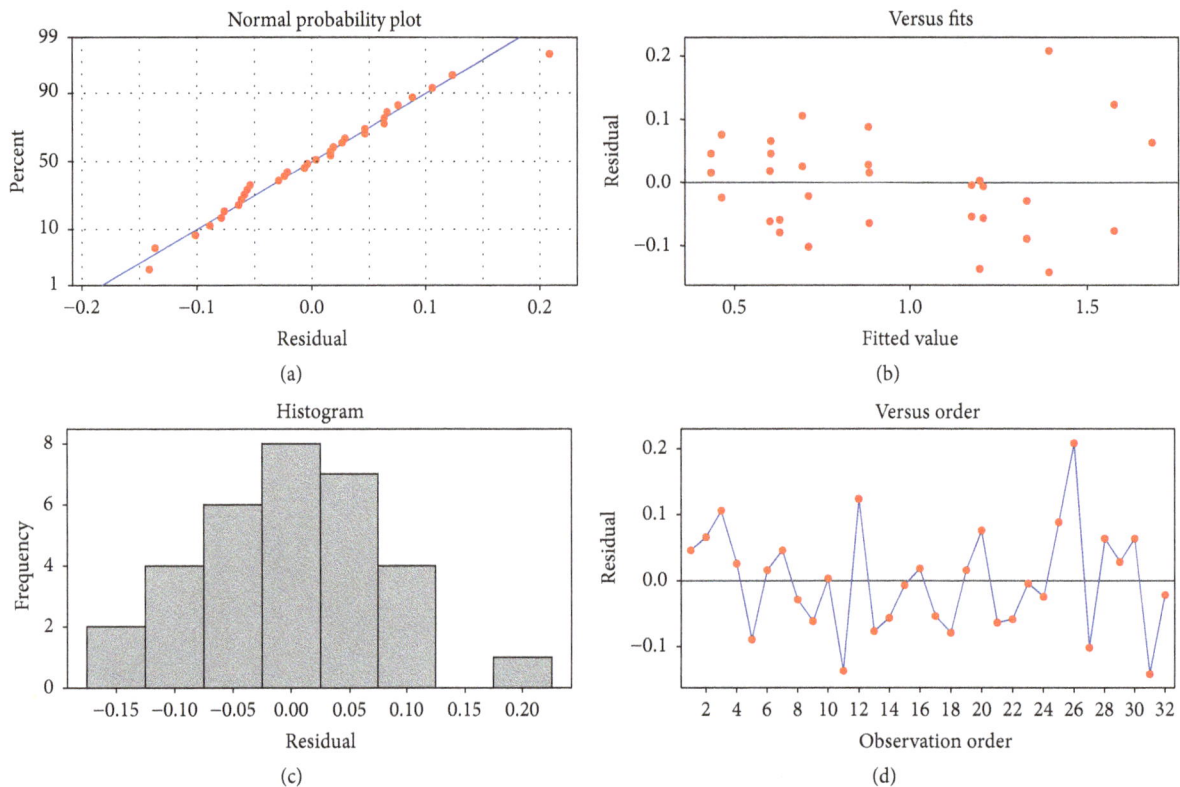

FIGURE 5: Residual plots for RC: fractional factorial design. Residual plots for RC_ms (m/s^2).

depicted in Table 10 confirm the fitment of model. PRESS and predicted R-squared values give an idea about the model's predictability and the value of PRESS here is smaller, that is, 0.0403, compared to that when three factors were considered. Similarly, predicted R-Squared is 91.98%. The higher value of predicted R-Squared shows better predictability and the negligible difference between adjusted R-Squared and R-Squared indicates that the model is not overfit.

The normal probability plot for RC, at Figure 5(a), indicates the presence of one outlier which can be neglected, as far as its impact on the model is concerned. In this analysis, the distribution of residuals over the zero line follows a bell shaped curve, but the presence of some outlier points on extreme left and right makes it slightly skewed. The versus-fit plots, at Figure 5(b), show a random and equal distribution without any usual patterns for both the responses which shows that the residuals possess a constant variance throughout the data. The graph at Figure 5(d) is the versus order plot which is the plot of residuals in time sequence. The plot is used to detect the correlation between the residuals. Ideally, the residuals on the plot should fall randomly around the center line. There is no strong pattern observed in the versus order plots for RC which indicate the random variation in experimental data. Hence, the residuals are deemed time independent and uncorrelated in the analysis.

3.2.3. *ANOVA*. ANOVA, that is, Analysis of Variance, explains the variability in the mean and variance of the residuals, taking into account the degree of freedom for each variable. The ANOVA table generated for RC is shown in Table 11. In ANOVA, the total sum of squares quantifies the total variation in the data and is divided into two parts, Seq SS (sequential sum of squares) and Adj SS (adjusted sum of squares). The equality of both these terms confirms the orthogonality of the design. The residual mean square (MS) of treatments is an unbiased estimator of variance.

Lower P values provide stronger evidence against the null hypothesis; that is, for P value less than 0.05, the effect of the factor is termed significant and the null hypothesis can be rejected. As suggested by the F-value and P value in Table 9, rotational speed, which was considered within a block, is significantly influential.

The lack of fit for the model is also tested by using F-test. To obtain the lack of fit estimation, replication of the model is mandatory. There are two parts of the error term in the model, pure error and the bias error. The error due to variations in the replications around their mean value is called pure error; on the other hand bias error is due to the variation of mean values around the model prediction. Ideally, the F-ratio of lack of fit should not be significant for the model to describe the functional relationship between the experimental factors and the response. For the model, in Table 11, the P value of lack of fit is 0.178, which is greater than the significance level of 0.05. This shows its insignificance which in turn depicts the fitment and credibility of the model generated. Also the variance of model error, that is, lack of fit and replicate error, that is,

TABLE 11: ANOVA analysis: fractional factorial design.

Source	DF	Seq SS	Adj SS	Adj MS	F	P
Blocks	1	1.1552	1.1552	1.1552	133.09	0
Main effects	6	2.84555	2.84555	0.47426	54.64	0
typ (psi)	1	0.00405	0.00405	0.00405	0.47	0.502
cma (degree)	1	0.3698	0.3698	0.3698	42.61	0
ks (N/m)	1	0.125	0.125	0.125	14.4	0.001
cs (N-s/m)	1	1.02245	1.02245	1.02245	117.8	0
toe (mm)	1	0.53045	0.53045	0.53045	61.11	0
m	1	0.7938	0.7938	0.7938	91.46	0
Two-way interactions	2	0.84725	0.84725	0.42363	48.81	0
typ (psi) ∗ cma (deg)	1	0.61605	0.61605	0.61605	70.98	0
typ (psi) ∗ m	1	0.2312	0.2312	0.2312	26.64	0
Residual error	22	0.19095	0.19095	0.00868		
Lack of fit	6	0.07515	0.07515	0.01253	1.73	0.178
Pure error	16	0.1158	0.1158	0.00724		
Total	31	5.03895				

pure error values for the RC model, is 0.01253 and 0.00724, respectively, which was very negligible and it indicates that the experimentation is having good reproducibility.

3.2.4. Model with Six Variables. Once all the assumptions were verified and consolidated and the fit of model is obtained, an experimental model explaining the behavioral relationship between all the selected suspension and steering geometry parameters is shown in

$$RC = -0.210200 \times typ - 2.18875 \times cma + 1.56250E$$
$$- 05 \times ks - 0.00140196 \times cs - 0.0257500 \times toe$$
$$- 0.0558750 \times m + 0.05550 \times typ \times cma \quad (2)$$
$$+ 0.001700 \times typ \times m - 0.1900 \times N + 8.61814.$$

4. Response Optimization

After the experimentation and the statistical analysis, the regression models for the desired responses are generated in terms of the influential parameters, which produce a value of response as per the given set of the factors. When the value of response is desired around a user-specified range and the combination of various influencing factors is required to yield the optimal response accordingly, the method of response optimization is applied. In this, the most desirable value of response possible under all the restrictions or conditions is achieved.

To serve the purpose, response optimizer, a software function in MINITAB, is used. It generates a combination of the variables which gives the optimal solution, based on the goal of optimization, range, weight, and relative importance as specified by the user. The model must fit all the responses separately, to be implemented in the response optimizer. The response can be minimized, maximized, or targeted as per the objective of optimization. RC is measured in terms of

RMS acceleration of sprung mass and improves as its value decreases; hence the response is to be minimized to obtain a high degree of comfort.

As recommended by ISO 2631-1:1997 [2] standards, RC should be less than 0.5 m/s^2, so the target value for RC was 0.3 m/s^2 and upper value was 0.6 m/s^2. An interactive optimizer plot in Figure 6 is also provided by the response optimizer which shows the optimum factor settings (values in red) and the response values (in blue) along with their respective desirabilities.

The predicted value(s) of response(s) can be calculated at a particular setting of interest for one or more factors and sensitivity of factors can be determined. When one or more input factors are changed to a new level, the graphs of desirabilities are redrawn and the predicted responses and desirabilities are recalculated. The model gives an optimum RMS acceleration of 0.2663 m/s^2 at the optimum combination of tire pressure of 35 psi, camber angle of 3 degree, toe of 10 mm, spring stiffness of 18000 N/m, and damper of damping coefficient 673 N-s/m with mass of 41 kg.

5. Results and Verification

The optimum setting for all the factors which minimizes the RMS acceleration, that is, improves the RC to the optimum level, is shown in Table 12. This optimum setting obtained is again verified by executing the combination on the test rig and the experimental value of RC was found to be 0.302 m/s^2 against the theoretical value of 0.2663 m/s^2.

6. Conclusion

In this work, the experimentation was accomplished by incorporating various combinations of the parameters on the test rig developed, for the simulation of real time behavior of suspension system of a vehicle. It can be concluded

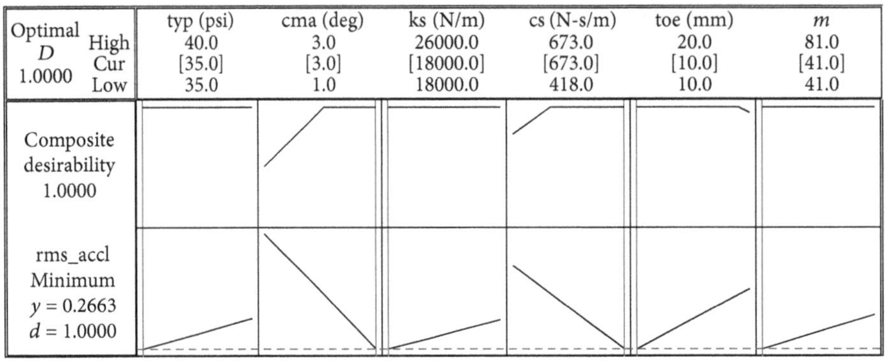

Optimal D 1.0000	High Cur Low	typ (psi) 40.0 [35.0] 35.0	cma (deg) 3.0 [3.0] 1.0	ks (N/m) 26000.0 [18000.0] 18000.0	cs (N-s/m) 673.0 [673.0] 418.0	toe (mm) 20.0 [10.0] 10.0	m 81.0 [41.0] 41.0
Composite desirability 1.0000							
rms_accl Minimum $y = 0.2663$ $d = 1.0000$							

FIGURE 6: Response optimization plot.

TABLE 12: Results and verification.

Factors	Optimum level
Tire pressure	35 psi
Camber	3 degrees
Spring stiffness	18000 N/m
Damping coefficient	673 Ns/m
toe	10 mm
Mass	41 kg
Verification	
RC (response optimization)	RC [experimental value (test rig)]
0.2663 m/s^2	0.302 m/s^2

from the aforementioned work that the RC, measured in terms of RMS acceleration of sprung mass in a vehicle, cannot be successfully optimized by controlling only spring stiffness, damping coefficient, and the sprung mass. The steering geometry angles, that is, camber and toe with the tire pressure, significantly affect the ride behavior of a vehicle along with the primary suspension parameters.

Also, it was found that the fractional factorial method of optimization reduces the number of runs significantly with the negligible compromise in the result, as the application of full factorial method is only limited to the experimentations where number of runs is small.

Later, the experimental model for RC and the optimized combination of six influential parameters were obtained after the regression analysis of data acquired, which was again verified on the test rig. As the responses have been treated with the road profile of a bump, it can be deemed that the optimized set of values will render a very comfortable ride as far as normal roads are considered.

Competing Interests

The authors declare that they have no competing interests.

Acknowledgments

The authors are thankful to the MESCOE-NI LabVIEW Academy Lab, Department of Mechanical Engineering, MES's College of Engineering, Pune, India, for providing the necessary testing facilities.

References

[1] SAE J670e, Vehicle Dynamics Terminology. Standard, Society of Automotive Engineers, 1976.

[2] ISO, "Mechanical vibration and shock evaluation of human exposure to whole body vibration. Part 1: general requirements," ISO Standard 2631-1:1997, International Organization for Standardization, Geneva, Switzerland, 1997.

[3] H. Damian, *Multibody Systems Approach to Vehicle Dynamics*, Butterworth-Heinemann, Oxford, UK, 2004.

[4] M. N. Khajavi, B. Notghi, and G. Paygane, "Multi objective optimization approach to optimize vehicle ride and handling characteristics," *World Academy of Science, Engineering and Technology*, vol. 62, pp. 580–584, 2010.

[5] J. Zhang, Y. Zhang, and R. Gao, "Genetic algorithms for optimal design of vehicle suspensions," in *Proceedings of the IEEE International Conference on Engineering of Intelligent Systems (ICEIS '06)*, pp. 1–6, IEEE, Islamabad, Pakistan, April 2006.

[6] T. M. Farid, A. Salah, and W. Abbas, "Design of optimal linear suspension for quarter car with human model using genetic algorithms," *Journal of Applied Sciences Research*, vol. 7, no. 11, pp. 1709–1720, 2011.

[7] S. Kilian, U. Zander, and F. E. Talke, "Suspension modeling and optimization using finite element analysis," *Tribology International*, vol. 36, no. 4–6, pp. 317–324, 2003.

[8] S. Roy and Z. Liu, "Road vehicle suspension and performance evaluation using a two-dimensional vehicle model," *International Journal of Vehicle Systems Modelling and Testing*, vol. 3, no. 1-2, pp. 68–93, 2008.

[9] J. P. C. Gonçalves and J. A. C. Ambrósio, "Road vehicle modeling requirements for optimization of ride and handling," *Multibody System Dynamics*, vol. 13, no. 1, pp. 3–23, 2005.

[10] A. F. Naudé and J. A. Snyman, "Optimisation of road vehicle passive suspension systems—part 1. Optimisation algorithm and vehicle model," *Applied Mathematical Modelling*, vol. 27, no. 4, pp. 249–261, 2003.

[11] A. F. Naudé and J. A. Snyman, "Optimisation of road vehicle passive suspension systems. Part 2. Qualification and case study," *Applied Mathematical Modelling*, vol. 27, no. 4, pp. 263–274, 2003.

[12] M. Sharifi and B. Shahriari, "Pareto optimization of vehicle suspension vibration for a nonlinear halfcar model using a

multi-objective genetic algorithm," *Research Journal of Recent Sciences*, vol. 1, no. 8, pp. 17–22, 2012.

[13] A. Bagheri, M. J. Mahmoodabadi, H. Rostami, and S. Kheybari, "Pareto optimization of a two-degree of freedom passive linear suspension using a new multiobjective genetic algorithm," *International Journal of Engineering, Transactions A: Basics*, vol. 24, no. 3, pp. 291–299, 2011.

[14] A. Kuznetsov, M. Mammadov, I. Sultan, and E. Hajilarov, "Optimization of a quarter-car suspension model coupled with the driver biomechanical effects," *Journal of Sound and Vibration*, vol. 330, no. 12, pp. 2937–2946, 2011.

[15] Z. Chi, Y. He, and G. F. Naterer, "Design optimization of vehicle suspensions with a quarter-vehicle model," *CSME Transactions*, vol. 32, no. 2, pp. 297–312, 2008.

[16] P. E. Uys, P. S. Els, and M. Thoresson, "Suspension settings for optimal ride comfort of off-road vehicles travelling on roads with different roughness and speeds," *Journal of Terramechanics*, vol. 44, no. 2, pp. 163–175, 2007.

[17] S. Mostaani, D. Singh, K. Firouzbakhsh, and M. T. Ahmadian, "Optimization of a passive vehicle suspension system for ride comfort enhancement with different speeds based on DOE method," in *Proceedings of the International Joint Colloquiums on Computer Electronics Electrical Mechanical and Civil (CEMC '11)*, Kerala, India, September 2011.

[18] J. Marzbanrad, M. Mohammadi, and S. Mostaani, "Optimization of a passive vehicle suspension system for ride comfort enhancement with different speeds based on design of experiment method (DOE) method," *Journal of Mechanical Engineering Research*, vol. 5, no. 3, pp. 50–59, 2013.

[19] P. Mohan, D. Marzougui, E. Arispe, and C. Story, "Component and full-scale tests of the 2007 chevrolet silverado suspension system," NCAC Test Report, The George Washington University, Washington, DC, USA, 2009.

[20] M. Arraigada and M. Partl, "Calculation of displacements of measured accelerations, analysis of two accelerometers and application in road engineering," in *Proceedings of the 6th Swiss Transport Research Conference*, vol. 1517, Ascona, Switzerland, 2006.

[21] D. C. Montgomery, *Design and Analysis of Experiments*, John Wiley & Sons, Hoboken, NJ, USA, 2009.

[22] P. S. Pillai, "Inflation pressure effect on whole tyre hysteresis ratio and radial spring constant," *Indian Journal of Engineering and Materials Sciences*, vol. 13, no. 2, pp. 110–116, 2006.

Permissions

List of Contributors

A. H. Ansari
Department of Mathematics, AlBaha University, Al Baha 1988, Saudi Arabia

Nabeel Alshabatat
Department of Mechanical Engineering, Tafila Technical University, Tafila 66110, Jordan

Rahul Dixit
Control Systems Laboratory, Research Center Imarat, Vigyanakancha, Hyderabad 500069, India

R. Prasanth Kumar
Department of Mechanical & Aerospace Engineering, Indian Institute of Technology Hyderabad, Kandi, Telangana 502285, India

A. Khnaijar and R. Benamar
Equipe des Etudes et Recherches en Simulation, Instrumentation et Mesures (ERSIM), Universit´e Mohammed V, Ecole Mohammadia des Ing´enieurs, Avenue Ibn Sina, BP 765, Rabat, Morocco

Ayush Raizada, Pravin Singru, Vishnuvardhan Krishnakumar and Varun Raj
Birla Institute of Technology and Science-Pilani, K.K. Birla Goa Campus, Goa 403726, India

I. P. G. Sopan Rahtika
Department of Mechanical Engineering, Bali State Polytechnics, Badung, Bali 80361, Indonesia
Department of Mechanical Engineering, Brawijaya University, Malang, East Java 65144, Indonesia

I. N. G. Wardana, A. A. Sonief and E. Siswanto
Department of Mechanical Engineering, Brawijaya University, Malang, East Java 65144, Indonesia

Meng Peng and Hans A. DeSmidt
Department of Mechanical, Aerospace and Biomedical Engineering, The University of Tennessee, 606 Dougherty Engineering Building, Knoxville, TN 37996-2210, USA

Edgar J. Gunter
RODYN Vibration Analysis, Inc., Charlottesville, VA, USA
Rotor Dynamics Laboratory, Mechanical and Aerospace Engineering Department, University of Virginia, Charlottesville, VA, USA

Brian K. Weaver
Rotating Machinery and Controls Laboratory, Mechanical and Aerospace Engineering Department, University of Virginia, Charlottesville, VA, USA

Piotr Krauze and Jerzy Kasprzyk
Institute of Automatic Control, Silesian University of Technology, Akademicka 16, Gliwice, Poland

AhmedM. Al-Jumaily and AtaMeshkinzar
Institute of Biomedical Technologies, Auckland University of Technology, Private Bag 92006, Auckland 1142, New Zealand

Vincent O. S. Olunloyo
Department of Systems Engineering, Faculty of Engineering, University of Lagos, Akoka-Yaba, Lagos 23401, Nigeria

Charles A. Osheku
Centre for Space Transport and Propulsion, National Space Research and Development Agency, Federal Ministry of Science and Technology, FCT, PMB 437, Abuja, Nigeria

Patrick S. Olayiwola
Department of Mechanical & Biomedical Engineering, College of Engineering, Bells University of Technology, Ota 234037, Ogun State, Nigeria

H. Larsson and K. Farhang
Department of Mechanical Engineering and Energy Processes, Southern Illinois University Carbondale, Carbondale, IL 62901, USA

Nader Sawalhi
Mechanical Engineering Department, Prince Mohammad Bin Fahd University, AL-Khobar, Saudi Arabia

Rahul Kalyankar and Nasim Uddin
Department of Civil Construction and Environmental Engineering, University of Alabama at Birmingham, Birmingham, AL, USA

Jhonatan Camacho-Navarro and Marco Flórez
Escuela de Ingeniería Electrónica, Universitaria de Investigación (UDI), Grupo de Investigación en Procesamiento de Señales (GPS), Bucaramanga, Colombia

R. Guzmán-López and Sergio Gómez
Universidad Pontificia Bolivariana, Bucaramanga, Colombia

Anirban C. Mitra, Tanushri Soni and G. R. Kiranchand
Department of Mechanical Engineering, MES's College of Engineering, Wadia College Campus, 19 Late Prin, V. K. Joag Path, Pune 411001, India

Index

www.ingramcontent.com/pod-product-compliance
Lightning Source LLC
Chambersburg PA
CBHW050457200326
41458CB00014B/5217